The Book of Prime Number Records

Paulo Ribenboim

The Book of
Prime Number Records

Springer-Verlag
New York Berlin Heidelberg
London Paris Tokyo

Paulo Ribenboim
Department of Mathematics and Statistics
Queen's University
Kingston, Ontario K7L 3N6
Canada

Mathematics Subject Classification (1980): 10H15, 10H20

Library of Congress Cataloging-in-Publication Data
Ribenboim, Paulo.
  The book of prime number records.
  Bibliography: p.
  Includes index.
  1. Numbers, Prime. I. Title.
QA246.R47   1988   512'.72   87-14811

Printed and bound by R.R. Donnelley and Sons, Harrisonburg, Virginia.
Printed in the United States of America.

9 8 7 6 5 4 3 2 1

ISBN 0-387-96573-4 Springer-Verlag New York Berlin Heidelberg
ISBN 3-540-96573-4 Springer-Verlag Berlin Heidelberg New York

Narrow road to
a far province.

      Bashō

À Jean Pierre Serre, qui
montre le chemin.

PREFACE

This text originated as a lecture delivered November 20, 1984, at Queen's University, in the undergraduate colloquium series established to honour Professors A. J. Coleman and H. W. Ellis and to acknowledge their long-lasting interest in the quality of teaching undergraduate students.

In another colloquium lecture, my colleague Morris Orzech, who had consulted the latest edition of the *Guinness Book of Records*, reminded me very gently that the most "innumerate" people of the world are of a certain tribe in Mato Grosso, Brazil. They do not even have a word to express the number "two" or the concept of plurality. "Yes Morris, I'm from Brazil, but my book will contain numbers different from 'one.' "

He added that the most boring 800-page book is by two Japanese mathematicians (whom I'll not name), and consists of about 16 million digits of the number π.

"I assure you Morris, that in spite of the beauty of the apparent randomness of the decimal digits of π, I'll be sure that my text will also include some words."

Acknowledgment. The manuscript of this book was prepared on the word processor by Linda Nuttall. I wish to express my appreciation for the great care, speed, and competence of her work.

Paulo Ribenboim

CONTENTS

GUIDING THE READER

If a notation, which is not self-explanatory, appears without explanation on, say page 103, look at the Index of Notations, which is organized by page number; the definition of the notation should appear before page 103.

If you wish to see where and how often your name is quoted in this book, turn to the Index of Names, at the end of the book. Should I say that there is no direct relation between achievement and number of quotes earned?

If, finally, you do not want to read the book but you just want to have some information about Knödel numbers – which is perfectly legitimate, if not laudable – go quickly to the Subject Index. Do not look under the heading "Numbers," but rather "Knödel." And for a subject like "Strong Lucas pseudoprimes," you have exactly three possibilities ... .

INDEX OF NOTATIONS

The following traditional notations are used in the text without explanation:

Notation	Explanation
$m \mid n$	the integer m divides the integer n
$m \nmid n$	the integer m does not divide the integer n
$p^e \parallel n$	p is a prime, $p^e \mid n$ but $p^{e+1} \nmid n$
$\gcd(m,n)$	greatest common divisor of the integers m, n
$\operatorname{lcm}(m,n)$	least common multiple of the integers m, n
$\log x$	natural logarithm of the real number $x > 0$
$\mathbf{Z}$	ring of integers
$\mathbf{Q}$	field of rational numbers
$\mathbf{R}$	field of real numbers
$\mathbf{C}$	field of complex numbers

The following notations are listed as they appear in the book:

Page	Notation	Explanation
6	U_n	nth Fibonacci number
18	g_p	smallest primitive root modulo p
21	$[x]$	the largest integer in x, that is, the only integer such that $[x] \leq x < [x] + 1$
25	$\phi(n)$	totient or Euler's function
27	$\lambda(n)$	Carmichael's function
27	$\omega(n)$	number of distinct prime factors of n
28	$L(x)$	number of composite n, such that $n \leq x$ and $\phi(n)$ divides $n - 1$
29	$N_\phi(m)$	$\#\{n \geq 1 \mid \phi(n) = m\}$
30	$E_\phi(k)$	$\#\{(n,m) \mid n > m \geq 1,\ n{-}m = k,\ \phi(n) = \phi(m)\}$
32	t_n^*	primitive part of $a^n - b^n$
32	$k(m)$	square-free kernel of m
33	$P(m)$	largest prime factor of m
33	S_κ	set of integers n with at most $\kappa \log \log n$ distinct prime factors
34	(a/p)	Legendre symbol
35	(a/b)	Jacobi symbol
41	$U_n=U_n(P,Q)$	nth term of the Lucas sequence with parameters (P,Q)
41	$V_n=V_n(P,Q)$	nth term of the companion Lucas sequence with parameters (P,Q)
47	$\rho(n)=\rho(n,U)$	smallest $r \geq 1$ such that $\rho(n)$ divides U_r
47	$\psi(p)$	$p - (D/p)$
49	$\left[\dfrac{\alpha,\beta}{p}\right]$	a symbol associated to the roots α, β of $X^2 - PX + Q$
49	$\lambda_{\alpha,\beta}(\Pi p^e)$	$\ell cm\{\psi_{\alpha,\beta}(p^e)\}$
53	$\mathbb{P}(U)$	set of primes p dividing some term U_n
53	$\mathbb{P}(V)$	set of primes p dividing some term V_n
55	U_n^*	primitive part of U_n
61	$U_n=U_n(\sqrt{R},Q)$	nth term of the Lehmer sequence with parameters $\sqrt{R},\ Q$
61	$V_n=V_n(\sqrt{R},Q)$	nth term of the companion Lehmer sequences with parameters $\sqrt{R},\ Q$

Page	Notation	Explanation

Page	Notation	Explanation
249	$SP\pi(x)$	number of strong pseudoprimes to base 2, less than or equal to x
249	$SP\pi_a(x)$	same, to base a
251	$\mathrm{psp}(d,a)$	smallest pseudoprime in the arithmetic progression $\{a+kd \mid k \geqslant 1\}$ with $\gcd(a,d) = 1$
252	$CN(x)$	$= \#\{n \mid n \leqslant x,\ n$ Carmichael number$\}$
253	$L\pi(x)$	number of Lucas pseudoprimes [with parameters (P,Q)] $n \leqslant x$
254	$SL\pi(x)$	number of strong Lucas pseudoprimes [with parameters (P,Q)] $n \leqslant x$
258	$ii(p)$	irregularity index of p
258	$\pi_{iis}(x)$	number of primes $p \leqslant x$ such that $ii(p) = s$
258	$\pi_{reg}(N)$	number of regular primes $p \leqslant x$
258	$\pi_{ii}(x)$	number of irregular primes $p \leqslant x$
259	K_n	$= Q(\zeta_{p^{n+1}})$
259	K_n^+	$= Q(\zeta_{p^{n+1}} + \zeta_{p^{n+1}}^{-1})$
259	h_n	class number of K_n
259	h_n^+	class number of K_n^+
262	$S_{d,a}(x)$	$= \#\{p$ prime $\mid p \leqslant x,\ a + pd$ is a prime$\}$
265	$q_p(a)$	$= \dfrac{a^{p-1} - 1}{p}$, Fermat quotient of p, with base a
266	$\mathfrak{W}_{\ell}^{(k)}$	$= \{p$ prime $\mid \ell^{p-1} \equiv 1 \pmod{p^k}\}$
267	$\mathfrak{N}_L$	$= \{p$ prime $\mid$ there exists c, not a multiple of p, such that $pc = u \pm v$, where all prime factors of uv are at most $L\}$
268	$\mathfrak{N}_{\ell}^{(k)}$	$= \{p$ prime $\mid$ there exists $s \geqslant 1$ such that p divides $\ell^s + 1$, but p^{k+1} does not divide $\ell^s + 1\}$
274	$\mathfrak{P}(F)$	$= \{p$ prime $\mid$ there exists n such that p divides $F_n\}$
274	$\mathfrak{P}(M)$	$= \{p$ prime $\mid$ there exists a prime q such that p divides $M_q\}$
274	$\mathfrak{P}^{(2)}(F)$	$= \{p$ prime $\mid$ there exists n such that p^2 divides $F_n\}$

INTRODUCTION

The *Guinness Book of Records* became famous as an authoritative source of information to settle amiable disputes between drinkers of, it was hoped, the Guinness brand of stout. Its immense success in recording all sorts of exploits, anomalies, endurance performances, and so on has in turn influenced and sparked these very same performances. So one sees couples dancing for countless hours or persons buried in coffins with snakes, for days and days — just for the purpose of having their name in this bible of trivia. There are also records of athletic performances, extreme facts about human size, longevity, procreation, etc.

Little is found in the scientific domain. Yet, scientists — mathematicians in particular — also like to chat while sipping wine or drinking a beer in a bar. And when the spirits mount, bets may be exchanged about the latest advances, for example, about recent discoveries concerning numbers.

Frankly, if I were to read in the *Whig-Standard* that a brawl in one of our pubs began with a heated dispute concerning which is the largest known pair of twin prime numbers, I would find this highly civilized.

But not everybody agrees that fights between people are desirable, even for such all-important reasons. So, maybe I should reveal some of these records. Anyone who knows better should not hesitate to pass me additional information.

I'll restrict my discussion to prime numbers: these are natural numbers, like 2, 3, 5, 7, 11, ... , which are not multiples of any smaller natural number (except 1). If a natural number is neither 1 nor a prime, it is called a composite number.

Prime numbers are important, since a fundamental theorem in arithmetic states that every natural number greater than 1 is a product of prime numbers, and moreover, in a unique way.

Without further ado it is easy to answer the following question: "Which is the oddest prime number?" It is 2, because it is the only even prime number!

There will be plenty of opportunities to encounter other prime numbers, like 1093 and 608,981,813,029, possessing interesting distinctive properties. Prime numbers are like cousins, members of the same family, resembling one another, but not quite alike.

Facing the task of presenting the records on prime numbers, I was led to think how to organize this volume. In other words, to classify the main lines of investigation and development of the theory of prime numbers.

It is quite natural, when studying a set of numbers — in this case the set of prime numbers — to ask the following questions, which I phrase informally as follows:

How many? How to decide whether an arbitrary given number is in the set? How to describe them? What is the distribution of these numbers, both at large and in short intervals? And, then, to focus attention on distinguished types of such numbers, as well as to experiment with these numbers and make predictions — just as in any science.

Thus, I have divided the presentation into the following topics:

(1)   How many prime numbers are there?
(2)   How to recognize whether a natural number is a prime?
(3)   Are there functions defining prime numbers?
(4)   How are the prime numbers distributed?
(5)   Which special kinds of primes have been considered?
(6)   Heuristic and probabilistic results about prime numbers.

The discussion of these topics will lead me to indicate the relevant records.

Chapter 1

HOW MANY PRIME NUMBERS ARE THERE?

The answer to the question of how many prime numbers exist is given by the fundamental theorem:

There exist infinitely many prime numbers.

I shall give nine and a half(!) proofs of this theorem, by famous, but also by forgotten, mathematicians. Some proofs suggest interesting developments; other proofs are just clever or curious. There are of course more (but not quite infinitely many) proofs of the existence of infinitely many primes.

I. Euclid's Proof

Euclid's Proof. Suppose that $p_1 = 2 < p_2 = 3 < \cdots < p_r$ are all the primes. Let $P = p_1 p_2 \cdots p_r + 1$ and let p be a prime dividing P; then p cannot be any of $p_1, p_2, \ldots, p_r$, otherwise p would divide the difference $P - p_1 p_2 \cdots p_r = 1$, which is impossible. So this prime p is still another prime, and $p_1, p_2, \ldots, p_r$ would not be all the primes. $\square$

I shall write the infinite increasing sequence of primes as

$$p_1 = 2, \quad p_2 = 3, \quad p_3 = 5, \quad p_4 = 7, \ldots, p_n, \ldots .$$

Euclid's proof is pretty simple; however, it does not give any information about the new prime, only that it is at most equal to the number P, but it may well be smaller.

For every prime p, let $p\#$ denote the product of all primes q, such that $q \leqslant p$. Following a suggestion of Dubner, $p\#$ may be called the primorial of p.

The answer to the following questions are unknown:

Are there infinitely many primes p for which $p\# + 1$ is prime?

Are there infinitely many primes p for which $p\# + 1$ is composite?

Record

$13649\# + 1$ is the largest known prime of the form $p\# + 1$; it has 5862 digits and was discovered by Dubner (1987), who also identified $p = 11549$, 4787, 4547, and 3229 with the same property (Dubner & Dubner, 1985).

Previous work by Borning (1972), Templer (1980), and Buhler, Crandall & Penk (1982) established that $p\# + 1$ is prime for $p = 2, 3, 5, 7$, 11, 31, 379, 1019, 1021, 2657, and $p\# + 1$ is composite for all other $p < 11213$.

A variant of Euclid's proof is the following: Consider the sequence $q_1 = 2$, $q_2 = 3$, $q_3 = 7$, $q_4 = 43$, $q_5 = 139$, $q_6 = 50{,}207$, $q_7 = 340{,}999$, $q_8 = 2{,}365{,}347{,}734{,}339$, ..., where q_{n+1} is the highest prime factor of $q_1 q_2 \cdots q_n + 1$ (so $q_{n+1} \neq q_1, q_2, ..., q_n$). In 1963, Mullin asked: Does the sequence $(q_n)_{n \geqslant 1}$ contain all the prime numbers? Does it exclude at most finitely many primes? Is the sequence monotonic?

Cox & van der Poorten found in 1968 congruences which are sufficient to decide whether a given prime is excluded; moreover, they have shown that the primes 5, 11, 13, 17, 19, 23, 29, 31, 37, 41, and 47 do not appear in the sequence. They have also conjectured that there exist infinitely many excluded primes.

II. Kummer's Proof

In 1878, Kummer gave the following:

Kummer's Proof. Suppose that there exist only finitely many primes $p_1 < p_2 < \cdots < p_r$. Let $N = p_1 p_2 \cdots p_r > 2$. The integer $N - 1$, being a product of primes, has a prime divisor p_i in common with N; so, p_i divides $N - (N-1) = 1$, which is absurd! □

This proof, by an eminent mathematician, is like a pearl, round, bright, and beautiful in its simplicity.

III. Pólya's Proof

Pólya's proof uses the following idea: It is enough to find an infinite sequence of natural numbers $1 < a_1 < a_2 < a_3 < \cdots$ that are pairwise relatively prime (i.e., without common prime factor). So, if p_1 is a prime dividing a_1, if p_2 is a prime dividing a_2, etc., then p_1, p_2, ..., are all different.

For this proof, the numbers a_n are chosen to be the Fermat numbers $F_n = 2^{2^n} + 1$ $(n \geq 0)$. Indeed, it is easy to see, by induction on m, that $F_m - 2 = F_0 F_1 \cdots F_{m-1}$; hence, if $n < m$, then F_n divides $F_m - 2$.

If a prime p would divide both F_n and F_m, then it would divide $F_m - 2$ and F_m, hence also 2, so $p = 2$. But F_n is odd, hence not divisible by 2. This shows that the Fermat numbers are pairwise relatively prime. $\square$

Explicitly, the first Fermat numbers are $F_0 = 3$, $F_1 = 5$, $F_2 = 17$, $F_3 = 257$, $F_4 = 65537$, and it is easy to see that they are prime numbers. F_5 already has 10 digits, and each subsequent Fermat number is about the square of the preceding one, so the sequence grows very quickly. A natural task is to determine explicitly whether F_n is a prime, or at least to find a prime factor of it. I shall return to this point in Chapter 2.

It would be desirable to find other infinite sequences of pairwise relatively prime integers, without already assuming the existence of infinitely many primes. In a paper of 1964, Edwards examined this question and indicated various sequences, defined recursively, having this property. For example, if S_0, a are relative prime integers, with $S_0 > a \geq 1$, the sequence defined by the recursive relation

(S1) $\qquad S_n - a = S_{n-1}(S_{n-1} - a)$ $\qquad$ (for $n \geq 1$)

consists of pairwise relatively prime natural numbers.

Similarly, if S_0 is odd and

(S2) $\qquad S_n = S_{n-1}^2 - 2$ $\qquad$ (for $n \geq 1$),

then, again, the integers S_n are pairwise relatively prime.

In the best situation, that is, when $S_0 = 3$, $a = 2$, the sequence (S1) is in fact the sequence of Fermat numbers: $S_n = F_n = 2^{2^n} + 1$.

The sequence (S2), which grows essentially just as quickly, has been considered by Lucas, and I shall return to it in Chapter 2.

If a sequence $(S_n)_{n \geqslant 1}$ of pairwise relatively prime natural numbers has been found, then for every exponent $k \geqslant 1$, the sequence of kth powers $(S_n^k)_{n \geqslant 1}$ consists also of pairwise relatively prime integers.

There is also another very simple method, using the Fibonacci numbers U_n, to produce infinitely many sequences of pairwise relatively prime integers.

Originally, Fibonacci numbers appeared in a problem in his book *Liber Abaci*, published in 1202, where the Arabic numbers were also first introduced in Europe. The problem, now reproduced in many elementary books, concerned rabbits having certain reproductive patterns. I would rather speak about rabbits cooked with noodles in a good wine sauce, as I had once the occasion of tasting at "L'Aubette" in Strasbourg, in 1952. I can remember it very well, much better than how many offspring a rabbit should have to make a Fibonacci sequence ... .

The Fibonacci numbers may be explicitly defined by $U_0 = 0$, $U_1 = 1$, and, for $n \geqslant 2$, $U_n = U_{n-1} + U_{n-2}$. Thus, the sequence begins with 0, 1, 1, 2, 3, 5, 8, 13, 21, 34, 55, 89, 144, ... I shall return to this and similar sequences, in Chapters 2 and 4 and again in Chapter 5, Section VIII. For the moment, I just note that it may be shown, in quite an elementary way, that if $d = \gcd(m,n)$, then $U_d = \gcd(U_m,U_n)$. In particular, if m,n are relatively prime, then so are U_m, U_n. Thus, once a sequence $(S_n)_{n \geqslant 1}$ of pairwise relatively prime natural numbers has been found, then the sequences

$$U_{S_1}, \ U_{S_2}, \ U_{S_3}, \ ...,$$

$$U_{U_{S_1}}, \ U_{U_{S_2}}, \ U_{U_{S_3}}, \ ...,$$

$$U_{U_{U_{S_1}}}, \ U_{U_{U_{S_2}}}, \ U_{U_{U_{S_3}}}, \ ...,$$

all consist of pairwise relatively prime integers. However, this construction goes exactly against what would be desirable.

IV. Euler's Proof

This is a rather indirect proof, which, in some sense is unnatural; but, on the other hand, as I shall indicate, it leads to the most important developments.

Euler showed that there must exist infinitely many primes because a certain expression formed with all the primes is infinite.

If p is any prime, then $1/p < 1$; hence, the sum of the geometric series is

$$\sum_{k=0}^{\infty} \frac{1}{p^k} = \frac{1}{1 - (1/p)} .$$

Similarly, if q is another prime, then

$$\sum_{k=0}^{\infty} \frac{1}{q^k} = \frac{1}{1 - (1/q)} .$$

Multiplying these equalities:

$$1 + \frac{1}{p} + \frac{1}{q} + \frac{1}{p^2} + \frac{1}{pq} + \frac{1}{q^2} + \cdots = \frac{1}{1 - (1/p)} \times \frac{1}{1 - (1/q)} .$$

Explicitly, the left-hand side is the sum of the inverses of all natural numbers of the form $p^h q^k$ ($h \geqslant 0$, $k \geqslant 0$), each counted only once, because every natural number has a unique factorization as a product of primes. This simple idea is the basis of the proof.

Euler's Proof. Suppose that $p_1, p_2, \ldots, p_n$ are all the primes. For each $i = 1, \ldots, n$

$$\sum_{k=0}^{\infty} \frac{1}{p_i^k} = \frac{1}{1 - (1/p_i)} .$$

Multiplying these n equalities, one obtains

$$\prod_{i=1}^{n} \left(\sum_{k=0}^{\infty} \frac{1}{p_i^k} \right) = \prod_{i=1}^{n} \frac{1}{1 - (1/p_i)} ,$$

and the left-hand side is the sum of the inverses of all natural numbers, each counted once — this follows from the fundamental theorem that every natural number is equal, in a unique way, to the product of primes.

But the series $\sum_{n=1}^{\infty} (1/n)$ is divergent; being a series of positive terms, the order of summation is irrelevant, so the left-hand side is infinite, while the right-hand side is clearly finite. This is absurd. $\square$

In Chapter 4 I'll return to developments along this line.

V. Thue's Proof

Thue's proof uses only the fundamental theorem of unique factorization of natural numbers as products of prime numbers.

Thue's Proof. First let $n,k \geqslant 1$ be integers such that $(1+n)^k < 2^n$. Let $p_1 = 2$, $p_2 = 3$, p_3, ..., p_r be all the primes satisfying $p \leqslant 2^n$. Suppose that $r \leqslant k$.

By the fundamental theorem, every integer m, $1 \leqslant m \leqslant 2^n$, may be written in a unique way in the form

$$m = 2^{e_1} 3^{e_2} \cdots p_r^{e_r},$$

where $0 \leqslant e_1 \leqslant n$, $0 \leqslant e_2 < n$, ..., $0 \leqslant e_r < n$.

Counting all the possibilities, it follows that $2^n \leqslant (n+1)n^{r-1} < (n+1)^r \leqslant (n+1)^k < 2^n$, and this is absurd. So $r \geqslant k + 1$.

Choose $n = 2k^2$. From $1 + 2k^2 < 2^{2k}$ for every $k \geqslant 1$, it follows that

$$(1 + 2k^2)^k < 2^{2k^2} = 4^{k^2}.$$

Thus, there exist at least $k + 1$ primes p such that $p < 4^{k^2}$. $\square$

Since k may be taken arbitrarily large, this shows that there are infinitely many primes, and actually that $k + 1$ is a lower bound for the number of primes less than 4^{k^2}. This is a quantitative result, which is, of course, very poor. In Chapter 4, I shall further investigate this kind of question.

VI. Two-and-a-Half Forgotten Proofs

The next proofs are by Perott, Auric, and Métrod. Who remembers these names? If it were not for Dickson's *History of the Theory of Numbers*, they would be totally forgotten. As I shall show, these proofs are very pleasant and ingenious; yet, they do not add new insights.

A. Perott's Proof

Perott's proof dates from 1881.

Perott's Proof. It is required to know that $\sum_{n=1}^{\infty}(1/n^2)$ is convergent with sum smaller than 2. (As a matter of fact, it is a famous result of Euler that the sum is exactly $\pi^2/6$, and I shall return to this point in Chapter 4.)

Indeed,

$$\sum_{n=1}^{\infty} \frac{1}{n^2} < 1 + \sum_{n=1}^{\infty} \frac{1}{n(n+1)} = 1 + \sum_{n=1}^{\infty} \left(\frac{1}{n} - \frac{1}{n+1} \right) = 1 + 1 = 2.$$

Let

$$\delta = 2 - \sum_{n=1}^{\infty} \frac{1}{n^2}.$$

Suppose that there exist only r prime numbers $p_1 < p_2 < \cdots < p_r$. Let N be any integer such that $p_1 p_2 \cdots p_r < N$. The number of integers $m \leqslant N$ that are not divisible by a square is therefore 2^r (which is the number of all possible sets of distinct primes), because every integer is, in a unique way, the product of primes. The number of integers $m \leqslant N$ divisible by p_i^2 is at most N/p_i^2; so the number of integers $m \leqslant N$ divisible by some square is at most $\sum_{i=1}^{r}(N/p_i^2)$. Hence,

$$N \leqslant 2^r + \sum_{i=1}^{r} \frac{N}{p_i^2} < 2^r + N\left[\sum_{n=1}^{\infty} \frac{1}{n^2} - 1 \right] = 2^r + N(1 - \delta).$$

By choosing N such that $N\delta \geqslant 2^r$, it follows a contradiction. $\square$

B. Auric's Proof

Auric's proof, which appeared in 1915, is very simple.

Auric's Proof. Suppose that there exist only r primes $p_1 < p_2 < \cdots < p_r$.

Let $t \geqslant 1$ be any integer and let $N = p_r^t$.

By the unique factorization theorem, each integer m, $1 \leqslant m \leqslant N$, is written $m = p_1^{f_1} p_2^{f_2} \cdots p_r^{f_r}$ and the sequence $(f_1, ..., f_r)$, with each $f_i \geqslant 0$, is uniquely defined. Note also that $p_1^{f_i} \leqslant m \leqslant N \leqslant p_r^t$. Letting $E = (\log p_r)/(\log p_1)$, then $f_i \leqslant tE$. Thus, the number N (of integers m, $1 \leqslant m \leqslant N$) is at most the number of sequences $(f_1, f_2, ..., f_r)$; hence, $p_r^t = N \leqslant (tE + 1)^r \leqslant t^r (E + 1)^r$. If t is sufficiently large, this inequality cannot hold, which shows that the number of primes must be infinite. $\square$

C. Métrod's Proof

Métrod's proof of 1917 is also very simple.

Métrod's Proof. Assume that there exist only r primes $p_1 < p_2 < \cdots < p_r$. Let $N = p_1 p_2 \cdots p_r$, and for each $i = 1, ..., r$, let $Q_i = N/p_i$. Note that p_i does not divide Q_i, while p_i divides Q_j, for all $j \neq i$. Let $S = \sum_{i=1}^r Q_i$. If q is any prime dividing S, then $q \neq p_i$ because p_i divides Q_j (for $j \neq i$) but p_i does not divide Q_i. Thus there exists yet another prime! $\square$

Actually, Métrod's proof is no more than a variant of a previous proof of Stieltjes (1890), which itself follows the same idea of Euclid. Therefore, I'm counting it as just "half a proof."

Stieltjes did as follows: let $N = mn$ be any factorization. Each p_i divides one of m, n, but not both. So $m + n$ is not divisible by any of the existing primes, which is impossible because $m + n \neq 1$. $\square$

Taking $n = 1$, this is just Euclid's proof.

VII. Washington's Proof

Washington's (1980) is a proof via commutative algebra. The ingredients are elementary facts of the theory of principal ideal domains, unique factorization domains, Dedekind domains, and algebraic numbers, and may be found in any textbook on the subject, such as Samuel's (1967) book: there is no mystery involved. First, I recall the needed facts:

1. In every number field (of finite degree) the ring of algebraic
    integers is a Dedekind domain: every ideal is, in a unique way,
    the product of prime ideals.
2. In every number field (of finite degree) there are only finitely
    many prime ideals that divide any given prime number p.
3. A Dedekind domain with only finitely many prime ideals is a
    principal ideal domain, and as such, every element is, up to
    units, the product of prime elements in a unique way.

Washington's Proof. Consider the field of all numbers of the form
$a + b\sqrt{-5}$, where a, b are rational numbers. The ring of algebraic
integers in this field consists of the numbers of the above form, with
a, b ordinary integers. It is easy to see that 2, 3, $1 + \sqrt{-5}$, $1 - \sqrt{-5}$
are prime elements of this ring, since they cannot be decomposed
into factors that are algebraic integers, unless one of the factors is a
"unit" 1 or −1. Note also that $(1 + \sqrt{-5})(1 - \sqrt{-5}) = 2 \times 3$, the decomposition of 6 into a product of primes is not unique up to units, so
this ring is not a unique factorization domain; hence, it is not a
principal ideal domain. So, it must have infinitely many prime
ideals (by fact 3 above) and (by fact 2 above) there exist infinitely
many prime numbers.   □

VIII. Fürstenberg's Proof

Fürstenberg's is an ingenious proof based on topological ideas. Since
it is so short, I cannot do any better than transcribe it verbatim; it
appeared in 1955:

> In this note we would like to offer an elementary "topological"
> proof of the infinitude of the prime numbers. We introduce a
> topology into the space of integers S, by using the arithmetic
> progressions (from $-\infty$ to $+\infty$) as a basis. It is not difficult to
> verify that this actually yields a topological space. In fact,
> under this topology, S may be shown to be normal and hence
> metrizable. Each arithmetic progression is closed as well as
> open, since its complement is the union of the other arith-
> metic progressions (having the same difference). As a result,

the union of any finite number of arithmetic progressions is closed. Consider now the set $A = \cup A_p$, where A_p consists of all multiples of p, and p runs through the set of primes ≥ 2. The only numbers not belonging to A are -1 and 1, and since the set $\{-1,1\}$ is clearly not an open set, A cannot be closed. Hence A is not a finite union of closed sets which proves that there are an infinity of primes.   □

Golomb developed further the idea of Fürstenberg, and wrote an interesting short paper in 1959.

Chapter 2

HOW TO RECOGNIZE WHETHER A NATURAL NUMBER IS A PRIME?

In the art. 329 of *Disquisitiones Arithmeticae*, Gauss (1801) wrote:

> The problem of distinguishing prime numbers from composite numbers and of resolving the latter into their prime factors is known to be one of the most important and useful in arithmetic. It has engaged the industry and wisdom of ancient and modern geometers to such an extent that it would be superfluous to discuss the problem at length... . Further, the dignity of the science itself seems to require that every possible means be explored for the solution of a problem so elegant and so celebrated.

Despite what Gauss wrote, to recognize the primality or to find the factors of large numbers may at first seem like a pointless exercise. However, this is an erroneous impression. Factorization and primality have become quite important in public key cryptography, and this represents, today, one of the most striking applications of number theory to communications. I shall return to these matters.

More may be said about factorization, and I wish to quote from a letter of Brillhart, who has been involved in this problem for a long time:

> It is relevant to number theorists who work from examples; being able to factor has always been one of the most important

analytic techniques, since the secret of a problem often is revealed when (and only when) the prime factors of the numbers in question have been discovered by their factorization. I know that many people — in fact the vast majority of people — do not realize this, because they have never been realistically confronted with needing to know the factors of some numbers; but I assure you, that when you do need them, you are terribly frustrated with not being able to get them, because the number is too large to be factored. It is not a matter of sport nor a passing fad.

The first observation concerning the problem of primality and factorization is clear: there is an algorithm for both problems. By this, I mean a procedure involving finitely many steps, which is applicable to every number N and which will indicate whether N is a prime, or, if N is composite, which are its prime factors. Namely, given the natural number N, try in succession every number $n = 2, 3,$ ... up to $[\sqrt{N}]$ (the largest integer not greater than $\sqrt{N}$) to see whether it divides N. If none does, then N is a prime. If, say, N_0 divides N, write $N = N_0 N_1$, so $N_1 < N$, and then repeat the same procedure with N_0 and with N_1. Eventually this gives the complete factorization into prime factors.

What I have just said is so evident as to be irrelevant. It should, however, be noted that for large numbers N, it may take a long time with this algorithm to decide whether N is a prime or composite.

This touches the most important practical aspect, the need to find an efficient algorithm — one which involves as few operations as possible, and therefore requires less time and costs less money to be performed.

It is my intention to divide this chapter into several sections in which I will examine various approaches, as well as explain the required theoretical results.

I. The Sieve of Eratosthenes

As I have already said, it is possible to find if N is a prime using trial division by every number n such that $n^2 \leqslant N$.

Since multiplication is an easier operation than division, Eratosthenes (in the 3rd century BC) had the idea of organizing the computations in the form of the well-known sieve. It serves to determine all the prime numbers, as well as the factorizations of composite numbers, up to any given number N. This is illustrated now for $N = 101$.

Do as follows: write all the numbers up to 101; cross out all the multiples of 2, bigger than 2; in each subsequent step, cross out all the multiples of the smallest remaining number p, which are bigger than p. It suffices to do it for $p^2 < 101$.

	2	3	~~4~~	5	~~6~~	7	~~8~~	~~9~~	~~10~~
11	~~12~~	13	~~14~~	~~15~~	~~16~~	17	~~18~~	19	~~20~~
~~21~~	~~22~~	23	~~24~~	~~25~~	~~26~~	~~27~~	~~28~~	29	~~30~~
31	~~32~~	~~33~~	~~34~~	~~35~~	~~36~~	37	~~38~~	~~39~~	~~40~~
41	~~42~~	43	~~44~~	~~45~~	46	47	~~48~~	~~49~~	~~50~~
~~51~~	~~52~~	53	~~54~~	~~55~~	~~56~~	~~57~~	~~58~~	59	~~60~~
61	~~62~~	~~63~~	~~64~~	~~65~~	~~66~~	67	~~68~~	~~69~~	~~70~~
71	~~72~~	73	~~74~~	~~75~~	~~76~~	~~77~~	~~78~~	79	~~80~~
~~81~~	~~82~~	83	~~84~~	~~85~~	~~86~~	~~87~~	~~88~~	89	~~90~~
~~91~~	~~92~~	~~93~~	~~94~~	~~95~~	~~96~~	97	~~98~~	~~99~~	~~100~~
101									

Thus, all the multiples of 2, 3, 5, 7 $< \sqrt{101}$ are sifted away. The number 53 is prime because it remained. Thus the primes up to 101 are 2, 3, 5, 7, 11, 13, 17, 19, 23, 29, 31, 37, 41, 43, 47, 53, 59, 61, 67, 71, 73, 79, 83, 89, 97, 101.

This procedure is the basis of sieve theory, which has been developed to provide estimates for the number of primes satisfying given conditions.

II. Some Fundamental Theorems on Congruences

In the next section, I intend to describe some classical methods to test primality and to find factors. They rely on theorems on congruences, especially Fermat's little theorem, the old theorem of Wilson, as well as Euler's generalization of Fermat's theorem. I shall also include a subsection on quadratic residues, a topic of central importance, which is also related with primality testing, as I shall have occasion to indicate.

A. Fermat's Little Theorem and Primitive Roots Modulo a Prime

Fermat's Little Theorem. *If p is a prime number and if a is an integer, then $a^p \equiv a$ (mod p). In particular, if p does not divide a, then $a^{p-1} \equiv 1$ (mod p).*

Euler published the first proof of Fermat's little theorem.

Proof. It is true for $a = 1$. Assuming that it is true for a, then, by induction, $(a+1)^p \equiv a^p + 1 \equiv a + 1$ (mod p). So the theorem is true for every natural number a.   □

The above proof required only the fact that if p is a prime number and if $1 \leqslant k \leqslant p - 1$, then the binomial coefficient $\binom{p}{k}$ is a multiple of p.

Note the following immediate consequence: if $p \nmid a$ and p^n is the highest power of p dividing $a^{p-1} - 1$, then p^{n+e} is the highest power of p dividing $a^{p^e(p-1)} - 1$ (where $e \geqslant 1$); in this statement, if $p = 2$, then n must be at least 2.

It follows from the theorem that for any integer a, which is not a multiple of the prime p, there exists the smallest exponent $h \geqslant 1$, such that $a^h \equiv 1$ (mod p). Moreover, $a^k \equiv 1$ (mod p) if and only if h divides k; in particular, h divides $p - 1$. This exponent h is called the order of a modulo p. Note that a mod p, a^2 mod p, ..., a^{h-1} mod p, and 1 mod p are all distinct.

It is a basic fact that for every prime p there exists at least one integer g, not a multiple of p, such that the order of g modulo p is equal to $p - 1$. Then, the set $\{g$ mod p, g^2 mod p, ..., g^{p-2} mod p, 1 mod $p\}$ is equal to the set $\{1$ mod p, 2 mod p, ..., $(p - 1)$ mod $p\}$.

Every integer g, $1 \leqslant g \leqslant p - 1$, such that g mod p has order $p - 1$, is called a primitive root modulo p.

Let p be any odd prime, $k \geqslant 1$, and $S = \sum_{j=1}^{p-1} j^k$. Then

$$S \equiv \begin{cases} -1 \bmod p, & \text{when } p-1 \mid k, \\ \\ 0 \bmod p, & \text{when } p-1 \nmid k. \end{cases}$$

Indeed, if $p-1$ divides k, then $j^k \equiv 1 \pmod{p}$ for $j = 1,2, ..., p-1$; so $S \equiv p - 1 \equiv -1 \pmod{p}$. If $p - 1$ does not divide k, let g be a primitive root modulo p. Then $g^k \not\equiv 1 \pmod{p}$. Since the sets of residue classes $\{1 \bmod p, 2 \bmod p, ... , (p-1) \bmod p\}$ and $\{g \bmod p, 2g \bmod p, ... , (p-1)g \bmod p\}$ are the same, then

$$g^k S \equiv \sum_{j=1}^{p-1} (gj)^k \equiv \sum_{j=1}^{p-1} j^k \equiv S \pmod{p}.$$

Hence $(g^k - 1)S \equiv 0 \pmod{p}$ and, since p does not divide $g^k - 1$, then $S \equiv 0 \pmod{p}$. $\square$

The determination of a primitive root modulo p may be effected by a simple method indicated by Gauss in articles 73, 74 of *Disquisitiones Arithmeticae*.

Proceed as follows:

Step 1. Choose any integer a, $1 < a < p$, for example, $a = 2$, and write the residues modulo p of a, a^2, a^3, ... Let t be the smallest exponent such that $a^t \equiv 1 \pmod{p}$. If $t = p - 1$, then a is a primitive root modulo p. Otherwise, proceed to the next step.

Step 2. Choose any number b, $1 < b < p$, such that $b \not\equiv a^i \pmod{p}$ for $i = 1, ..., t$; let u be the smallest exponent such that $b^u \equiv 1 \pmod{p}$. It is simple to see that u cannot be a factor of t, otherwise $b^t \equiv 1 \pmod{p}$; but $1, a, a^2, ..., a^{t-1}$ are t pairwise incongruent solutions of the congruence $X^t \equiv 1 \pmod{p}$; so they are all the possible solutions, and therefore $b \equiv a^m \pmod{p}$, for some m, $0 \leqslant m \leqslant t - 1$, which is contrary to the hypothesis. If $u = p - 1$, then b is a primitive root modulo p. If $u \neq p - 1$, let v be the least common multiple of t, u; so $v = mn$ with m dividing t, n dividing u, and $\gcd(m,n) = 1$. Let $a' \equiv a^{t/m} \pmod{p}$, $b' \equiv b^{u/n} \pmod{p}$ so $c = a'b'$ has order $mn = v$ modulo p. If $v = p - 1$, then c is a primitive root modulo p. Otherwise, proceed to the next step, which is similar to step 2.

Note that $v > t$, so in each step either one reaches a primitive root modulo p, or one constructs an integer with a bigger order modulo p.

The process must stop; one eventually reaches an integer with order $p - 1$ modulo p, that is, a primitive root modulo p.

Gauss also illustrated the procedure with the example $p = 73$, and found that $g = 5$ is a primitive root modulo 73.

The above construction leads to a primitive root modulo p, but not necessarily to the smallest integer g_p, $1 < g_p < p$, which is a primitive root modulo p.

The determination of g_p is done by trying successively the various integers $a = 2, 3, \ldots$ and computing their orders modulo p. There is no uniform way of predicting, for all primes p, which is the smallest primitive root modulo p. However, several results were known about the size of g_p. In 1944, Pillai proved that there exist infinitely many primes p, such that $g_p > C \log \log p$ (where C is a positive constant). In particular $\lim \sup_{p \to \infty} g_p = \infty$. A few years later, using a very deep theorem of Linnik (see Chapter 4) on primes in arithmetic progressions, Fridlender (1949), and independently Salié (1950), proved that $g_p > C \log p$ (for some constant C) and infinitely many primes p. On the other hand, g_p does not grow too fast, as proved by Burgess in 1962:

$$g_p \leqslant C p^{\frac{1}{4} + \epsilon}$$

(for $\epsilon > 0$, a constant $C > 0$, and p sufficiently large).

Grosswald made Burgess' result explicit in 1981: if $p > e^{e^{24}}$ then $g(p) < p^{0.499}$.

The proof of the weaker result (with $\frac{1}{2}$ in place of $\frac{1}{4}$), attributed to Vinogradov, is in Landau's *Vorlesungen über Zahlentheorie*, Part VII, Chapter 14 (see General References).

The proof of the following result is elementary (problem proposed by Powell in 1983, solution by Kearnes in 1984):

For any positive integer M, there exist infinitely many primes p such that $M < g_p < p - M$.

On the next page, as an illustration, is a table of the smallest primitive roots g_p, for each prime $p < 200$.

A simple glance at the table suggests the following question: Is 2 a primitive root for infinitely many primes? More generally, if the integer $a \neq \pm 1$ is not a square, is it a primitive root modulo infinitely difficult problem.

p	g_p	p	g_r
2	1	89	3
3	2	97	5
5	2	101	2
7	3	103	5
11	2	107	2
13	2	109	6
17	3	113	3
19	2	127	3
23	5	131	2
29	2	137	3
31	3	139	2
37	2	149	2
41	6	151	6
43	3	157	5
47	5	163	2
53	2	167	5
59	2	173	2
61	2	179	2
67	2	181	2
71	7	191	19
73	5	193	5
79	3	197	2
83	2	199	3

B. The Theorem of Wilson

Wilson's Theorem. *If p is a prime number, then*

$$(p - 1)! \equiv - 1 \pmod{p}.$$

This is just a corollary of Fermat's little theorem. Indeed, 1, 2, ..., $p - 1$ are roots of the congruence $X^{p-1} - 1 \equiv 0 \pmod{p}$. But a congruence modulo p cannot have more roots than its degree. Hence,

$$X^{p-1} - 1 \equiv (X - 1)(X - 2) \cdots (X - (p-1)) \pmod{p}.$$

Comparing the constant terms, $-1 \equiv (-1)^{p-1}(p-1)! = (p-1)! \pmod{p}$. (This is also true if $p = 2$.) ☐

Wilson's theorem gives a characterization of prime numbers. Indeed, if $N > 1$ is a natural number that is not a prime, then $N = mn$, with

$1 < m, n < N - 1$, so m divides N and $(N-1)!$, and therefore $(n-1)! \not\equiv -1 \pmod{N}$.

However, Wilson's characterization of the prime numbers is not of practical value to test the primality of N, since there is no known algorithm to rapidly compute $N!$, say, in log N steps.

C. The Properties of Giuga, Wolstenholme and Mann & Shanks

Now, I shall consider other properties that are satisfied by prime numbers.

First, I note that if p is a prime, then by Fermat's little theorem (as already indicated)

$$1^{p-1} + 2^{p-1} + \cdots + (p - 1)^{p-1} \equiv -1 \pmod{p}.$$

In 1950, Giuga asked whether the converse is true: If $n > 1$ and n divides $1^{n-1} + 2^{n-1} + \cdots + (n-1)^{n-1} + 1$, then is n a prime number? It is easy to show that n satisfies Giuga's condition if and only if, for every prime p dividing n, $p^2(p - 1)$ divides $n - p$.

Indeed, writing $n = pt$, Giuga's condition becomes:

$$A = 1 + \sum_{j=1}^{pt-1} j^{pt-1} \equiv 0 \pmod{p},$$

while the condition that $p^2(p - 1)$ divides $pt - p$ is equivalent to the conjunction of both conditions: $p|t - 1$ and $p - 1|t - 1$. But $pt - 1 = (p-1)t + (t-1)$; hence, by Fermat's little theorem,

$$A \equiv 1 + \sum_{j=1}^{pt-1} j^{t-1} \equiv 1 + tS \pmod{p},$$

where $S = \sum_{j=1}^{p-1} j^{t-1}$. Hence,

$$A \equiv \begin{cases} 1 - t \pmod{p} & \text{when } p - 1 \mid t - 1, \\ 1 \pmod{p} & \text{when } p - 1 \nmid t - 1. \end{cases}$$

Thus, if $A \equiv 0 \pmod{p}$, then $p - 1|t - 1$ and $p|t - 1$. But, conversely, these latter conditions imply that $p \nmid t$, so n is square-free, and therefore $A \equiv 0 \pmod{p}$.   □

It follows at once that $n \equiv p \equiv 1 \pmod{p - 1}$; so, if $p|n$, then $p - 1|n - 1$.

In Section IX, I shall indicate that this condition is severely restrictive. At any rate, it is now known that if there exists a composite integer n satisfying Giuga's condition, then n must have at least 1700 digits (see Bedocchi, 1985).

Now, I shall discuss another property of prime numbers.

In 1862, Wolstenholme proved the following interesting result: If p is a prime, $p \geqslant 5$, then the numerator of

$$1 + \frac{1}{2} + \frac{1}{3} + \cdots + \frac{1}{p - 1}$$

is divisible by p^2, and the numerator of

$$1 + \frac{1}{2^2} + \frac{1}{3^2} + \cdots + \frac{1}{(1 - p)^2}$$

is divisible by p.

For a proof, see Hardy & Wright (1938, p. 88, General References). Based on this property, it is not difficult to deduce that if n is a prime number, then

$$\binom{2n - 1}{n - 1} \equiv 1 \pmod{n^3}.$$

Is the converse true? This question, still unanswered today, has been asked by Jones for a few years. An affirmative reply would provide an interesting and formally simple characterization of prime numbers.

Still in the same frame of ideas, Mann & Shanks (1972) characterized the prime numbers by the divisibility of certain binomial coefficients: p is a prime number if and only if, for every

$$k = \left[\frac{p+2}{3}\right], \left[\frac{p+2}{3}\right] + 1, ..., \left[\frac{p}{2}\right], \quad k \text{ divides } \binom{k}{p - 2k}.$$

Here the notation $[x]$, for any real number x, indicates the only integer such that $[x] \leqslant x < [x] + 1$; thus $[x]$ is called the largest integer in x.

D. The Power of a Prime Dividing a Factorial

In 1808, Legendre determined the exact power p^m of the prime p that divides a factorial $a!$ (so p^{m+1} does not divide $a!$).

There is a very nice expression of m in terms of the p-adic development of a:

$$a = a_k p^k + a_{k-1} p^{k-1} + \cdots + a_1 p + a_0$$

where $p^k \leq a < p^{k+1}$ and $0 \leq a_i \leq p - 1$ (for $i = 0,1, \ldots, k$). The integers $a_0, a_1, \ldots, a_k$ are the digits of a in base p.

For example, in base 5, $328 = 2 \times 5^3 + 3 \times 5^2 + 3$, so the digits of 328 in base 5 are 2, 3, 0, 3. Using the above notation,

$$m = \sum_{i=0}^{\infty} \left[\frac{a}{p^i} \right] = \frac{a - (a_0 + a_1 + \cdots + a_k)}{p - 1}.$$

Proof. By definition $a! = p^m b$, where $p \nmid b$.

Let $a = q_1 p + r_1$ with $0 \leq q_1$, $0 \leq r_1 < p$; so $q_1 = [a/p]$. The multiples of p, not bigger than a are $p, 2p, \ldots, q_1 p \leq a$. So $p^{q_1}(q_1!) = p^m b'$, where $p \nmid b'$. Thus $q_1 + m_1 = m$, where p^{m_1} is the exact power of p which divides $q_1!$. Since $q_1 < a$, by induction,

$$m_1 = \left[\frac{q_1}{p} \right] + \left[\frac{q_1}{p^2} \right] + \left[\frac{q_1}{p^3} \right] + \cdots .$$

But

$$\left[\frac{q_1}{p^i} \right] = \left[\frac{[a/p]}{p^i} \right] = \left[\frac{a}{p^{i+1}} \right],$$

as may be easily verified. So

$$m = \left[\frac{a}{p} \right] + \left[\frac{a}{p^2} \right] + \left[\frac{a}{p^3} \right] + \cdots .$$

Now, I derive the second expression, involving the p-adic digits of $a = a_k p^k + \cdots + a_1 p + a_0$. Then

$$\left[\frac{a}{p} \right] = a_k p^{k-1} + \cdots + a_1,$$

$$\left[\frac{a}{p^2} \right] = a_k p^{k-2} + \cdots + a_2,$$

$$\cdots$$

$$\left[\frac{a}{p^k}\right] = a_k .$$

So

$$\sum_{i=0}^{\infty} \left[\frac{a}{p^i}\right] = a_1 + a_2(p+1) + a_3(p^2+p+1) + \cdots +$$

$$a_k(p^{k-1} + p^{k-2} + \cdots + p + 1)$$

$$= \frac{1}{p-1}\{ a_1(p-1) + a_2(p^2-1) + \cdots + a_k(p^k - 1)\}$$

$$= \frac{1}{p-1}\{a - (a_0 + a_1 + \cdots + a_k)\}. \quad \square$$

In 1852, Kummer used Legendre's result to determine the exact power p^m of p dividing a binomial coefficient

$$\left[\begin{array}{c} a + b \\ a \end{array}\right] = \frac{(a + b)!}{a!b!},$$

where $a \geqslant 1$, $b \geqslant 1$.

Let

$$a = a_0 + a_1 p + \cdots + a_t p^t,$$
$$b = b_0 + b_1 p + \cdots + b_t p^t,$$

where $0 \leqslant a_i \leqslant p - 1$, $0 \leqslant b_i \leqslant p - 1$, and either $a_t \neq 0$ or $b_t \neq 0$. Let $S_a = \sum_{i=0}^{t} a_i$, $S_b = \sum_{i=0}^{t} b_i$ be the sums of p-adic digits of a, b. Let c_i, $0 \leqslant c_i \leqslant p - 1$, and $\epsilon_i = 0$ or 1, be defined successively as follows:

$$a_0 + b_0 = \epsilon_0 p + c_0 ,$$
$$\epsilon_0 + a_1 + b_1 = \epsilon_1 p + c_1 ,$$
$$\epsilon_1 + a_2 + b_2 = \epsilon_2 p + c_2 ,$$

$$\cdots$$

$$\epsilon_{t-1} + a_t + b_t = \epsilon_t p + c_t .$$

Multiplying these equations successively by 1, p, p^2, ... and adding them:

$$a + b + \epsilon_0 p + \epsilon_1 p^2 + \cdots + \epsilon_{t-1} p^t = \epsilon_0 p + \epsilon_1 p^2 + \cdots$$
$$+ \epsilon_{t-1} p^t + \epsilon_t p^{t+1} + c_0 + c_1 p + \cdots + c_t p^t.$$

So, $a + b = c_0 + c_1 p + \cdots + c_t p^t + \epsilon_t p^{t+1}$, and this is the expression of $a + b$ in the base p. Similarly, by adding those equations:

$$S_a + S_b + (\epsilon_0 + \epsilon_1 + \cdots + \epsilon_{t-1}) = (\epsilon_0 + \epsilon_1 + \cdots + \epsilon_t)p + S_{a+b} - \epsilon_t.$$

By Legendre's result

$$(p - 1)m = (a + b) - S_{a+b} - a + S_a - b + S_b$$
$$= (p - 1)(\epsilon_0 + \epsilon_1 + \cdots + \epsilon_t).$$

Hence, the result of Kummer:

The exact power of p dividing $\binom{a+b}{a}$ is equal to $\epsilon_0 + \epsilon_1 + \cdots + \epsilon_t$, which is the number of "carry-overs" when performing the addition of a, b, written in the base p.

The results of Legendre and Kummer have found many applications, in p-adic analysis, and also, for example, in Chapter 3, Section III.

E. The Chinese Remainder Theorem

Even though my paramount interest is on prime numbers, there is no way to escape also dealing with arbitrary integers — which essentially amounts, in many questions, to the simultaneous consideration of several primes, because of the decomposition of integers into the product of prime powers.

One of the keys connecting results for integers n and for their prime power factors is very old; indeed, it was known to the ancient Chinese and, therefore, it is called the Chinese remainder theorem. I'm sure that every one of my readers knows it already:

If $n_1, n_2, \ldots, n_k$ are pairwise relatively prime positive integers, and if $a_1, a_2, \ldots, a_k$ are any integers, then there exists an integer a such that

$$\begin{cases} a \equiv a_1 \pmod{n_1} \\ a \equiv a_2 \pmod{n_2} \\ \quad \cdots \\ a \equiv a_k \pmod{n_k}. \end{cases}$$

Another integer a' satisfies also the same congruences as a if and only if $a \equiv a'$ (mod $n_1 n_2 \cdots n_k$). *So, there exists a unique integer a, as above, with* $0 \leqslant a < n_1 n_2 \cdots n_k$.

The proof is indeed very simple; it is in many books and also in a short note by Mozzochi (1967).

There are many uses for the Chinese remainder theorem. It was in fact in this way that Chinese generals used to count the number of their soldiers:

> Line up 7 by 7!   (Not factorial of 7, but a SCREAMED military command.)

> Line up 11 by 11!

> Line up 13 by 13!

> Line up 17 by 17!

and they could go in this way, just counting the remainders.

Here is another application: if $n = p_1 p_2 \cdots p_t$ is a product of distinct primes, if g_i is a primitive root modulo p_i, if g is such that $1 \leqslant g \leqslant n - 1$ and $g \equiv g_i$ (mod p_i) for every $i = 1, ..., t$, then g is a common primitive root modulo every p_i.

F. Euler's Function

Euler generalized Fermat's little theorem by introducing the totient or Euler's function.

For every $n \geqslant 1$, let $\phi(n)$ denote the number of integers a, $1 \leqslant a < n$, such that $\gcd(a,n) = 1$. Thus, if $n = p$ is a prime, then $\phi(p) = p - 1$; also

$$\phi(p^k) = p^{k-1}(p - 1) = p^k \left(1 - \frac{1}{p} \right).$$

Moreover, if $m,n \geqslant 1$ and $\gcd(m,n) = 1$, then $\phi(mn) = \phi(m)\phi(n)$, that is, ϕ is a multiplicative function. Hence, for any integer $n = \prod_p p^k$ (product for all primes p dividing n, and $k \geqslant 1$), then

$$\phi(n) = \prod_p p^{k-1}(p - 1) = n \prod_p \left(1 - \frac{1}{p} \right).$$

Another simple property is: $n = \sum_{d|n} \phi(d)$.

Euler proved the following:

If $\gcd(a,n) = 1$, *then* $a^{\phi(n)} \equiv 1 \pmod{n}$.

Proof. Let $r = \phi(n)$ and let $b_1, ..., b_r$ be integers, pairwise incongruent modulo n, such that $\gcd(b_i,n) = 1$ for $i = 1, ..., r$.

Then $ab_1, ..., ab_r$ are again pairwise incongruent modulo n and $\gcd(ab_i,n) = 1$ for $i = 1, ..., r$. Therefore, the sets $\{b_1 \bmod n, ..., b_r \bmod n\}$ and $\{ab_1 \bmod n, ..., ab_r \bmod n\}$ are equal. Now,

$$a^r \prod_{i=1}^{r} b_i \equiv \prod_{i=1}^{r} ab_i \equiv \prod_{i=1}^{r} b_i \pmod{n}.$$

Hence,

$$(a^r - 1) \prod_{i=1}^{r} b_i \equiv 0 \pmod{n} \quad \text{and so} \quad a^r \equiv 1 \pmod{n}. \quad \square$$

Just like Fermat's little theorem, it follows also from Euler's theorem that there exists the smallest positive exponent e such that $a^e \equiv 1 \pmod{n}$. It is called the order of a modulo n. If n is a prime number, this definition coincides with the previous one. Note also that $a^m \equiv 1 \pmod{n}$ if and only if m is a multiple of the order e of $a \bmod n$; thus, in particular, e divides $\phi(n)$.

Once again, it is natural to ask: Given $n > 2$ does there always exist an integer a, relatively prime to n, such that the order of a mod n is equal to $\phi(n)$? Recall that when $n = p$ is a prime, such numbers exist, namely, the primitive roots modulo p. If $n = p^e$, a power of an odd prime, it is also true. More precisely, the following assertions are equivalent:

(i) g is a primitive root modulo p and $g^{p-1} \not\equiv 1 \pmod{p^2}$;

(ii) g is a primitive root modulo p^2;

(iii) for every $e \geqslant 2$, g is a primitive root modulo p^e.

Note that 10 is a primitive root modulo 487, but $10^{486} \equiv 1 \pmod{487^2}$, so 10 is not a primitive root modulo 487^2. This is the smallest example illustrating this possibility, when the base is 10. Another example is 14 modulo 29.

However, if n is divisible by $4p$, or pq, where p, q are distinct odd primes, then there is no number a, relatively prime to n, with order equal to $\phi(n)$. Indeed, it is easy to see that the order of $a \bmod n$ is at most equal to $\lambda(n)$, where $\lambda(n)$ is the following function, defined by Carmichael in 1912:

$\lambda(1) = 1$, $\lambda(2) = 1$, $\lambda(4) = 2$, $\lambda(2^r) = 2^{r-2}$ (for $r \geq 3$), $\lambda(p^r) = p^{r-1}(p - 1) = \phi(p^r)$ for any odd prime p and $r \geq 1$,

$$\lambda(2^r p_1^{r_1} p_2^{r_2} \cdots p_s^{r_s}) = \ell\text{cm}\{\lambda(2^r), \lambda(p_1^{r_1}), ..., \lambda(p_s^{r_s})\}$$

(ℓcm denotes the least common multiple).

Note that $\lambda(n)$ divides $\phi(n)$, but may be smaller, and that there is an integer a, relatively prime to n, with order of $a \bmod n$ equal to $\lambda(n)$.

I shall use this opportunity to study Euler's function in more detail. First I shall consider Lehmer's problem, and thereafter the values of ϕ, the valence, the values avoided, the average of the function, etc.

Recall that if p is a prime, then $\phi(p) = p - 1$. In 1932, Lehmer asked whether there exists any composite integer n such that $\phi(n)$ divides $n - 1$. This question remains open and its solution "seems as remote today as it was when Lehmer raised it a half century ago." (I am quoting from a review by my friend Charlie Small.) If the answer is negative, it will provide a characterization of prime numbers.

What can one say, anyway, when it is not possible to solve the problem? Only that the existence of composite integers n, for which $\phi(n)$ divides $n - 1$, is unlikely, for various reasons:

(a) any such number must be very large (if it exists at all);
(b) any such number must have many prime factors (if it exists at all);
(c) the number of such composite numbers, smaller than any given real number x, is bounded by a very small function of x.

Thus, Lehmer showed in 1932 that if n is composite and $\phi(n)$ divides $n - 1$, then it is odd and square-free, and the number of its distinct prime factors is $\omega(n) \geq 7$. Subsequent work by Schuh (1944) gave $\omega(n) \geq 11$. In 1970, Lieuwens showed: if $3|n$, then $\omega(n) \geq 213$ and $n > 5.5. \times 10^{570}$; if $30 \nmid n$, then $\omega(n) \geq 13$.

To date, the best results are $n > 10^{20}$, $\omega(n) \geq 14$ by Cohen & Hagis

(1980) and if $30 \nmid n$, then $\omega(n) \geqslant 26$ by Wall (1980), while if $3|n$, then Lieuwens' result is still the best.

Moreover, in 1977 Pomerance showed that for every sufficiently large positive real number x, the number $L(x)$ of composite n such that $\phi(n)|n - 1$ and $n \leqslant x$, satisfies

$$L(x) \leqslant x^{1/2} (\log x)^{3/4}.$$

Moreover, if $\omega(n) = k$, then $n < k^{2^k}$.

After Lehmer's problem, I shall discuss the values of Euler's function. Not every even integer $n \geqslant 1$ is a value of Euler's function – a fact which is not difficult to establish. For example, Schinzel showed in 1956 that, for every $k \geqslant 1, 2 \times 7^k$ is not a value of Euler's function. In 1976, Mendelsohn showed that there exist infinitely many primes p such that, for every $k \geqslant 1, 2^k p$ is not a value of the function ϕ. Moreover, for every $k \geqslant 1$ there exists n such that $\phi(n) = k!$; this was proved by Gupta in 1950.

Aigner showed in 1977 that if $k > 1$ is a square-free natural number (i.e., no square greater than 1 divides k), then there exist infinitely many integers n such that $\phi(n) = kh^2$ (with $h \geqslant 1$), and also infinitely many integers n such that $\phi(n^2) = kg^2$ (with $g \geqslant 1$).

The next results tell how erratic is the behaviour of Euler's function.

Thus, in 1950, Somayajulu showed that

$$\lim_{n \to \infty} \sup \frac{\phi(n + 1)}{\phi(n)} = \infty \quad \text{and} \quad \lim_{n \to \infty} \inf \frac{\phi(n + 1)}{\phi(n)} = 0.$$

This result was improved by Sierpinski and Schinzel (see Schinzel, 1954):

The set of all numbers $\phi(n+1)/\phi(n)$ is dense in the set of all real positive numbers.

Schinzel & Sierpiński (1954) and Schinzel (1954) also proved the following: For every $m, k \geqslant 1$, there exist $n, h \geqslant 1$ such that

$$\frac{\phi(n + i)}{\phi(n + i - 1)} > m \quad \text{and} \quad \frac{\phi(h + i - 1)}{\phi(h + i)} > m$$

for $i = 1, 2, ..., k$. It is also true that the set of all numbers $\phi(n)/n$ is dense in the interval $(0,1)$.

Now I shall examine the "valence" of Euler's function; in other words, how often a value $\phi(n)$ is assumed. In order to explain the results in a systematic way, it is better to introduce some notation. If $m \geqslant 1$, let

$$N_\phi(m) = \#\{n \geqslant 1 \mid \phi(n) = m\}.$$

What are the possible values of $N_\phi(m)$? I have already said that there are infinitely many even integers n for which $N_\phi(m) = 0$. It is also true that if $m = 2 \times 3^{6k+1}$ $(k \geqslant 1)$, then $\phi(n) = m$ exactly when $n = 3^{6k+2}$ or $n = 2 \times 3^{6k+2}$. Hence, there are infinitely many integers m such that $N_\phi(m) = 2$.

It is not difficult to show that $N_\phi(m) \neq \infty$ for every $m \geqslant 1$.

Schinzel gave a simpler proof (in 1956 and in 1958) of the following result of Pillai (1929):

$$\sup\{N_\phi(m)\} = \infty \, .$$

In other words, for every $k \geqslant 1$ there exists an integer m_k such that there exists at least k integers n with $\phi(n) = m_k$.

The conjecture that dominates the study of the valence of ϕ was proposed by Carmichael in 1922: N_ϕ does not assume the value 1. In other words, given $n \geqslant 1$, there exists $n' \geqslant 1$, $n' \neq n$, such that $\phi(n') = \phi(n)$.

A concise update on Carmichael's conjecture, written by Wagon, has just appeared in *The Mathematical Intelligencer*, 1986.

This conjecture was studied by Klee, who showed in 1947 that it holds for every integer n such that $\phi(n) < 10^{400}$. Klee's method was improved by Masai & Valette (1982), and now the statement is known to be true for all n such that $\phi(n) < 10^{10000}$.

Pomerance has shown (1974) the following: Suppose that m is a natural number such that if p is any prime and $p - 1$ divides $\phi(m)$, then p^2 divides m. Then $N_\phi(\phi(m)) = 1$.

Of course, if there exists a number m satisfying the above condition, then Carmichael's conjecture would be false. However, the existence of such a number m is far from established, and perhaps unlikely.

Finally, at variance with Carmichael's conjecture, it is reasonable to expect that every $s \neq 1$ is a value of N_ϕ (this was conjectured by Sierpiński). As a matter of fact, this will be shown in Chapter 6, Section II, under the assumption of an unproved hypothesis.

And how about the valence of the valence function N_ϕ? I have already said that there exist infinitely many m that are not values of ϕ, for which $N_\phi(m) = 0$. So N_ϕ assumes the value 0 infinitely often.

This was generalized by Erdös in 1958: If $s \geq 1$ is a value of N_ϕ, then it is assumed infinitely often. (Try to phrase this statement directly using Euler's function, to see whether you are understanding my notation.)

In Chapter 4, Section I, I shall return to Euler's function to study its properties of distribution.

The studies on the function ϕ have not stopped at this point! For example, another question investigated is the following: How different may be n, m when $\phi(n) = \phi(m)$?

To explain the results clearly, I introduce the following notation. For every $k \geq 1$ let $E_\phi(k) = \#\{(n,m) \mid n > m \geq 1, n - m = k, \phi(n) = \phi(m)\}$. For example, since $\phi(3) = \phi(4) = \phi(6) = 2$, then $E_\phi(1) \geq 1$, $E_\phi(2) \geq 1$; also $\phi(5) = \phi(8) = 4$, so $E_\phi(3) \geq 1$. What is the range of E_ϕ?

Sierpiński showed in 1956 that for every $k \geq 1$ there exists $n_k \geq 1$ such that $\phi(n_k + k) = \phi(n_k)$; in other words, 0 is not in the range of E_ϕ. On the other hand, Schinzel proved in 1958 that $\sup\{E_\phi(k)\} = \infty$; that is, for every $m \geq 1$ there exists $k \geq 1$ such that there exists at least m integers $n_1, \ldots, n_m$ such that $\phi(n_i + k) = \phi(n_i)$ $(i = 1, \ldots, m)$.

I could not locate in the literature more results about the exact range of E_ϕ. Anyway, the set $\{k \mid E_\phi(k) \geq 2\}$ is quite large, as shown by Schinzel & Wakulicz in 1959: it contains all integers $k \leq 2 \times 10^{58}$, and also, for every integer k there exists $a \geq 1$ such that $E(k^a) \geq 2$ (shown by Schinzel in 1958).

As an aside, here is a curious result that was conjectured by Recaman Santos (1976) and proved in 1978 by Hausman & Shapiro: $n = 12$ is the largest integer such that $n + a$ is prime, for every a, $1 \leq a < n$, $\gcd(a,n) = 1$.

I haven't yet considered the growth of the function ϕ. Since $\phi(p) = p - 1$ for every prime p, then $\limsup \phi(n) = \infty$. Similarly, from

$\phi(p) = p - 1$, lim sup $\phi(n)/n = 1$.

I shall postpone the indication of other results about the growth of ϕ until Chapter 4: they depend on methods that will be discussed in that chapter.

G. Sequences of Binomials

The preceding considerations referred to congruences modulo a given integer $n > 1$, and a was any positive integer relatively prime to n.

Another point of view is very illuminating. This time, let $a > 1$ be given, and consider the sequence of integers $a^n - 1$ (for $n \geqslant 1$), as well as the companion sequence of integers $a^n + 1$ (for $n \geqslant 1$). More generally, if $a > b \geqslant 1$ with $\gcd(a,b) = 1$, one may consider the sequences $a^n - b^n$ ($n \geqslant 1$) and $a^n + b^n$ ($n \geqslant 1$).

A first natural question, with an immediate answer, is the following: to determine all primes p, such that there exists $n \geqslant 1$ for which p divides $a^n - b^n$ (respectively, $a^n + b^n$). These are the primes p not dividing ab because a, b are relatively prime, and if $p \mid ab$, $bb' \equiv 1 \pmod{p}$ and n is the order of ab' mod p (resp. of $-ab'$ mod p), then p divides $a^n - b^n$ (resp. $a^n + b^n$).

If $n \geqslant 1$ is the smallest integer such that p divides $a^n - b^n$ (resp. $a^n + b^n$), then p is called a primitive prime factor of the sequence of binomials in question. In this case, by Fermat's little theorem, n divides $p - 1$; this was explicitly observed by Legendre.

So, every prime $p \nmid ab$ appears as a primitive factor of some binomial. Does, conversely, every binomial have a primitive factor?

In 1892, Zsigmondy proved the following theorem, which is very interesting and has many applications:

If $a > b \geqslant 1$ and $\gcd(a,b) = 1$, then every number $a^n - b^n$ has a primitive prime factor — the only exception being $2^6 - 1 = 63$.

Equally, if $a > b \geqslant 1$, then every number $a^n + b^n$ has a primitive prime factor — with the exception of $2^3 + 1 = 9$.

The special case, where $b = 1$, had been proved by Bang in 1886.

Later, this result was proved again, unknowingly, by a long list of mathematicians: Birkhoff & Vandiver (1904), Carmichael (1913), Kanold (1950), Artin (1955), Lüneburg (1981), and probably others.

The proof is definitely not so obvious; however, it is very easy to write up such sequences and watch the successive appearance of new primitive prime factors.

It is interesting to consider the primitive part t_n^* of $a^n - b^n$; namely, write $a^n - b^n = t_n^* t_n'$ with $\gcd(t_n^*, t_n') = 1$ and a prime p divides t_n^* if and only if p is a primitive factor of $a^n - b^n$.

By experimenting numerically with sequences $a^n - b^n$, it is observed that, apart from a few initial terms, t_n^* is composite.

In fact, Schinzel indicated in 1962 the theorem below.

Let $k(m)$ denote the square-free kernel of m, that is, m divided by its largest square factor. Let

$$
e = \begin{cases} 1 & \text{if } k(ab) \equiv 1 \pmod 4, \\ 2 & \text{if } k(ab) \equiv 2 \text{ or } 3 \pmod 4. \end{cases}
$$

If $n/ek(ab)$ is integral and odd, then $a^n - b^n$ has at least two primitive prime factors, with some exceptions (of which the largest possible is $n = 20$). In particular, if $b = 1$ the exceptions are

if $a = 2$: $n = 4, 12, 20$;
if $a = 3$: $n = 3, 6, 12$;
if $a = 4$: $n = 3, 6$.

Therefore, there are infinitely many n such that the primitive part of $a^n - b^n$ is composite.

Schinzel also proved that if $ab = c^h$ with $h \geqslant 3$, or $h = 2$ and $k(c)$ odd, then there are infinitely many n such that the primitive part of $a^n - b^n$ has at least three prime factors.

For the sequence of binomials $a^n + b^n$, it follows at once: if $k(ab)$ is even, $n/k(ab)$ is odd, and $n > 10$, then the primitive part of $a^n + b^n$ is composite. Just note that each primitive prime factor of $a^{2n} - b^{2n}$ is also a primitive prime factor of $a^n + b^n$.

Here are some questions that are very difficult to answer:

Are there infinitely many n such that the primitive part of $a^n - b^n$ is prime?

Are there infinitely many n such that the primitive part of $a^n - b^n$ is square-free?

And how about the seemingly easier question:

Are there infinitely many n such that the primitive part t_n^* of $a^n - b^n$ has a prime factor p such that p_n^2 does not divide $a^n - b^n$?

Are there infinitely many n such that t_n^* has a square-free kernel $k(t_n^*) \neq 1$?

In Chapter 5, Section III, I shall discuss this question for the special case when $b = 1$. I will show how it is ultimately related, in a very surprising way, to Fermat's last theorem!

It is also an interesting problem to estimate the size of the largest prime factor of $a^n - b^n$, where $a > b \geqslant 1$. The following notation will be used: $P(m)$ = largest prime factor of m.

It is not difficult to show, using Zsigmondy's theorem, that $P(a^n - b^n) \geqslant n + 1$ when $n > 2$.

In 1962, Schinzel showed that $P(a^n - b^n) \geqslant 2n + 1$ in the following cases, with $n > 2$:

$4 \nmid n$, with exclusion $a = 2$, $b = 1$, $n = 6$;

$k(ab)|n$ or $k(ab) = 2$, with exclusions $a = 2$, $b = 1$, $n = 4, 6$, or 12.

Erdös conjectured in 1965 that $\lim_{n \to \infty} P(2^n - 1)/n = \infty$. Despite very interesting work, this conjecture has not yet been settled completely; but there are very good partial results, which I report now.

In 1975, using Baker's inequalities for linear forms of logarithms, Stewart showed the following. Let $0 < \kappa < 1/\log 2$, and let S_κ be the set of integers n having at most $\kappa \log \log n$ distinct prime factors (the set S_κ has density 1); then

$$\lim_{\substack{n \to \infty \\ n \in S_\kappa}} \frac{P(a^n - b^n)}{n} = \infty.$$

How fast does the expression increase? This was answered by Stewart in 1977, with sharper inequalities of Baker's type:

$$\frac{P(a^n - b^n)}{n} > C \frac{(\log n)^\lambda}{\log \log \log n},$$

$\lambda = 1 - \kappa \log 2$, $C > 0$ constant.

Stewart showed also that, for every sufficiently large prime p, $P(a^p - b^p)/p > C \log p$ (with $C > 0$). The special case of Mersenne numbers $2^p - 1$ had been established in 1976 by Erdös & Shorey.

There is also a close connection between the numbers $a^n - 1$, the values of the cyclotomic polynomials, and primes in certain arithmetic progressions. But I cannot explain everything at the same time — so be patient and wait until I reconsider this matter again in Chapter 4, Section IV.

H. Quadratic Residues

In the study of quadratic diophantine equations, developed by Fermat, Euler, Legendre, Gauss, it was very important to determine when an integer a is a square modulo a prime $p > 2$.

If $p > 2$ does not divide a and if there exists an integer b such that $a \equiv b^2 \pmod{p}$, then a is called a quadratic residue modulo p; otherwise, it is a nonquadratic residue modulo p.

Legendre introduced the following practical notation:

$$\left(\frac{a}{p}\right) = (a/p) = \begin{cases} +1 & \text{if } a \text{ is a quadratic residue modulo } p, \\ -1 & \text{otherwise.} \end{cases}$$

It is also convenient to define $(a/p) = 0$ when p divides a.

I shall now indicate the most important properties of the Legendre symbol. References are plentiful — practically every book in elementary number theory.

If $a \equiv a' \pmod{p}$, then

$$\left(\frac{a}{p}\right) = \left(\frac{a'}{p}\right).$$

For any integers a, a':

$$\left(\frac{aa'}{p}\right) = \left(\frac{a}{p}\right)\left(\frac{a'}{p}\right).$$

So, for the computation of the Legendre symbol, it suffices to calculate (q/p), where $q = -1$, 2 or any odd prime different from p.

Euler proved the following congruence:

$$\left[\frac{a}{p}\right] \equiv a^{(p-1)/2} \pmod{p}.$$

In particular,

$$\left[\frac{-1}{p}\right] = \begin{cases} +1 & \text{when } p \equiv 1 \pmod{4}, \\ -1 & \text{when } p \equiv -1 \pmod{4}, \end{cases}$$

and

$$\left[\frac{2}{p}\right] = \begin{cases} +1 & \text{when } p \equiv \pm 1 \pmod{8}, \\ -1 & \text{when } p \equiv \pm 3 \pmod{8}. \end{cases}$$

The computation of the Legendre symbol (q/p), for any odd prime $q \neq p$, can be performed with an easy, explicit and fast algorithm (needing only Euclidean division), by using Gauss' reciprocity law:

$$\left[\frac{p}{q}\right] = \left[\frac{q}{p}\right](-1)^{\frac{p-1}{2} \times \frac{q-1}{2}}.$$

The importance of Legendre's symbol was such that it prompted Jacobi to consider the following generalization, now called Jacobi symbol. Again, references are abundant, for example, Grosswald's book (1966, second edition 1984), or, why not?, my own book (1972).

Let a be a nonzero integer, let b be an odd integer, such that $\gcd(a,b) = 1$. The Jacobi symbol (a/b) is defined as an extension of Legendre's symbol, in the following manner. Let $|b| = \prod_{p|b} p^{e_p}$ (with $e_p \geq 1$). Then

$$\left[\frac{a}{b}\right] = \left[\frac{a}{-b}\right] = \prod_{p|b}\left[\frac{a}{p}\right]^{e_p}.$$

Therefore, (a/b) is equal to $+1$ or -1. Note that

$$\left[\frac{a}{1}\right] = \left[\frac{a}{-1}\right] = +1.$$

Here are some of the properties of the Jacobi symbol (under the assumptions of its definition):

$$\left[\frac{aa'}{b}\right] = \left[\frac{a}{b}\right]\left[\frac{a'}{b}\right],$$

$$\left[\frac{a}{bb'}\right] = \left[\frac{a}{b}\right]\left[\frac{a}{b'}\right].$$

$$\left[\frac{-1}{b}\right] = (-1)^{(b-1)/2} = \begin{cases} +1 & \text{if } b \equiv 1 \pmod{4}, \\ -1 & \text{if } b \equiv -1 \pmod{4}, \end{cases}$$

$$\left[\frac{2}{b}\right] = (-1)^{(b^2-1)/8} = \begin{cases} +1 & \text{if } b \equiv \pm 1 \pmod{8}, \\ -1 & \text{if } b \equiv \pm 3 \pmod{8}. \end{cases}$$

For the calculation of the Jacobi symbol, the key result is the reciprocity law, which follows easily from Gauss' reciprocity law for the Legendre symbol:

If a, b are relatively prime odd integers and $b \geqslant 3$, then

$$\left[\frac{a}{b}\right] = (-1)^{\frac{a-1}{2} \times \frac{b-1}{2}} \left[\frac{b}{a}\right].$$

If $b \geqslant 3$ and a is a square modulo b, then

$$\left[\frac{a}{b}\right] = +1.$$

III. Classical Primality Tests Based on Congruences

After the discussion of the theorems of Fermat, Wilson, and Euler, I am ready. For me, the classical primality tests based on congruences are those indicated by Lehmer, extending or using previous tests by Lucas, Pocklington, and Proth. Yes, Professor Lehmer, even though your tests are not so old, they have become classical and are quite appropriate for numbers of special forms, as I will soon indicate. I reserve another section for your classical tests based on recurring sequences.

Wilson's theorem, which characterizes prime numbers, might seem very promising. But, it has to be discarded as a practical test, since the computation of factorials is very time consuming.

Fermat's little theorem says that if p is a prime and a is any natural number not a multiple of p, then $a^{p-1} \equiv 1 \pmod{p}$. However, I note right away that a crude converse of this theorem is not true — because there exist composite integers N, and $a \geqslant 2$, such that $a^{N-1} \equiv 1 \pmod{N}$. I shall devote Section VIII to the study of these numbers, which are very important in primality questions.

Nevertheless, a useful and true converse of Fermat's little theorem was discovered by Lucas in 1876. It says:

Test 1. Let $N > 1$. Assume that there exists an integer $a > 1$ such that:

(i) $a^{N-1} \equiv 1 \pmod{N}$,
(ii) $a^m \not\equiv 1 \pmod{N}$, for $m = 1, 2, ..., N - 2$.

Then N is a prime.

Defect of this test: it might seem perfect, but it requires $N - 2$ successive multiplications by a, and finding residues modulo N — too many operations.

Proof. It suffices to show that every integer m, $1 \leqslant m < N$, is prime to N, that is, $\phi(N) = N - 1$. For this purpose, it suffices to show that there exists a, $1 \leqslant a < N$, $\gcd(a,N) = 1$, such that the order of a mod N is $N - 1$. This is exactly spelled out in the hypothesis. $\square$

In 1891, Lucas gave the following test:

Test 2. Let $N > 1$. Assume that there exists an integer $a > 1$ such that:

(i) $a^{N-1} \equiv 1 \pmod{N}$,
(ii) $a^m \not\equiv 1 \pmod{N}$ for every $m < N$, such that m divides $N - 1$.

Then N is a prime.

Defect of this test: it requires the knowledge of all factors of $N - 1$, thus it is only easily applicable when N has a special form, like $N = 2^n + 1$, or $N = 3 \times 2^n + 1$.

The proof of Test 2 is, of course, the same as that of Test 1.

In 1927, Lehmer made Lucas' test more practical; this was made even more flexible by Brillhart, Lehmer & Selfridge, in 1975:

Test 3. Let $N > 1$. Assume that for every prime factor q of $N - 1$ there exists an integer $a = a(q) > 1$ such that

(i) $a^{N-1} \equiv 1 \pmod{N}$,

(ii) $a^{(N-1)/q} \not\equiv 1 \pmod{N}$.

Then N is a prime.

Defect of this test: once again, it is necessary to know the prime factors of $N - 1$, but fewer congruences have to be satisfied.

If the reader is observant, he or she should now note that, after all, to verify that $a^{N-1} \equiv 1 \pmod{N}$ it is necessary in particular to calculate, as one goes, the residue of a^n modulo N (for every $n \leqslant N - 1$), and so the first Lucas' criterion could have been used. The point is that there is a fast algorithm to find the power a^n, hence also $a^n \bmod N$, without computing all the preceding powers. It runs as follows.

Write the exponent n in base 2:

$$n = n_0 2^k + n_1 2^{k-1} + \cdots + n_{k-1} 2 + n_k,$$

where each n_i is equal to 0 or 1, and $n_0 = 1$.

Put $s_0 = n_0 = 1$ and if s_j has been calculated, let $s_{j+1} = 2s_j + n_{j+1}$.

Let $r_j = a^{s_j}$. Then $r_{j+1} = r_j^2 a^{n_{j+1}}$; so $r_{j+1} = r_j^2$ when $n_{j+1} = 0$, or $r_{j+1} = a r_j^2$ when $n_{j+1} = 1$. Noting that $r_k = a^n$, then it is only necessary to perform $2k$ operations, which are either a squaring or a multiplication by a. If the computation is of $a^n \bmod N$, then it is even easier; at each stage r_j is to be replaced by its residue modulo N. Now, k is equal to

$$\left[\frac{\log n}{\log 2} \right].$$

So, if $n = N - 1$, then only about

$$2\left[\frac{\log n}{\log 2} \right]$$

operations are needed to find $a^{N-1} \bmod N$, and there is no requirement of computing all powers $a^n \bmod N$.

Why don't you try calculating $2^{1092} \bmod 1093^2$ in this way? You should find $2^{1092} \equiv 1 \pmod{1093^2}$ – if you really succeed! This has nothing to do directly with primality – but it will appear much later, in Chapter 5.

I return to Lehmer's Test 3 and give its proof.

Proof of Test 3. It is enough to show that $\phi(N) = N - 1$, and since $\phi(N) \leqslant N - 1$, it suffices to show that $N - 1$ divides $\phi(N)$. If this is false, there exists a prime q and $r \geqslant 1$ such that q^r divides $N - 1$, but q^r does not divide $\phi(N)$. Let $a = a(q)$ and let e be the order of a mod N. Thus e divides $N - 1$ and e does not divide $(N-1)/q$, so q^r divides e. Since $a^{\phi(N)} \equiv 1 \pmod{N}$, then e divides $\phi(N)$, so $q^r \mid \phi(N)$, which is a contradiction, and concludes the proof.   □

In the section on Fermat numbers, I will derive Pepin's primality test for Fermat numbers, as a consequence of Lehmer's Test 3.

To make the primality tests more efficient, it is desirable to avoid the need to find all prime factors of $N - 1$. So there are tests that only require a partial factorization of $N - 1$. The basic result was proved by Pocklington in 1916, and it is indeed very simple:

Let $N - 1 = q^n R$, where q is a prime, $n \geqslant 1$, and q does not divide R. Assume that there exists an integer $a > 1$ such that:

(i) $a^{N-1} \equiv 1 \pmod{N}$,
(ii) $\gcd(a^{(N-1)/q} - 1, N) = 1$.

Then each prime factor of N is of the form $mq^n + 1$, with $m \geqslant 1$.

Proof. Let p be a prime factor of N, and let e be the order of a mod p, so e divides $p - 1$; by condition (ii), e cannot divide $(N-1)/q$, because p divides N; hence, q does not divide $(N-1)/e$; so q^n divides e, and a fortiori, q^n divides $p - 1$.   □

The above statement looks more like a result on factors than a primality test. However, if it may be verified that each prime factor $p = mq^n + 1$ is greater than $\sqrt{N}$, then N is a prime. When q^n is fairly large, this verification is not too time consuming.

In 1928, Lehmer refined the result of Pocklington by using the Chinese remainder theorem:

Let $N - 1 = FR$, where $\gcd(F,R) = 1$ and the factorization of F is known. Assume that for every prime q dividing F there exists an integer $a = a(q) > 1$ such that

(i) $a^{N-1} \equiv 1 \pmod{N}$,

(ii) $\gcd(a^{(N-1)/q} - 1, N) = 1$.

Then each prime factor of N is of the form $mF + 1$, with $m \geqslant 1$.

The same comments to Pocklington's criterion apply here. So, if $F > \sqrt{N}$, then N is a prime.

This result is very apt to prove the primality of numbers of special form. The old criterion of Proth (1878) is easily deduced:

Test 4. Let $N = 2^n h + 1$ with h odd and $2^n > h$. Assume that there exists an integer $a > 1$ such that $a^{(N-1)/2} \equiv -1 \pmod{N}$. Then N is prime.

Proof. $N - 1 = 2^n h$, with $\gcd(2^n,h) = 1$ and $a^{N-1} \equiv 1 \pmod{N}$. Since N is odd, then $\gcd(a^{(N-1)/2} - 1, N) = 1$. By Lehmer's result above, each prime factor p of N is of the form $p = 2^n m + 1 > 2^n$. But $N = 2^n h + 1 < 2^{2n}$, hence $\sqrt{N} < 2^n < p$ and so N is prime. $\square$

In the following test (using the same notation) it is required to know that R (the nonfactored part of $N - 1$) has no prime factor less than a given bound B. Precisely:

Test 5. Let $N - 1 = FR$, where $\gcd(F,R) = 1$, the factorization of F is known, B is such that $FB > \sqrt{N}$, and R has no prime factors less than B. Assume:

(i)   For each prime q dividing F there exists an integer $a = a(q) > 1$ such that $a^{N-1} \equiv 1 \pmod{N}$ and $\gcd(a^{(N-1)/q} - 1, N) = 1$.

(ii) There exists an integer $b > 1$ such that $b^{N-1} \equiv 1 \pmod{N}$ and $\gcd(b^F - 1, N) = 1$.

Then N is a prime.

Proof. Let p be any prime factor of N, let e be the order of b modulo N, so e divides $p - 1$ and also e divides $N - 1 = FR$. Since e does not divide F, then $\gcd(e,R) \neq 1$, so there exists a prime q such that $q|e$, $q|R$; hence, $q|p - 1$. But, by the previous result of Lehmer, F divides $p - 1$; since $\gcd(F,R) = 1$, then qF divides $p - 1$. So $p - 1 \geqslant qF \geqslant BF > \sqrt{N}$. This implies that $p = N$, so N is a prime.   □

The paper of Brillhart, Lehmer & Selfridge (1975) contains other variants of these tests, which have been put to good use to determine the primality of numbers of the form $2^r + 1$, $2^{2r} \pm 2^r + 1$, $2^{2r-1} \pm 2^r + 1$.

I have already said enough and will make only one comment: these tests require prime factors of $N - 1$. Later, using linear recurring sequences, other tests will be presented, requiring prime factors of $N + 1$.

IV. Lucas Sequences

Let P, Q be nonzero integers.

Consider the polynomial $X^2 - PX + Q$; its discriminant is $D = P^2 - 4Q$ and the roots are

$$\left.\begin{array}{c} \alpha \\ \beta \end{array}\right\} = \frac{P \pm \sqrt{D}}{2} .$$

So

$$\left\{\begin{array}{c} \alpha + \beta = P , \\ \alpha\beta = Q , \\ \alpha - \beta = \sqrt{D} . \end{array}\right.$$

I shall assume that $D \neq 0$. Note that $D \equiv 0 \pmod 4$ or $D \equiv 1 \pmod 4$. Define the sequences of numbers

$$U_n(P,Q) = \frac{\alpha^n - \beta^n}{\alpha - \beta} \quad \text{and} \quad V_n(P,Q) = \alpha^n + \beta^n, \quad \text{for } n \geqslant 0.$$

In particular, $U_0(P,Q) = 0$, $U_1(P,Q) = 1$, while $V_0(P,Q) = 2$, $V_1(P,Q) = P$.

The sequences $U(P,Q) = (U_n(P,Q))_{n \geqslant 1}$ and $V(P,Q) = (V_n(P,Q))_{n \geqslant 1}$ are called the Lucas sequences associated to the pair (P,Q). Special cases had been considered by Fibonacci, Fermat, and Pell, among others. Many particular facts were known about these sequences; however, the general theory was first developed by Lucas in a seminal paper, which appeared in Volume I of the *American Journal of Mathematics*, 1878. It is a long memoir with a rich content, relating Lucas sequences to many interesting topics, like trigonometric functions, continued fractions, the number of divisions in the algorithm of the greatest common divisor, and, also, primality tests. It is for this latter reason that I discuss Lucas sequences. If you are curious about the other connections that I have mentioned, look at the references at the end of the book and/or consult the paper in the library.

I should, however, warn that despite the importance of the paper, the methods employed are often indirect and cumbersome, so it is advisable to read Carmichael's long article of 1913, where he corrected errors and generalized results.

The first thing to note is that, for every $n \geqslant 2$,

$$U_n(P,Q) = PU_{n-1}(P,Q) - QU_{n-2}(P,Q),$$

$$V_n(P,Q) = PV_{n-1}(P,Q) - QV_{n-2}(P,Q)$$

(just check it). So, these sequences deserve to be called linear recurring sequences of order 2 (each term depends linearly on the two preceding terms). Conversely, if P, Q are as indicated, and $D = P^2 - 4Q \neq 0$, if $W_0 = 0$ (resp. 2), $W_1 = 1$ (resp. P), if $W_n = PW_{n-1} - QW_{n-2}$ for $n \geqslant 2$, then Binet showed (in 1843) that

$$W_n = \frac{\alpha^n - \beta^n}{\alpha - \beta} \quad (\text{resp.,} \quad W_n = \alpha^n + \beta^n) \quad \text{for } n \geqslant 0;$$

here α, β are the roots of the polynomial $X^2 - PX + Q$. This is trivial, because the sequences of numbers

$$(W_n)_{n \geqslant 0} \quad \text{and} \quad \left[\frac{\alpha^n - \beta^n}{\alpha - \beta} \right]_{n \geqslant 0} \quad (\text{resp.,} \ (\alpha^n + \beta^n)_{n \geqslant 0}),$$

have the first two terms equal and both have the same linear second-order recurrence definition.

Before I continue, here are the main special cases that had been considered before the full theory was developed.

The sequence corresponding to $P = 1$, $Q = -1$, $U_0 = U_0(1,-1) = 0$, and $U_1 = U_1(1,-1) = 1$ was first considered by Fibonacci, and it begins as follows:

$$0 \quad 1 \quad 1 \quad 2 \quad 3 \quad 5 \quad 8 \quad 13 \quad 21 \quad 34 \quad 55 \quad 89 \quad 144 \quad 233$$
$$377 \quad 610 \quad 987 \quad 1597 \quad 2584 \quad 4181 \quad 6765 \quad \dots .$$

Remember that I have mentioned the Fibonacci numbers in Chapter 1, as examples of sequences $(S_n)_{n \geqslant 0}$ with the property that $S_1 = 1$, and if $n,m \geqslant 1$, $\gcd(n,m) = 1$, then $\gcd(S_n, S_m) = 1$.

The companion sequence of Fibonacci numbers, still with $P = 1$, $Q = -1$, is the sequence of Lucas numbers: $V_0 = V_0(1,-1) = 2$, $V_1 = V_1(1,-1) = 1$, and it begins as follows:

$$2 \quad 1 \quad 3 \quad 4 \quad 7 \quad 11 \quad 18 \quad 29 \quad 47 \quad 76 \quad 123 \quad 199 \quad 322$$
$$521 \quad 843 \quad 1364 \quad 2207 \quad 3571 \quad 5778 \quad 9349 \quad 15127 \quad \dots .$$

If $P = 3$, $Q = 2$, then the sequences obtained are

$$U_n(3,2) = 2^n - 1 \quad \text{and} \quad V_n(3,2) = 2^n + 1, \quad \text{for} \quad n \geqslant 0.$$

These sequences were the reason of many sleepless nights for Fermat (see details in Sections VI and VII). The sequences associated to $P = 2$, $Q = -1$, are called the Pell sequences; they begin as follows:

$$U_n(2,-1): \quad 0 \quad 1 \quad 2 \quad 5 \quad 12 \quad 29 \quad 70 \quad 169 \quad 408$$
$$985 \quad 2378 \quad 5741 \quad 17223 \quad \dots ,$$
$$V_n(2,-1): \quad 2 \quad 2 \quad 6 \quad 14 \quad 34 \quad 82 \quad 198 \quad 478 \quad 1154$$
$$2786 \quad 6726 \quad 16238 \quad 39202 \quad \dots .$$

Lucas noted a great similarity between the sequences of numbers $U_n(P,Q)$ (resp. $V_n(P,Q)$) and $(a^n-b^n)/(a-b)$ (resp. $a^n + b^n$), where a, b are given, $a > b \geqslant 1$, $\gcd(a,b) = 1$ and $n \geqslant 0$. No wonder, one is a special case of the other. Just observe that for the pair $(a+b, ab)$, $D = (a - b)^2 \neq 0$, $\alpha = a$, $\beta = b$, so

$$U_n(a+b, ab) = \frac{a^n - b^n}{a - b}, \quad V_n(a+b, ab) = a^n + b^n.$$

It is clearly desirable to extend the main results about the sequence of numbers $(a^n-b^n)/(a-b)$, $a^n + b^n$ (in what relates to divisibility and primality) for the wider class of Lucas sequences.

I shall therefore present the generalizations of Fermat's little theorem, Euler's theorem, etc., to Lucas sequences. There is no essential difficulty, but the development requires a surprising number of steps — true enough, all at an elementary level. In what follows, I shall record, one after the other, the facts needed to prove the main results. If you wish, work out the details. But I'm also explicitly giving the beginning of several Lucas sequences, so you may be happy just to check my statements numerically (see tables at the end of the section).

First, the algebraic facts, then the divisibility facts. To simplify the notations, I write only $U_n = U_n(P,Q)$, $V_n = V_n(P,Q)$.

I repeat:

(IV.1) $U_n = PU_{n-1} - QU_{n-2}$ $(n \geq 2)$, $U_0 = 0$, $U_1 = 1$,
$V_n = PV_{n-1} - QV_{n-2}$ $(n \geq 2)$, $V_0 = 2$, $V_1 = P$.

(IV.2) $U_{2n} = U_n V_n$,
$V_{2n} = V_n^2 - 2Q^n$.

(IV.3) $U_{m+n} = U_m V_n - Q^n U_{m-n}$,
$V_{m+n} = V_m V_n - Q^n V_{m-n}$ (for $m \geq n$).

(IV.4) $U_{m+n} = U_m U_{n+1} - QU_{m-1}U_n$,
$2V_{m+n} = V_m V_n + DU_m U_n$.

(IV.5) $DU_n = 2V_{n+1} - PV_n$,
$V_n = 2U_{n+1} - PU_n$.

(IV.6) $U_n^2 = U_{n-1}U_{n+1} + Q^{n-1}$,
$V_n^2 = DU_n^2 + 4Q^n$.

(IV.7) $U_m V_n - U_n V_m = 2Q^n U_{m-n}$ (for $m \geq n$),
$U_m V_n + U_n V_m = 2U_{m+n}$.

(IV.8) $2^{n-1}U_n = \begin{pmatrix} n \\ 1 \end{pmatrix} P^{n-1} + \begin{pmatrix} n \\ 3 \end{pmatrix} P^{n-3}D + \begin{pmatrix} n \\ 5 \end{pmatrix} P^{n-5}D^2 + \cdots$,

$2^{n-1}V_n = P^n + \begin{pmatrix} n \\ 2 \end{pmatrix} P^{n-2}D + \begin{pmatrix} n \\ 4 \end{pmatrix} P^{n-4}D^2 + \cdots$.

(IV.9)   If m is odd and $k \geq 1$, then

$$D^{(m-1)/2}U_k^m = U_{km} - \begin{pmatrix} m \\ 1 \end{pmatrix} Q^k U_{k(m-2)} + \begin{pmatrix} m \\ 2 \end{pmatrix} Q^{2k} U_{k(m-4)}$$

$$- \cdots \pm \begin{pmatrix} m \\ (m-1)/2 \end{pmatrix} Q^{\frac{m-1}{2}k} U_k ,$$

$$V_k^m = V_{km} + \begin{bmatrix} m \\ 1 \end{bmatrix} Q^k V_{k(m-2)} + \begin{bmatrix} m \\ 2 \end{bmatrix} Q^{2k} V_{k(m-4)}$$

$$+ \cdots + \begin{bmatrix} m \\ (m-1)/2 \end{bmatrix} Q^{\frac{m-1}{2}k} V_k .$$

If m is even and $k \geqslant 1$, then

$$D^{m/2} U_k^m = V_{km} - \begin{bmatrix} m \\ 1 \end{bmatrix} Q^k V_{k(m-2)} + \begin{bmatrix} m \\ 2 \end{bmatrix} Q^{2k} V_{k(m-4)}$$

$$+ \cdots + \begin{bmatrix} m \\ m/2 \end{bmatrix} Q^{(m/2)k} \times 2,$$

$$V_k^m = V_{km} + \begin{bmatrix} m \\ 1 \end{bmatrix} Q^k V_{k(m-2)} + \begin{bmatrix} m \\ 2 \end{bmatrix} Q^{2k} V_{k(m-4)}$$

$$+ \cdots + \begin{bmatrix} m \\ m/2 \end{bmatrix} Q^{(m/2)k} \times 2.$$

(IV.10) $U_m = V_{m-1} + QV_{m-3} + Q^2 V_{m-5} + \cdots + \text{(last summand)},$
where

$$\text{last summand} = \begin{cases} Q^{(m-2)/2} P & \text{if } m \text{ is even,} \\ Q^{(m-1)/2} & \text{if } m \text{ is odd.} \end{cases}$$

$$P^m = V_m + \begin{bmatrix} m \\ 1 \end{bmatrix} QV_{m-2} + \begin{bmatrix} m \\ 2 \end{bmatrix} Q^2 V_{m-4} + \cdots + \text{(last summand)},$$

where

$$\text{last summand} = \begin{cases} \begin{bmatrix} m \\ m/2 \end{bmatrix} Q^{m/2} & \text{if } m \text{ is even,} \\ \\ \begin{bmatrix} m \\ (m-1)/2 \end{bmatrix} Q^{(m-1)/2} P & \text{if } m \text{ is odd.} \end{cases}$$

The following identity of Lagrange, dating from 1741, is required for the next property:

$$X^n + Y^n = (X+Y)^n - \frac{n}{1} XY(X+Y)^{n-2} + \frac{n}{2} \begin{bmatrix} n-3 \\ 1 \end{bmatrix} X^2 Y^2 (X+Y)^{n-4}$$

$$- \frac{n}{3} \begin{bmatrix} n-4 \\ 2 \end{bmatrix} X^3 Y^3 (X+Y)^{n-6} + \cdots + (-1)^r \frac{n}{r} \begin{bmatrix} n-r-1 \\ r-1 \end{bmatrix} X^r Y^r (X+Y)^{n-2r}$$

$$\pm \cdots ,$$

where the sum is extended for $2r \leqslant n$. Note that each coefficient is an integer.

(IV.11)   If $m \geqslant 1$ and q is odd,

$$U_{mq} = D^{(q-1)/2}U_m^q + \frac{q}{1} Q^m D^{(q-3)/2}U_m^{q-2} + \frac{q}{2}\left[\begin{array}{c}q-3\\1\end{array}\right]Q^{2m}D^{(q-5)/2}U_m^{q-4}$$

$$+ \cdots + \frac{q}{r}\left[\begin{array}{c}q-r-1\\r-1\end{array}\right]Q^{mr}D^{(q-2r-1)/2}U_m^{q-2r} + \cdots + \text{(last summand)},$$

where the last summand is

$$\frac{q}{(q-1)/2}\left[\begin{array}{c}(q-1)/2\\(q-3)/2\end{array}\right]Q^{\frac{q-1}{2}m}U_m = qQ^{\frac{q-1}{2}m}U_m.$$

Now, I begin to indicate, one after the other, the divisibility properties, in the order in which they may be proved.

(IV.12) $U_n \equiv V_{n-1} \pmod{Q}$
            $V_n \equiv P^n \pmod{Q}$.

Hint: use (IV.10).

(IV.13)   Let p be an odd prime, then

$$U_{kp} \equiv D^{(p-1)/2} U_k \pmod{p}$$

and, for $e \geqslant 1$,

$$U_{p^e} \equiv D^{\frac{p-1}{2}e} \pmod{p}.$$

In particular,

$$U_p \equiv \left(\frac{D}{p}\right) \pmod{p}.$$

Hint: Use (IV.9).

(IV.14) $V_p \equiv P \pmod{p}$.
Hint: use (IV.10).

(IV.15)   If $n,k \geqslant 1$, then U_n divides U_{kn}.
Hint: use (IV.3).

(IV.16)   If $n,k \geqslant 1$ and k is odd, then V_n divides V_{kn}.

Hint: use (IV.9).

Notation. If $n \geqslant 2$ and if there exists $r \geqslant 1$ such that n divides U_r, denote by $\rho(n) = \rho(n,U)$ the smallest such r.

(IV.17) Assume that $\rho(n)$ exists and $\gcd(n,2Q) = 1$. Then $n|U_k$ if and only if $\rho(n)|k$.
Hint: use (IV.15) and (IV.7).

It will be seen that $\rho(n)$ exists, for many — not for all — values of n, such that $\gcd(n,2Q) = 1$.

(IV.18) If Q is even and P is even, then U_n is even (for $n \geqslant 2$) and V_n is even (for $n \geqslant 1$).
   If Q is even and P is odd, then U_n, V_n are odd (for $n \geqslant 1$).
   If Q is odd and P is even, then $U_n \equiv n \pmod 2$ and V_n is even.
   If Q is odd and P is odd, then U_n, V_n are even if 3 divides n, while U_n, V_n are odd, otherwise.
   In particular, if U_n is even, then V_n is even.
   Hint: use (IV.12), (IV.5), (IV.2), (IV.6), and (IV.1).

Here is the first main result, which is a companion of (IV.18) and generalizes Fermat's little theorem:

(IV.19) Let p be an odd prime.
   If $p \mid P$ and $p \mid Q$, then $p \mid U_k$ for every $k > 1$.
   If $p \mid P$ and $p \nmid Q$, then $p \mid U_k$ exactly when k is even.
   If $p \nmid P$ and $p \mid Q$, then $p \nmid U_n$ for every $n \geqslant 1$.
   If $p \nmid P$, $p \nmid Q$, and $p \mid D$, then $p \mid U_k$ exactly when $p \mid k$.
   If $p \nmid PQD$, then $p \mid U_{\psi(p)}$, where $\psi(p) = p - (D/p)$, and (D/p) denotes the Legendre symbol.

Proof. If $p \mid P$ and $p \mid Q$, by (IV.1) $p \mid U_k$ for every $k > 1$.
   If $p \mid P = U_2$, by (IV.15) $p \mid U_{2k}$ for every $k \geqslant 1$. Since $p \nmid Q$, and $U_{2k+1} = PU_{2k} - QU_{2k-1}$, by induction, $p \nmid U_{2k+1}$.
   If $p \nmid P$ and $p \mid Q$, by induction and (IV.1), $p \nmid U_n$ for every $n \geqslant 1$.
   If $p \nmid PD$ and $p \mid D$, by (IV.8), $2^{p-1}U_p \equiv 0 \pmod p$ so $p \mid U_p$. On the other hand, if $p \nmid n$, then by (IV.8) $2^{n-1}U_n \equiv nP^{n-1} \not\equiv 0 \pmod p$, so $p \nmid U_n$.

Finally the more interesting case: assume $p \nmid PQD$.
If $(D/p) = -1$, then by (IV.8)

$$2^P U_{p+1} = \binom{p+1}{1} P^p + \binom{p+1}{3} P^{p-2} D + \cdots + \binom{p+1}{p} P D^{(p-1)/2}$$

$$\equiv P + PD^{(p-1)/2} \equiv 0 \pmod{p}, \text{ so } p \mid U_{p+1}.$$

If $(D/p) = 1$, there exists C such that $P^2 - 4Q = D \equiv C^2 \pmod{p}$; hence, $P^2 \not\equiv C^2 \pmod{p}$ and $p \nmid C$. By (IV.8), noting that

$$\binom{p-1}{1} \equiv -1 \pmod{p}, \quad \binom{p-1}{3} \equiv -1 \pmod{p}, \dots :$$

$$2^{p-2} U_{p-1} = \binom{p-1}{1} P^{p-2} + \binom{p-1}{3} P^{p-4} D + \binom{p-1}{5} P^{p-6} D^2 + \cdots +$$

$$\binom{p-1}{p-2} PD^{(p-3)/2} \equiv -[P^{p-2} + P^{p-4} D + P^{p-6} D^2 + \cdots + PD^{(p-3)/2}]$$

$$\equiv -P\left[\frac{P^{p-1} - D^{(p-1)/2}}{P^2 - D} \right] \equiv -P\, \frac{P^{p-1} - C^{p-1}}{P^2 - C^2} \equiv 0 \pmod{p}.$$

So $p \mid U_{p-1}$.   □

If I want to use the notation $\rho(p)$ introduced before, some of the assertions of (IV.19) may be restated as follows:
If p is an odd prime and $p \nmid Q$.
If $p \mid P$, then $\rho(p) = 2$.
If $p \nmid P$, $p \mid D$, then $\rho(p) = p$.
If $p \nmid PD$, then $\rho(p)$ divides $\psi(p)$.
Don't conclude hastily that, in this latter case, $\rho(p) = \psi(p)$. I shall return to this point, after I list the main properties of the Lucas sequences.

For the special Lucas sequence $U_n(a+1,a)$, the discriminant is $D = (a-1)^2$; so if $p \nmid a(a^2-1)$, then

$$\left(\frac{D}{p} \right) = 1 \quad \text{and} \quad p \mid U_{p-1} = \frac{a^{p-1} - 1}{a - 1},$$

so $p \mid a^{p-1} - 1$ (this is trivial if $p \mid a^2 - 1$) — which is Fermat's little theorem.

(IV.20) Let $e \geq 1$, and let p^e be the exact power of p dividing U_m.

If $p \nmid k$ and $f \geq 1$, then p^{e+f} divides U_{mkp^f}.

Moreover, if $p \nmid Q$ and $p^e \neq 2$, then p^{e+f} is the exact power of p dividing U_{mkp^f}, while if $p^e = 2$ then $U_{mk/2}$ is odd.

Hint: use (IV.19), (IV.18), (IV.11), and (IV.6).

And now the generalization of Euler's theorem:

If α, β are roots of $X^2 - PX + Q$, define the symbol:

$$\left[\frac{\alpha, \beta}{2} \right] = \begin{cases} 1 & \text{if } Q \text{ is even,} \\ 0 & \text{if } Q \text{ is odd, } P \text{ even,} \\ -1 & \text{if } Q \text{ is odd, } P \text{ odd.} \end{cases}$$

and $p \neq 2$:

$$\left[\frac{\alpha, \beta}{p} \right] = \left[\frac{D}{p} \right]$$

(so it is 0 if $p \mid D$).

Put

$$\psi_{\alpha,\beta}(p) = p - \left[\frac{\alpha, \beta}{p} \right]$$

for every prime p, also

$$\psi_{\alpha,\beta}(p^e) = p^{e-1}\psi_{\alpha,\beta}(p) \qquad \text{for } e \geq 1.$$

If $n = \Pi_{p|n} p^e$, define the Carmichael function

$$\lambda_{\alpha,\beta}(n) = \ell\text{cm}\{\psi_{\alpha,\beta}(p^e)\}$$

(where ℓcm denotes the least common multiple), and define the generalized Euler function

$$\psi_{\alpha,\beta}(n) = \Pi_{p|n} \psi_{\alpha,\beta}(p^e).$$

So $\lambda_{\alpha,\beta}(n)$ divides $\psi_{\alpha,\beta}(n)$.

It is easy to check that $\psi_{a,1}(p) = p - 1 = \phi(p)$ for every prime p not dividing a; so if $\gcd(a,n) = 1$, then $\psi_{a,1}(n) = \phi(n)$ and also $\lambda_{a,1}(n) = \lambda(n)$, where $\lambda(n)$ is the function, also defined by Carmichael, and considered in Section II.

And here is the extension of Euler's theorem:

(IV.21)   If $\gcd(n,Q) = 1$, then n divides $U_{\lambda_{\alpha,\beta}(n)}$; hence, also n divides $U_{\psi_{\alpha,\beta}(n)}$.

Hint: use (IV.19) and (IV.20).

It should be said that the divisibility properties of the companion sequence $(V_n)_{n \geqslant 1}$ are not so simple to describe. Note, for example,

(IV.22) If $p \nmid 2QD$, then $V_{p-(D/p)} \equiv 2Q^{\frac{1}{2}[1-(D/p)]} \pmod{p}$.

Hint: use (IV.5), (IV.13), (IV.19), and (IV.14).

This may be applied to give divisibility results for $U_{\psi(p)/2}$ and $V_{\psi(p)/2}$.

(IV.23)   Assume that $p \nmid 2QD$. Then

$P | U_{\psi(p)/2}$ if and only if $(Q/P) = 1$;

$P | V_{\psi(p)/2}$ if and only if $(Q/p) = -1$.

Hint: for the first assertion, use (IV.2), (IV.6), (IV.22) and the congruence $(Q/p) \equiv p^{(Q-1)/2} \pmod{p}$.

For the second assertion, use (IV.2), (IV.19), the first assertion, and also (IV.6).

For the next results, I shall assume that $\gcd(P,Q) = 1$.

(IV.24) $\gcd(U_n,Q) = 1$, $\gcd(V_n,Q) = 1$.

Hint: use (IV.12).

(IV.25) $\gcd(U_n,V_n) = 1$ or 2.

Hint: use (IV.16), and (IV.24).

(IV.26)   If $d = \gcd(m,n)$, then $U_d = \gcd(U_m,U_n)$.

Hint: use (IV.15), (IV.7), (IV.24), (IV.18), and (IV.6).

This proof is actually not so easy, and requires the use of the Lucas sequence $(U_n(V_d,Q^d))_{n \geqslant 0}$.

(IV.27)   If $\gcd(m,n) = 1$, then $\gcd(U_m,U_n) = 1$.

No hint for this one.

(IV.28)   If $d = \gcd(m,n)$ and m/d, n/d are odd, then $V_d = \gcd(V_m,V_n)$.

Hint: use the same proof as for (IV.26).

And here is a result similar to (IV.17), but with the assumption that $\gcd(P,Q) = 1$:

(IV.29) Assume that $\rho(n)$ exists. Then $n|U_k$ if and only if $\rho(n)|k$.
Hint: use (IV.15), (IV.24), and (IV.3).

I pause to write explicitly what happens for the Fibonacci numbers U_n and Lucas numbers V_n; now $P = 1$, $Q = -1$, $D = 5$.

Property (IV.18) becomes the law of appearance of p; even though
I am writing this text on Halloween's evening, it would hurt me to
call it the "apparition law" (as it was badly translated from the
French *loi d'apparition*; in all English dictionaries "apparition" means
"ghost"). Law of apparition (oops!, appearance) of p:

$$p|U_{p-1} \text{ if } (5/p) = 1, \quad \text{that is, } p \equiv \pm 1 \pmod{10}$$
$$p|U_{p+1} \text{ if } (5/p) = -1, \quad \text{that is, } p \equiv \pm 3 \pmod{10}.$$

Property (IV.19) is the law of repetition.

For the Lucas numbers, the following properties hold:

$$p|V_{p-1} - 2 \text{ if } (5/p) = 1, \quad \text{that is, } p \equiv \pm 1 \pmod{10}$$
$$p|V_{p+1} + 2 \text{ if } (5/p) = -1, \quad \text{that is, } p \equiv \pm 3 \pmod{10}.$$

Jarden showed in 1958 that, for the Fibonacci sequence, the function

$$\frac{\psi(p)}{\rho(p)} = \frac{p - (5/p)}{\rho(p)}$$

is unbounded (when the prime p tends to infinity).

This result was generalized by Kiss & Phong in 1978: there exists
$C > 0$ (depending only on P, Q) such that $\psi(p)/\rho(p)$ is unbounded, but
still $\psi(p)/\rho(p) < C[p/(\log p)]$ (when the prime p tends to infinity).

Now I shall indicate the behaviour of Lucas sequences modulo a
prime p.

If $p = 2$, this is as described in (IV.18). For example, if P, Q are
odd, then the sequences U_n mod 2, V_n mod 2 are equal to

$$1 \quad 1 \quad 0 \quad 1 \quad 1 \quad 0 \quad 1 \quad 1 \quad 0 \quad \dots .$$

It is more interesting when p is an odd prime.

(IV.30) If $p \nmid 2QD$ and $(D/p) = 1$, then

$$V_{n+p-1} \equiv U_n \pmod{p},$$
$$\dot{V}_{n+p-1} \equiv V_n \pmod{p}.$$

Thus, the sequences U_n mod p, V_n mod p have period $p - 1$.

Proof. By (D4), $U_{n+p-1} = U_n U_p - Q U_{n-1} U_{p-1}$; by (IV.19), $\rho_p(U)$ divides $p - (D/p) = p - 1$; by (IV.15), $p|U_{p-1}$; this is also true if $p|P$, $p \nmid Q$, because then $p - 1$ is even, so $p|U_{p-1}$, by (IV.19). By (IV.13),

$$U_p \equiv (D/p) \equiv 1 \pmod{p}.$$

So $U_{n+p-1} \equiv U_n \pmod{p}$.

Now, by (IV.5), $V_{n+p-1} = 2U_{n+p} - PU_n \equiv 2U_{n+1} - PU_n \equiv V_n \pmod{p}$.

$\square$

The companion result is the following:

(IV.31)   Let $p \nmid 2QD$, let e be the order of Q mod p. If $(D/p) = -1$, then

$$U_{n+e(p+1)} \equiv U_n \pmod{p},$$
$$V_{n+e(p-1)} \equiv V_n \pmod{p}.$$

Thus, the sequences U_n mod p, V_n mod p have period $e(p + 1)$.

Proof. By (IV.19), (IV.15),

$$p|U_{p-(D/p)} = U_{p+1}.$$

By (IV.22), $V_{p+1} \equiv 2Q \pmod{p}$. Now I show, by induction on $r \geqslant 1$, that $V_{r(p+1)} \equiv 2Q^r \pmod{p}$.
If this is true for $r \geqslant 1$, then by (IV.4)

$$2V_{(r+1)(p+1)} = V_{r(p+1)}V_{p+1} + DU_{r(p+1)}U_{p+1} \equiv 4Q^{r+1} \pmod{p},$$

so $V_{(r+1)(p+1)} \equiv 2Q^{r+1} \pmod{p}$. In particular, $V_{e(p+1)} \equiv 2Q^e \equiv 2 \pmod{p}$.
By (IV.7),

$$U_{n+e(p+1)}V_{e(p+1)} - U_{e(p+1)}V_{n+e(p+1)} = 2Q^{e(p+1)}U_n,$$

hence $2U_{n+e(p+1)} \equiv 2U_n \pmod{p}$ and the first congruence is established.

The second congruence follows using (IV.5). $\square$

It is good to summarize some of the preceding results, by writing explicitly the sets

$$P(U) = \{p \text{ prime} \mid \text{there exists } n \text{ such that } p|U_n\},$$
$$P(V) = \{p \text{ prime} \mid \text{there exists } n \text{ such that } p|V_n\}.$$

These are the prime divisors of the sequence $U = (U_n)_{n \geq 1}$, $V = (V_n)_{n \geq 1}$.

The parameters (P,Q) are assumed to be nonzero relatively prime integers and the discriminant is $D = P^2 - 4Q \neq 0$.

A first case arises if there exists $n > 1$ such that $U_n = 0$; equivalently, $\alpha^n = \beta^n$, that is α/β is a root of unity. If n is the smallest such index, then $U_r \neq 0$ for $r = 1, ..., n - 1$ and $U_{nk+r} = \alpha^{nk}U_r$ (for every $k \geq 1$), so $P(U)$ consists of the prime divisors of $U_2 \cdots U_{n-1}$. Similarly, $P(V)$ consists of the same prime numbers.

The more interesting case is when α/β is not a root of unity, so $U_n \neq 0$, $V_n \neq 0$ for every $n \geq 1$. Then $P(U) = \{p \text{ prime} \mid p \text{ does not divide } Q\}$.

This follows from (IV.18) and (IV.19). In particular, for the sequence of Fibonacci numbers, $P(U)$ is the set of all primes.

Nothing so precise may be said about the companion Lucas sequence $V = (V_n)_{n \geq 1}$. From $U_{2n} = U_n V_n$ ($n \geq 1$) it follows that $P(V)$ is a subset of $P(U)$. From (IV.18), $2 \in P(V)$ if and only if Q is odd. Also, from (IV.24) and (IV.6), if $p \neq 2$ and if $p|DQ$, then $p \notin P(V)$, while if $p \nmid 2DQ$ and $(Q/p) = -1$, then $p \in P(V)$ [see (IV.23)]; on the other hand, if $p \nmid 2DQ$, $(Q/p) = 1$, and $(D/p) = -(-1/p)$, then $p \notin P(V)$. This does not determine, without a further analysis, whether a prime p, such that $p \nmid 2DQ$, $(Q/p) = 1$, and $(D/p) = (-1/p)$ belongs, or does not belong, to $P(V)$.

At any rate, it shows that $P(V)$ is also an infinite set.

For the sequence of Lucas numbers, with $P = 1$, $Q = -1$, $D = 5$, the preceding facts may be explicitly stated as follows:

if $p \equiv 3, 7, 11, 19 \pmod{20}$, then $p \in P(V)$;

if $p \equiv 13, 17 \pmod{20}$, then $p \notin P(V)$.

For $p \equiv 1, 9 \pmod{20}$, no decision may be obtained without a careful study, as, for example, that done by Ward in 1961. Already in 1958 Jarden had shown that there exist infinitely many primes p, $p \equiv 1 \pmod{20}$, such that $p \notin \mathcal{P}(V)$, and, on the other hand, there exist also infinitely many primes p, $p \equiv 1 \pmod{40}$, such that $p \in \mathcal{P}(V)$.

Later, in Chapter 5, Section VIII, I shall return to the study of the sets $\mathcal{P}(U)$, $\mathcal{P}(V)$, asking for their density in the set of all primes.

In analogy with the theorem of Bang and Zsigmondy, Carmichael also considered the primitive prime factors of the Lucas sequences, with parameters (P,Q): p is a primitive prime factor of U_k (resp. V_k) if $p|U_k$ (resp. $p|V_k$), but p does not divide any preceding number in the sequence in question.

The proof of Zsigmondy's theorem is not too simple; here it is somewhat more delicate.

Carmichael showed that if the discriminant D is positive, then for every $n \neq 1, 2, 6$, U_n has a primitive prime factor, except if $n = 12$ and $P = \pm 1$, $Q = -1$.

Moreover, if D is a square, then it is better: for every n, U_n has a primitive prime factor, except if $n = 6$, $P = \pm 3$, $Q = 2$.

Do you recognize that this second statement includes Zsigmondy's theorem?   Also, if $P = 1$, $Q = -1$ the exception is the Fibonacci number $U_{12} = 144$.

For the companion sequence, if $D > 0$, then for every $n \neq 1, 3$, V_n has a primitive prime factor, except if $n = 6$, $P = \pm 1$, $Q = -1$ (the Lucas number $V_6 = 18$). Moreover, if D is a square, then the only exception is $n = 3$, $P = \pm 3$, $Q = 2$, also contained in Zsigmondy's theorem.

If, however, $D < 0$, the result indicated is no longer true. Thus, as Carmichael already noted, if $P = 1$, $Q = 2$, then for $n = 1, 2, 3, 5, 7, 8, 12, 13, 18$, $U_n = \pm 1$, so it has no primitive prime factors.

Schinzel showed the following in 1962:

Let $(U_n)_{n \geqslant 0}$ be the Lucas sequence with relatively prime parameters (P,Q) and assume that the discriminant is $D < 0$. Assume that α/β is not a root of unity.   Then there exists n_0 (depending on P, Q), effectively computable, such that if $n > n_0$, then U_n has a primitive prime factor.

Later, in 1974, Schinzel proved the same result with an absolute constant n_0 — independent of the Lucas sequence.

Making use of the methods of Baker, Stewart determined in 1977 that if $n > e^{452}2^{67}$, then U_n has a primitive prime factor. Moreover, Stewart also showed that if n is given ($n \neq 6$, $n > 4$), there are only finitely many Lucas sequences, which may be determined explicitly (so says Stewart, without doing it), for which U_n has no primitive prime factor.

It is interesting to consider the primitive part U_n^* of U_n:

$$U_n = U_n^* U_n' \quad \text{with} \quad \gcd(U_n^*, U_n') = 1$$

and p divides U_n^* if and only if p is a primitive prime factor of U_n.

In 1963, Schinzel indicated conditions for the existence of two (or even $e > 2$) distinct primitive prime factors. It follows that if $D > 0$ or $D < 0$ and α/β is not a root of unity, there exist infinitely many n such that the primitive part U_n^* is composite.

Can one say anything about U_n^* being square-free? This is a very deep question. Just think of the special case where $P = 3$, $Q = 2$, which gives the sequence $2^n - 1$ (see my comments in Section II).

Just as in the case of the binomials $a^n - b^n$ (described in Section II), there are results about the size of the largest prime factor $P(U_n)$, $P(V_n)$ of U_n, resp. V_n. If the discriminant is positive, it follows from Carmichael's work that $P(U_n) \geq n - 1$, $P(V_n) \geq 2n - 1$. With his more powerful methods, Stewart obtained the same estimates, indicated in Section II for the binomials $a^n - b^n$; namely,

$$P(U_n) > \frac{Cn(\log n)^{\lambda}}{\log \log \log n},$$

where $C > 0$, $\lambda = 1 - \kappa \log \log n$, $\kappa < 1/\log 2$, and the number $\omega(n)$ of distinct prime divisors of n satisfies $\omega(n) < \kappa \log \log n$.

$$P = 1, \quad Q = -1$$

Fibonacci Numbers	Lucas Numbers
$U(0) = 0 \quad U(1) = 1$	$V(0) = 2 \quad V(1) = 1$
$U(2) = 1$	$V(2) = 3$
$U(3) = 2$	$V(3) = 4$
$U(4) = 3$	$V(4) = 7$
$U(5) = 5$	$V(5) = 11$
$U(6) = 8$	$V(6) = 18$
$U(7) = 13$	$V(7) = 29$
$U(8) = 21$	$V(8) = 47$
$U(9) = 34$	$V(9) = 76$
$U(10) = 55$	$V(10) = 123$
$U(11) = 89$	$V(11) = 199$
$U(12) = 144$	$V(12) = 322$
$U(13) = 233$	$V(13) = 521$
$U(14) = 377$	$V(14) = 843$
$U(15) = 610$	$V(15) = 1364$
$U(16) = 987$	$V(16) = 2207$
$U(17) = 1597$	$V(17) = 3571$
$U(18) = 2584$	$V(18) = 5778$
$U(19) = 4181$	$V(19) = 9349$
$U(20) = 6765$	$V(20) = 15127$
$U(21) = 10946$	$V(21) = 24476$
$U(22) = 17711$	$V(22) = 39603$
$U(23) = 28657$	$V(23) = 64079$
$U(24) = 46368$	$V(24) = 103682$
$U(25) = 75025$	$V(25) = 167761$
$U(26) = 121393$	$V(26) = 271443$
$U(27) = 196418$	$V(27) = 439204$
$U(28) = 317811$	$V(28) = 710647$
$U(29) = 514229$	$V(29) = 1149851$
$U(30) = 832040$	$V(30) = 1860498$
$U(31) = 1346269$	$V(31) = 3010349$
$U(32) = 2178309$	$V(32) = 4870847$
$U(33) = 3524578$	$V(33) = 7881196$
$U(34) = 5702887$	$V(34) = 12752043$
$U(35) = 9227465$	$V(35) = 20633239$
$U(36) = 14930352$	$V(36) = 33385282$
$U(37) = 24157817$	$V(37) = 54018521$
$U(38) = 39088169$	$V(38) = 87403803$
$U(39) = 63245986$	$V(39) = 141422324$
$U(40) = 102334155$	$V(40) = 228826127$
$U(41) = 165580141$	$V(41) = 370248451$
$U(42) = 267914296$	$V(42) = 599074578$
$U(43) = 433494437$	$V(43) = 969323029$
$U(44) = 701408733$	$V(44) = 1568397607$
$U(45) = 1134903170$	$V(45) = 2537720636$
$U(46) = 1836311903$	$V(46) = 4106118243$

$P = 1$, $Q = -1$ (cont.)

$U(47) = 2971215073$	$V(47) = 6643838879$
$U(48) = 4807526976$	$V(48) = 10749957122$
$U(49) = 7778742049$	$V(49) = 17393796001$
$U(50) = 12586269025$	$V(50) = 28143753123$

$$P = 3, \quad Q = 2$$

Numbers $2^n - 1$	Numbers $2^n + 1$
$U(0) = 0 \quad U(1) = 1$	$V(0) = 2 \quad V(1) = 3$
$U(2) = 3$	$V(2) = 5$
$U(3) = 7$	$V(3) = 9$
$U(4) = 15$	$V(4) = 17$
$U(5) = 31$	$V(5) = 33$
$U(6) = 63$	$V(6) = 65$
$U(7) = 127$	$V(7) = 129$
$U(8) = 255$	$V(8) = 257$
$U(9) = 511$	$V(9) = 513$
$U(10) = 1023$	$V(10) = 1025$
$U(11) = 2047$	$V(11) = 2049$
$U(12) = 4095$	$V(12) = 4097$
$U(13) = 8191$	$V(13) = 8193$
$U(14) = 16383$	$V(14) = 16385$
$U(15) = 32767$	$V(15) = 32769$
$U(16) = 65535$	$V(16) = 65537$
$U(17) = 131071$	$V(17) = 131073$
$U(18) = 262143$	$V(18) = 262145$
$U(19) = 524287$	$V(19) = 524289$
$U(20) = 1048575$	$V(20) = 1048577$
$U(21) = 2097151$	$V(21) = 2097153$
$U(22) = 4194303$	$V(22) = 4194305$
$U(23) = 8388607$	$V(23) = 8388609$
$U(24) = 16777215$	$V(24) = 16777217$
$U(25) = 33554431$	$V(25) = 33554433$
$U(26) = 67108863$	$V(26) = 67108865$
$U(27) = 134217727$	$V(27) = 134217729$
$U(28) = 268435455$	$V(28) = 268435457$
$U(29) = 536870911$	$V(29) = 536870913$
$U(30) = 1073741823$	$V(30) = 1073741825$
$U(31) = 2147483647$	$V(31) = 2147483649$
$U(32) = 4294967295$	$V(32) = 4294967297$
$U(33) = 8589934591$	$V(33) = 8589934593$
$U(34) = 17179869183$	$V(34) = 17179869185$
$U(35) = 34359738367$	$V(35) = 34359738369$
$U(36) = 68719476735$	$V(36) = 68719476637$
$U(37) = 137438953471$	$V(37) = 137438953473$
$U(38) = 274877906943$	$V(38) = 274877906945$

$P = 3,$ $Q = 2$ (cont.)

$U(39) = 549755813887$	$V(39) = 549755813889$
$U(40) = 1099511627775$	$V(40) = 1099511627777$
$U(41) = 2199023255551$	$V(41) = 2199023255553$
$U(42) = 4398046511103$	$V(42) = 4398046511105$
$U(43) = 8796093022207$	$V(43) = 8796093022209$
$U(44) = 17592186044415$	$V(44) = 17592186044417$
$U(45) = 35184372088831$	$V(45) = 35184372088833$
$U(46) = 70368744177663$	$V(46) = 70368744177665$
$U(47) = 140737488355327$	$V(47) = 140737488355329$
$U(48) = 281474976710655$	$V(48) = 281474976710657$
$U(49) = 562949953421311$	$V(49) = 562949953421313$
$U(50) = 1125899906842623$	$V(50) = 1125899906842625$

$$P = 2, \qquad Q = -1$$

Pell Numbers	Companion Pell Numbers
$U(0) = 0 \quad U(1) = 1$	$V(0) = 2 \quad V(1) = 2$
$U(2) = 2$	$V(2) = 6$
$U(3) = 5$	$V(3) = 14$
$U(4) = 12$	$V(4) = 34$
$U(5) = 29$	$V(5) = 82$
$U(6) = 70$	$V(6) = 198$
$U(7) = 169$	$V(7) = 478$
$U(8) = 408$	$V(8) = 1154$
$U(9) = 985$	$V(9) = 2786$
$U(10) = 2378$	$V(10) = 6726$
$U(11) = 5741$	$V(11) = 16238$
$U(12) = 13860$	$V(12) = 39202$
$U(13) = 33461$	$V(13) = 94642$
$U(14) = 80782$	$V(14) = 228486$
$U(15) = 195025$	$V(15) = 551614$
$U(16) = 470832$	$V(16) = 1331714$
$U(17) = 1136689$	$V(17) = 3215042$
$U(18) = 2744210$	$V(18) = 7761798$
$U(19) = 6625109$	$V(19) = 18738638$
$U(20) = 15994428$	$V(20) = 45239074$
$U(21) = 38613965$	$V(21) = 109216786$
$U(22) = 93222358$	$V(22) = 263672646$
$U(23) = 225058681$	$V(23) = 636562078$
$U(24) = 543339720$	$V(24) = 1536796802$
$U(25) = 1311738121$	$V(25) = 3710155682$
$U(26) = 3166815962$	$V(26) = 8957108166$
$U(27) = 7645370045$	$V(27) = 21624372014$
$U(28) = 18457556052$	$V(28) = 52205852194$
$U(29) = 44560482149$	$V(29) = 126036076402$
$U(30) = 107578520350$	$V(30) = 304278004998$
$U(31) = 259717522849$	$V(31) = 734592086398$
$U(32) = 627013566048$	$V(32) = 1773462177794$
$U(33) = 1513744654945$	$V(33) = 4281516441986$
$U(34) = 3654502875938$	$V(34) = 10336495061766$
$U(35) = 8822750406821$	$V(35) = 24954506565518$
$U(36) = 21300003689580$	$V(36) = 60245508192802$
$U(37) = 51422757785981$	$V(37) = 145445522951122$
$U(38) = 124145519261542$	$V(38) = 351136554095046$
$U(39) = 299713796309065$	$V(39) = 847718631141214$
$U(40) = 723573111879672$	$V(40) = 2046573816377474$
$U(41) = 1746860020068409$	$V(41) = 4940866263896162$

$$P = 4, \quad Q = 3$$

$U(0) = 0 \qquad U(1) = 1$	$V(0) = 2 \qquad V(1) = 4$
$U(2) = 4$	$V(2) = 10$
$U(3) = 13$	$V(3) = 28$
$U(4) = 40$	$V(4) = 82$
$U(5) = 121$	$V(5) = 244$
$U(6) = 364$	$V(6) = 730$
$U(7) = 1093$	$V(7) = 2188$
$U(8) = 3280$	$V(8) = 6562$
$U(9) = 9841$	$V(9) = 19684$
$U(10) = 29524$	$V(10) = 59050$
$U(11) = 88573$	$V(11) = 177148$
$U(12) = 265720$	$V(12) = 531442$
$U(13) = 797161$	$V(13) = 1594324$
$U(14) = 2391484$	$V(14) = 4782970$
$U(15) = 7174453$	$V(15) = 14348908$
$U(16) = 21523360$	$V(16) = 43046722$
$U(17) = 64570081$	$V(17) = 129140164$
$U(18) = 193710244$	$V(18) = 387420490$
$U(19) = 581130733$	$V(19) = 1162261468$
$U(20) = 1743392200$	$V(20) = 3486784402$
$U(21) = 5230176601$	$V(21) = 10460353204$
$U(22) = 15690529804$	$V(22) = 31381059610$
$U(23) = 47071589413$	$V(23) = 94143178828$
$U(24) = 141214768240$	$V(24) = 282429536482$
$U(25) = 423644304721$	$V(25) = 847288609444$
$U(26) = 1270932914164$	$V(26) = 2541865828330$
$U(27) = 3812798742493$	$V(27) = 7625597484988$
$U(28) = 11438396227480$	$V(28) = 22876792454962$
$U(29) = 34315188682441$	$V(29) = 68630377364884$
$U(30) = 102945566047324$	$V(30) = 205891132094650$
$U(31) = 308836698141973$	$V(31) = 617673396283948$
$U(32) = 926510094425920$	$V(32) = 1853020188851842$
$U(33) = 2779530283277761$	$V(33) = 5559060566555524$

Addendum on Lehmer Numbers

In his Ph.D. thesis, Lehmer developed an extended theory of Lucas'
functions. These are now called the Lehmer numbers and constitute
a generalization of Lucas numbers.

Let α, β be complex numbers such that $\alpha + \beta = \sqrt{R}$, $\alpha\beta = Q$, with
R, Q relatively prime nonzero integers and α/β is not a root of unity.
The Lehmer numbers are

$$U_n = U_n(\sqrt{R},Q) = \begin{cases} \dfrac{\alpha^n - \beta^n}{\alpha - \beta} & \text{when } n \text{ is odd,} \\[2mm] \dfrac{\alpha^n - \beta^n}{\alpha - \beta} & \text{when } n \text{ is even,} \end{cases}$$

and the companion Lehmer numbers are

$$V_n = V_n(\sqrt{R},Q) = \begin{cases} \dfrac{\alpha^n + \beta^n}{\alpha + \beta} & \text{when } n \text{ is odd,} \\[2mm] \alpha^n + \beta^n & \text{when } n \text{ is even.} \end{cases}$$

This notation might be confusing, since it has been used for the
Lucas sequences. However, I shall not develop the theory of Lehmer
numbers, being content to indicate the main papers.

For the general theory: Lehmer (1930).

For primitive factors of Lehmer numbers: Ward (1955), Durst
(1959), Schinzel (1962, 1963, 1968), Stewart (1977, 1977), and Shorey
& Stewart (1981).

Lehmer numbers have found applications in primality testing, in
the same spirit as Lucas numbers have been used (see Section V).

V. Classical Primality Tests Based on Lucas Sequences

Lucas began, Lehmer continued, others refined. The primality tests
of N, to be presented now, require the knowledge of prime factors of
$N + 1$, and they complement the tests indicated in Section III, which
needed the prime factors of $N - 1$. Now, the tool will be the Lucas

sequences. By (IV.18), if N is an odd prime, if $U = (U_n)_{n\geqslant0}$ is a Lucas sequence with discriminant D, and the Jacobi symbol $(D/N) = -1$, then N divides $U_{N-(D/N)} = U_{N+1}$.

However, I note right away (as I did in Section III) that a crude converse does not hold, because there exist composite integers N, and Lucas sequences $(U_n)_{n\geqslant0}$ with discriminant D, such that N divides $U_{N-(D/N)}$. Such numbers will be studied in Section X.

It will be convenient to introduce for every integer $D > 1$ the function ψ_D, defined as follows:

If $N = \prod_{i=1}^{s}p_i^{e_i}$, let

$$\psi_D(N) = \frac{1}{2^{s-1}} \prod_{i=1}^{s} p_i^{e_i-1}\left[p_i - \left(\frac{D}{p_i}\right)\right].$$

Note that if $(U_n)_{n\geqslant0}$ is a Lucas sequence with discriminant D, if α, β are the roots of the associated polynomial, then the function $\psi_{\alpha,\beta}$ considered in Section IV is related to ψ_D as follows: $\psi_{\alpha,\beta}(N) = 2^{s-1}\psi_D(N)$.

As it will be necessary to consider simultaneously several Lucas sequences with the same discriminant D, it is preferable to work with ψ_D, and not with the functions $\psi_{\alpha,\beta}$ corresponding to the various sequences.

Note, for example, that if $U(P,Q)$ has discriminant D, if $P' = P + 2$, $Q' = P + Q + 1$, then also $U(P',Q')$ has discriminant D.

It is good to start with some preparatory and easy results.

(V.1) If N is odd, $\gcd(N,D) = 1$, then $\psi_D(N) = N - (D/N)$ if and only if N is a prime.

Proof. If N is a prime, by definition $\psi_D(N) = N - (D/N)$.

If $N = p^e$ with p prime, $e \geqslant 2$, then $\psi_D(N)$ is a multiple of p, while $N - (D/N)$ is not.

If $N = \prod_{i=1}^{s}p_i^{e_i}$, with $s \geqslant 2$, then

$$\psi_D(N) \leqslant \frac{1}{2^{s-1}} \prod_{i=1}^{s} p_i^{e_i-1}(p_i + 1) = 2N \prod_{i=1}^{s} \frac{1}{2}\left[1 + \frac{1}{p_i}\right] \leqslant 2N \times \frac{2}{3} \times \frac{3}{5} \times \cdots$$

$$= \frac{4N}{5} < N - 1,$$

since $N > 5$. □

(V.2) If N is odd, $\gcd(N,D) = 1$, and $N - (D/N)$ divides $\psi_D(N)$, then N is a prime.

Proof. Assume that N is composite. First, let $N = p^e$, with p prime, $e \geqslant 2$; then $\psi_D(N) = p^e - p^{e-1}(D/p)$. Hence,

$$p^e - p^{e-1} < p^e - 1 \leqslant p^e - (D/N) \leqslant p^e - p^{e-1}(D/p),$$

so $(D/p) = -1$ and $p^e \pm 1$ divides $p^e + p^{e-1} < 2p^e - 2$, which is impossible.

If N has at least two distinct prime factors, it was seen in (V.1) that $\psi_D(N) < N - 1 \leqslant N - (D/N)$, which is contrary to the hypothesis. So N must be a prime.   □

(V.3) If N is odd, $U = U(P,Q)$ is a Lucas sequence with discriminant D, and $\gcd(N,QD) = 1$, then $N|U_{\psi_D(N)}$.

Proof. Since $\gcd(N,Q) = 1$, then by (IV.12) N divides $\lambda_{\alpha,\beta}(N)$, where α, β are the roots of $X^2 - PX + Q$. If $N = \prod_{i=1}^{s} p_i^{e_i}$, then

$$\lambda_{\alpha,\beta}(N) = \ell cm\left\{ p_i^{e_i-1}\left[p_i - \left(\frac{D}{p_i}\right)\right]\right\} = 2\,\ell cm\left\{ \frac{1}{2} p_i^{e_i-1}\left[p_i - \left(\frac{D}{p_i}\right)\right]\right\}$$

and $\lambda_{\alpha,\beta}(N)$ divides

$$2\prod_{i=1}^{s} \frac{1}{2} p^{e_i-1}\left[p_i - \left(\frac{D}{p_i}\right)\right] = \psi_D(N).$$

By (IV.15), N divides $U_{\psi_D(N)}$.   □

(V.4) If N is odd, $U = U(P,Q)$ is a Lucas sequence with discriminant D such that $(D/N) = -1$, and N divides U_{N+1}, then $\gcd(N,QD) = 1$.

Proof. Since $(D/N) \neq 0$, then $\gcd(N,D) = 1$.

If there exists a prime p such that $p|N$ and $p|Q$, since $p\nmid D = P^2 - 4Q$, then $p\nmid P$. By (IV.18) $p\nmid U_n$ for every $n \geqslant 1$, which is contrary to the hypothesis. So $\gcd(N,Q) = 1$.   □

One more result which will be needed is the following:

(V.5) Let N be odd and q be any prime factor of $N + 1$. Assume that $U = U(P,Q)$ and $V = V(P,Q)$ are the Lucas sequences associated with the integers P, Q, having discriminant $D \neq 0$. Assume $\gcd(P,Q) = 1$ or $\gcd(N,Q) = 1$. If N divides $U_{(N+1)/q}$ and $V_{(N+1)/2}$ then N divides $V_{(N+1)/2q}$.

Proof.

$$\frac{N + 1}{2} = \frac{N + 1}{2q} + \frac{N + 1}{q} u \quad \text{with} \quad u = \frac{q - 1}{2}.$$

By (IV.4):

$$2V_{(N+1)/2} = V_{(N+1)/2q} V_{[(N+1)/q]u} + DU_{(N+1)/2q} U_{[(U+1)/q]u}.$$

By (IV.15), N divides $U_{[(N+1)/q]u}$ so N divides $V_{(N+1)/2q} V_{[(N+1)/q]u}$.

If $\gcd(P,Q) = 1$, by (IV.21) $\gcd(U_{[(N+1)/q]u}, V_{[(N+1)/q]u}) = 1$ or 2, hence $\gcd(N, V_{[(N+1)/q]u}) = 1$, so N divides $V_{(N+1)/2q}$.

If $\gcd(N,Q) = 1$ and if there exists a prime p dividing N and $V_{[(N+1)/q]u}$, then by (IV.6) p divides also $4Q$; since p is odd, then $p|Q$, which is a contradiction. $\square$

Before indicating primality tests, it is easy to give sufficient conditions for a number to be composite:

Let $N > 1$ be an odd integer. Assume that there exists a Lucas sequence $(U_n)_{n \geqslant 0}$ with parameters (P,Q), discriminant D, such that $\gcd(N,QD) = 1$, $(Q/N) = 1$, and $N \nmid U_{(1/2)(N-(D/N))}$. Then N is composite.

Similarly, let $N > 1$ be an odd integer. Assume that there exists a companion Lucas sequence $(V_n)_{n \geqslant 0}$, with parameters (P,Q), discriminant D, such that $N \nmid QD$, $(Q/N) = -1$ and $N \nmid V_{(1/2)(N-(D/N))}$. Then N is composite.

Indeed, if $N = p$ is an odd prime, not dividing QD and if $(Q/p) = 1$, then $p|U_{\psi(p)/2}$, and similarly, if $(Q/p) = -1$ then $p|V_{\psi(p)/2}$, as stated in (IV.23). In both cases there is a contradiction. $\square$

And now I'm ready to present several tests; each one better than the preceding one.

Test 1. Let $N > 1$ be an odd integer and $N + 1 = \prod_{i=1}^{s} q_i^{\,i}$. Assume that there exists an integer D such that $(D/N) = -1$, and for every prime factor q_i of $N + 1$, there exists a Lucas sequence $(U_n^{(i)})_{n \geqslant 0}$ with discriminant $D = P_i^2 - 4Q_i$, where $\gcd(P_i, Q_i) = 1$, or $\gcd(N, Q_i) = 1$ and such that $N | U_{N+1}^{(i)}$ and $N \nmid U_{(N+1)/q_i}^{(i)}$. Then N is a prime.

Proof. By (V.3), (V.4), $N | U_{\psi_D(N)}^{(i)}$ for every $i = 1, ..., s$. Let $\rho^{(i)}(N)$ be the smallest integer r such that $N | U_r^{(i)}$. By (IV.29) or (IV.22) and the hypothesis, $\rho^{(i)}(N) | (N+1)$, $\rho^{(i)}(N) \nmid (N+1)/q_i$, and also $\rho^{(i)}(N) | \psi_D(N)$. Hence $q_i^{f_i} | \rho^{(i)}(N)$ for every $i = 1, ..., s$. Therefore, $(N+1) | \psi_D(N)$ and by (V.2), N is a prime. $\square$

Defect of this test: it requires the knowledge of all the prime factors of $N + 1$ and the calculation of $U_n^{(i)}$ for $n = 1, 2, ..., N + 1$.

The following test needs only half of the computations:

Test 2. Let $N > 1$ be an odd integer and $N + 1 = \prod_{i=1}^{s} q_i^{f_i}$. Assume that there exists an integer D such that $(D/N) = -1$, and for every prime factor q_i of $N + 1$, there exists a Lucas sequence $(V_n^{(i)})_{n \geqslant 0}$ with discriminant $D = P_i^2 - 4Q_i$, where $\gcd(P_i, Q_i) = 1$ or $\gcd(N, Q_i) = 1$, and such that $N | V_{(N+1)/2}^{(i)}$ and $N \nmid V_{(N+1)/2q_i}^{(i)}$. Then N is a prime.

Proof. By (IV.2), $N | U_{N+1}^{(i)}$. By (V.5), $N \mid U_{(N+1)/q_i}^{(i)}$. By the test 1, N is a prime. $\square$

The following tests will require only a partial factorization of $N + 1$.

Test 3. Let $N > 1$ be an odd integer, let q be a prime factor of $N+1$ such that $2q > \sqrt{N} + 1$. Assume that there exists a Lucas sequence $(V_n)_{n \geqslant 0}$, with discriminant $D = P^2 - 4Q$, where $\gcd(P, Q) = 1$ or $\gcd(N, Q) = 1$, and such that $(D/N) = -1$, and $N | V_{(N+1)/2}$, $N \nmid V_{(N+1)/2q}$. Then N is a prime.

Proof. Let $N = \Pi_{i=1}^{s}p_i^{e_i}$. By (IV.2), $N|U_{N+1}$, so by (IV.29) or (IV.22), $\rho(N)|(N+1)$. By (V.5), $N \nmid U_{(N+1)/q}$; hence, $\rho(N) \nmid (N+1)/q$, therefore $q|\rho(N)$. By (V.4) and (V.3), $N|U_{\psi_D(N)}$, so $\rho(N)$ divides $\psi_D(N)$, which in turn divides $N \Pi_{i=1}^{s}(p_i - (D/p_i))$.

Since $q \nmid N$, then there exists p_i such that q divides $p_i - (D/p_i)$, thus $p_i \equiv (D/p_i) \pmod{2q}$. In conclusion, $p_i \geqslant 2q - 1 > \sqrt{N}$ and $1 \leqslant N/p_i < \sqrt{N} < 2q - 1$, and this implies that $N/p_i = 1$, that is N is a prime.   □

Defect of this test: it needs the knowledge of a fairly large prime factor of $N + 1$.

The next test, which was proposed by Morrison in 1975, may be viewed as the analogue of Pocklington–Lehmer's test indicated in Section III:

Test 4. Let $N > 1$ be an odd integer and $N + 1 = FR$, where $\gcd(F,R) = 1$ and the factorization of F is known. Assume that there exists D such that $(D/N) = -1$ and, for every prime q_i dividing F, there exists a Lucas sequence $(U_n^{(i)})_{n \geqslant 0}$ with discriminant $D = P_i^2 - 4Q_i$, where $\gcd(P_i,Q_i) = 1$, or $\gcd(N,Q_i) = 1$ and such that $N|U_{N+1}^{(i)}$ and $\gcd(U_{(N+1)/q_i}^{(i)},N) = 1$. Then each prime factor p of N satisfies $p \equiv (D/p) \pmod{F}$. If, moreover, $F > \sqrt{N} + 1$, then N is a prime.

Proof. From the hypothesis, $\rho^{(i)}(N)|N+1$; a fortiori, $\rho^{(i)}(p)|N+1$. But $p \nmid U_{(N+1)/q}^{(i)}$, so $\rho^{(i)}(p) \nmid (N+1)/q_i$, by (IV.29) or (IV.22). If $q_i^{f_i}$ is the exact power of q_i dividing F, then $q_i^{f_i}|\rho^{(i)}(p)$, so by (IV.18), $q_i^{f_i}$ divides $p - (D/p)$, and this implies that F divides $p - (D/p)$.

Finally, if $F > \sqrt{N} + 1$, then

$$p + 1 \geqslant p - (D/p) \geqslant F > \sqrt{N} + 1;$$

hence, $p > \sqrt{N}$. This implies that N itself is a prime.   □

The next result tells more about the possible prime factors of N.

Let N be an odd integer, $N + 1 = FR$, where $\gcd(F,R) = 1$ and the factorization of F is known. Assume that there exists a Lucas sequence $(U_n)_{n \geqslant 0}$ with discriminant $D = P^2 - 4Q$, where $\gcd(P,Q) = 1$ or $\gcd(N,Q) = 1$ and such that $(D/N) = -1$, $N|U_{N+1}$, and $\gcd(U_F,N) = 1$. If p is a prime factor of N, then there exists a prime factor q of R such that $p \equiv (D/p) \pmod{q}$.

Proof. $\rho(p) \mid (p - (D/p))$ by (IV.18) and $\rho(p)|(N+1)$. But $p \nmid U_F$, so $\rho(p) \nmid F$. Hence, $\gcd(\rho(p),R) \neq 1$ and there exists a prime q such that $q|R$ and $q|\rho(p)$; in particular, $p \equiv (D/q) \pmod{q}$. $\square$

This result is used in the following test:

Test 5. Let $N > 1$ be an odd integer and $N + 1 = FR$, where $\gcd(F,R) = 1$, the factorization of F is known, R has no prime factor less than B, where $BF > \sqrt{N} + 1$. Assume that there exists D such that $(D/N) = -1$ and the following conditions are satisfied:

(i)   For every prime q_i dividing F there exists a Lucas sequence $(U_n^{(i)})_{n \geqslant 0}$, with discriminant $D = P_i^2 - 4Q_i$, where $\gcd(P_i,Q_i) = 1$ or $\gcd(N,Q_i) = 1$ and such that $N|U_{N+1}^{(i)}$ and $\gcd(U_{(N+1)/q_i}^{(i)},N) = 1$.

(ii)   There exists a Lucas sequence $(U_n')_{n \geqslant 0}$, with discriminant $D = P'^2 - 4Q'$, where $\gcd(P',Q') = 1$ or $\gcd(N,Q') = 1$ and such that $N|U_{N+1}'$ and $\gcd(U_F',N) = 1$.

Then N is a prime.

Proof. Let p be a prime factor of N. By Test 4, $p \equiv (D/p) \pmod{F}$ and by the preceding result, there exists a prime factor q of R such that $p \equiv (D/p) \pmod{q}$. Hence, $p \equiv (D/p) \pmod{qF}$ and so,

$$p + 1 \geqslant (D/p) \geqslant qF \geqslant BF > \sqrt{N} + 1.$$

Therefore, $p > \sqrt{N}$ and N is a prime number. $\square$

The preceding test is more flexible than the others, since it requires only a partial factorization of $N + 1$ up to a point where it may be assured that the nonfactored part of $N + 1$ has no factors less than B.

I now want to indicate, in a very succinct way, how to quickly calculate the terms of Lucas sequences with large indices. One of the methods is similar to that used in the calculations of high powers, which was indicated in Section III.

Write $n = n_0 2^k + n_1 2^{k-1} + \cdots + n_k$, with $n_i = 0$ or 1 and $n_0 = 1$; so $k = [(\log n)/(\log 2)]$. To calculate U_n (or V_n) it is necessary to perform the simultaneous calculation of U_m, V_m for various values of m. The following formulas are needed:

$$\begin{cases} U_{2j} = U_j V_j \,, \\ V_{2j} = V_j^2 - 2Q^j \,, \end{cases} \qquad \text{[see formula (IV.2)]}$$

$$\begin{cases} 2U_{2j+1} = V_{2j} + PU_{2j} \,, \\ 2V_{2j+1} = PV_{2j} + DU_{2j} \,. \end{cases} \qquad \text{[see formula (IV.5)]}$$

Put $s_0 = n_0 = 1$, and $s_{j+1} = 2s_j + n_{j+1}$. Then $s_k = n$. So, it suffices to calculate U_{s_j}, V_{s_j} for $j \leqslant k$; note that

$$U_{s_{j+1}} = U_{2s_j+n_{j+1}} = \begin{cases} U_{2s_j} & \text{or} \\ U_{2s_j+1} \,, \end{cases}$$

$$V_{s_{j+1}} = V_{2s_j+n_{j+1}} = \begin{cases} V_{2s_j} & \text{or} \\ V_{2s_j+1} \,. \end{cases}$$

Thus, it is sufficient to compute $2k$ numbers U_i and $2k$ numbers V_i, that is, only $4k$ numbers.

If it is needed to know U_n modulo N, then in all steps the numbers may be replaced by their least positive residues modulo N.

The second method is also very quick. For $j \geqslant 1$,

$$\begin{pmatrix} U_{j+1} & V_{j+1} \\ U_j & V_j \end{pmatrix} = \begin{pmatrix} P & -Q \\ 1 & 0 \end{pmatrix} \begin{pmatrix} U_j & V_j \\ U_{j-1} & V_{j-1} \end{pmatrix}.$$

If

$$M = \begin{bmatrix} P & -Q \\ 1 & 0 \end{bmatrix},$$

then

$$\begin{bmatrix} U_n & V_n \\ U_{n-1} & V_{n-1} \end{bmatrix} = M^{n-1} \begin{bmatrix} U_1 & V_1 \\ 0 & 2 \end{bmatrix}.$$

To find the powers of M, say M^m, write m in binary form and proceed in the manner followed to calculate a power of a number.

If U_n modulo N is to be determined, all the numbers appearing in the above calculation should be replaced by their least positive residues modulo N.

To conclude this section, I like to stress that there are many other primality tests of the same family, which are appropriate for numbers of certain forms, and use either Lucas sequences or other similar sequences.

I note, for example, the paper of Ward (1959), where he gave a primality test involving the Sylvester cyclotomic numbers Q_n, associated to the Lucas sequences $(U_n)_{n \geqslant 0}$, with parameters P, Q, discriminant $D \neq 0$, roots α, β. These numbers are

$$Q_n = \prod_r (\alpha - \zeta^r \beta),$$

where $\zeta = \cos 2\pi/n + i \sin 2\pi/n$ is a primitive nth root of 1, and the product is extended to all exponents r, $1 \leqslant r < n$, $\gcd(r,n) = 1$. The numbers Q_n are integers and satisfy the relation

$$U_n = \prod_{d|n} Q_d.$$

Ward's test is the following:

Ward's Test. Let N be an odd integer. Assume that there exists a Lucas sequence $(U_n)_{n \geqslant 0}$, with associated Sylvester cyclotomic numbers $(Q_n)_{n \geqslant 0}$, such that there exists $n > \sqrt{N}$, $\gcd(n,N) = 1$, for which N divides Q_n. Then N is a prime, unless it is of one of the following two forms:

(1) $N = (n - 1)^2$ with $n - 1$ prime, $n > 4$.
(2) $N = n^2 - 1$ with $n - 1$ and $n + 1$ primes.

Sometimes it is practical to combine tests involving Lucas sequences with the tests discussed in Section III; see the paper of Brillhart, Lehmer & Selfridge (1975). As a comment, I add (half-jokingly) the following rule of thumb: the longer the statement of the testing procedure, the quicker it leads to a decision about the primality.

The tests indicated so far are applicable to numbers of the form $2^n - 1$ (see Section VII on Mersenne numbers, where the test will be given explicitly), but also to numbers of the form $k \times 2^n - 1$ [see, for example, Inkeri's paper of 1960 or Riesel's book, (1985)].

The success of Lucas sequences in primality tests prompted Lehmer to introduce more general sequences where the parameters P, Q of the Lucas sequences are replaced by $\sqrt{R}$, Q, with R, Q integers, and $\gcd(R,Q) = 1$. As I already mentioned, these numbers are called Lehmer numbers (see Section IV, Addendum for their definition). The reader may wish to consult the original paper, which was in fact Lehmer's thesis (1930), dutifully dedicated to his father D. N. Lehmer, who must have been — I suspect — a little responsible for his son's interest in prime numbers. The thesis contains the theory of extended Lucas sequences, modeled after the classical theory; this includes evidently the laws of apparition (stubborn word! ... appearance) and of repetition of primes. Lehmer applied his sequences to test the primality of numbers of the form $k \times 2^n - 1$ or $k \times 2^n + 1$, but his proof relied on properties of certain quadratic number fields.

More recently, in 1972, Williams used Lucas sequences to give a necessary and sufficient condition for the primality of numbers $k \times 3^n - 1$, where (k is even and) $k < 3^n$. Ten years later, Williams (again) gave a primality test for numbers of the form $kb^{2n} + hb^n - 1$, in particular, for numbers $10^{2n} \pm 10^n - 1$. In the same paper, Williams spelled out a whole class of primality tests involving sequences like Lucas' or Lehmer's, or even more general sequences.

VI. Fermat Numbers

For numbers having a special form, there are more suitable methods to test whether they are prime or composite.

The numbers of the form $2^m + 1$ were considered long ago.

If $2^m + 1$ is a prime, then m must be of the form $m = 2^n$, so it is a Fermat number $F_n = 2^{2^n} + 1$.

The Fermat numbers $F_0 = 3$, $F_1 = 5$, $F_2 = 17$, $F_3 = 257$, $F_4 = 65537$ are primes. Fermat believed and tried to prove that all Fermat numbers are primes. Since F_5 has 10 digits, in order to test its primality, it would be necessary to have a table of primes up to 100,000 (which was unavailable to him) or to derive and use some criterion for a number to be a factor of a Fermat number.

Euler showed that every factor of F_n (with $n \geqslant 2$) must be of the form $k \times 2^{n+2} + 1$ and thus he discovered that 641 divides F_5:

$$F_5 = 641 \times 6700417.$$

Proof. It suffices to show that every prime factor p of F_n is of the form indicated. Since $2^{2^n} \equiv -1 \pmod{p}$, then $2^{2^{n+1}} \equiv 1 \pmod{p}$, so 2^{n+1} is the order of 2 modulo p. By Fermat's little theorem 2^{n+1} divides $p - 1$; in particular, 8 divides $p - 1$. Therefore the Legendre symbol is $2^{(p-1)/2} \equiv (2/p) \equiv 1 \pmod{p}$, and so 2^{n+1} divides $(p-1)/2$; this shows that $p = k \times 2^{n+2} + 1$. $\square$

Since the numbers F_n increase very rapidly with n, it becomes laborious to check their primality.

Using the converse of Fermat's little theorem, as given by Lucas, Pepin obtained in 1877 a test for the primality of Fermat numbers. Namely:

Pepin's Test. Let $F_n = 2^{2^n} + 1$ (with $n \geqslant 2$) and $k \geqslant 2$. Then, the following conditions are equivalent:

(i) F_n is prime and $(k/F_n) = -1$.

(ii) $k^{(F_n-1)/2} \equiv -1 \pmod{F_n}$.

Proof. If (i) is assumed, then by Euler's criterion for the Legendre symbol

$$k^{(F_n-1)/2} \equiv \left(\frac{k}{F_n} \right) \equiv -1 \pmod{F_n}.$$

If, conversely, (ii) is supposed true, let a, $1 \leqslant a < F_n$, be such that $a \equiv k \pmod{F_n}$. Since $a^{(F_n-1)/2} \equiv -1 \pmod{F_n}$, then $a^{F_n-1} \equiv 1 \pmod{F_n}$. By Lehmer's Test 3 in Section III, F_n is prime. Hence

$$\left(\frac{k}{F_n} \right) \equiv k^{(F_n-1)/2} \equiv -1 \pmod{F_n}. \qquad \square$$

Possible choices of k are $k = 3, 5, 10$, because $F_n \equiv 2 \pmod 3$, $F_n \equiv 2 \pmod 5$, $F_n \equiv 1 \pmod 8$; hence, by Jacobi's reciprocity law

$$\left(\frac{3}{F_n} \right) = \left(\frac{F_n}{3} \right) = \left(\frac{2}{3} \right) = -1,$$

$$\left(\frac{5}{F_n} \right) = \left(\frac{F_n}{5} \right) = \left(\frac{2}{5} \right) = -1,$$

$$\left(\frac{10}{F_n} \right) = \left(\frac{2}{F_n} \right)\left(\frac{5}{F_n} \right) = -1.$$

This test is very practical in application. However, if F_n is composite, the test does not indicate any factor of F_n.

Lucas used it to show that F_6 is composite, and, in 1880, Landry showed that

$$F_6 = 274177 \times 67280421310721.$$

The factorization of F_7 was first performed by Morrison & Brillhart (1970, published in 1971), that of F_8 by Brent & Pollard (1981); see also Brent (1982).

For the numbers F_n (with $n \geqslant 9$) no complete factorization is known. In many cases, a factor $k \times 2^{n+2} + 1$ has been determined. For example, Selfridge showed in 1953 that $3150 \times 2^{18} + 1$ divides F_{16}. Using Pepin's primality testing procedure, Selfridge & Hurwitz concluded in 1963 that F_{14} is composite, without determining any of its factors.

In this respect, a table of what is known appears in the 1983 book by Brillhart et al., as well as in Riesel's recent book (1985, pp. 377-379), which is very well written and rich in information concerning matters of primality and factorization. I recommend it to everyone wanting to study these questions further.

Record

The largest known Fermat prime is $F_4 = 65537$.

The largest known composite Fermat number is F_{23471} (Keller, 1984, announced in 1985), which has the factor $5 \times 2^{23473} + 1$ and more than 10^{7000} digits. Keller has also shown in 1980 that F_{9448} is composite, having the factor $19 \times 2^{9450} + 1$.

The smallest Fermat numbers which are not known to be prime or composite are F_{22}, F_{24}, F_{28}. I have last minute information that F_{20} is composite (established by Young & Buell, 1987).

Here are some open problems:

(1) Are there infinitely many prime Fermat numbers?

This question became significant with the famous result of Gauss (see *Disquisitiones Arithmeticae* articles 365, 366 — the last ones in the book — as a crowning result for much of the theory previously developed). He showed that if $n \geqslant 3$ is an integer, the regular polygon with n sides may be constructed by ruler and compass, if and only if $n = 2^k p_1 p_2 \cdots p_h$, where $k \geqslant 0$, $h \geqslant 0$ and $p_1, \ldots, p_h$ are distinct odd primes, each being a Fermat number.

In 1844, Eisenstein proposed, as a problem, to prove that there are indeed infinitely many prime Fermat numbers. I should add, that already in 1828, an anonymous writer stated that

$$2 + 1, \quad 2^2 + 1, \quad 2^{2^2} + 1, \quad 2^{2^{2^2}} + 1, \quad 2^{2^{2^{2^2}}} + 1, \; \ldots$$

are all primes, and in fact, they are the only prime Fermat numbers (apart from $2^{2^3} + 1$). However, as Selfridge showed, F_{16} is not a prime, and this fact disproved that conjecture.

(2) Are there infinitely many composite Fermat numbers?

Questions (1) and (2) seem beyond the reach of present-day

methods and, side by side, they show how little is known on this
matter.

(3) Is every Fermat number square-free (i.e. without square factors)?

It has been conjectured, for example by Lehmer and by Schinzel,
that there exist infinitely many square-free Fermat numbers.

It is not difficult to show that if p is a prime number and p^2
divides some Fermat number, then $2^{p-1} \equiv 1 \pmod{p^2}$ — this will be
proved in detail in Chapter 5, Section III.   Since Fermat numbers
are pairwise relatively prime, if there exist infinitely many Fermat
numbers with a square factor, then there exist infinitely many
primes p satisfying the above congruence.

I shall discuss this congruence in Chapter 5. Let it be said here
that it is very rarely satisfied. In particular, it is not known
whether it holds infinitely often.

Sierpiński considered in 1958 the numbers of form $S_n = n^n + 1$,
with $n \geqslant 2$.

He proved that if S_n is a prime, then there exists $m \geqslant 0$ such that
$n = 2^{2^m}$, so S_n is a Fermat number $S_n = F_{m+2^m}$. It follows that the
only numbers S_n which are primes and have less than 30,000,000,000
digits, are 5 and 257. Indeed, if $m = 0, 1$, one has $F_1 = 5$, $F_3 = 257$;
if $m = 2$, one obtains F_6 and if $m = 3$, one gets F_{11}, both composite.
For $m = 4$, one has F_{20}, which is now known to be composite. For
$m = 5$, it is F_{37}, which is not known to be prime or composite. Since
$2^{10} > 10^3$ then

$$F_{37} > 2^{2^{37}} > 2^{10^{11}} = (2^{10})^{10^{10}} > 10^{3\times10^{10}} ,$$

so F_{37} has more than 30,000,000,000 digits.

The primes of the form $n^n + 1$ are very rare. Are there only
finitely many such primes? If so, there are infinitely many composite Fermat numbers. But all this is pure speculation, with no
basis for any reasonable conjecture.

VII. Mersenne Numbers

If a number of the form $2^m - 1$ is a prime, then $m = q$ is a prime. Even more, it is not a difficult exercise to show that if $2^m - 1$ is a prime power, it must be a prime, and so m is a prime. [If you cannot do it alone, look at the paper of Ligh & Neal (1974).]

The numbers $M_q = 2^q - 1$ (with q prime) are called Mersenne numbers, and their consideration was motivated by the study of perfect numbers (see the addendum to this section).

From Mersenne's time, it was known that some Mersenne numbers are prime, others composite. For example, $M_2 = 3$, $M_3 = 7$, $M_5 = 31$, $M_7 = 127$ are primes, while $M_{11} = 23 \times 89$. In 1640, Mersenne stated that M_q is also a prime for $q = 13, 17, 19, 31, 67, 127, 257$; he was wrong about 67 and 257, and he did not include 61, 89, 107 (among those less than 257) which also produce Mersenne primes. Yet, his statement was quite astonishing, in view of the size of the numbers involved.

The obvious problem is to recognize if a Mersenne number is a prime, and if not, to determine its factors.

A classical result about factors was stated by Euler in 1750 and proved by Lagrange (1775) and again by Lucas (1878):

If q is a prime $q \equiv 3 \pmod 4$, then $2q + 1$ divides M_q if and only if $2q + 1$ is a prime; in this case, if $q > 3$, then M_q is composite.

Proof. Let $n = 2q + 1$ be a factor of M_q. Since $2^2 \not\equiv 1 \pmod n$, $2^q \not\equiv 1 \pmod n$, $2^{2q} - 1 = (2^q + 1)M_q \equiv 0 \pmod n$, then by Lucas test 2 (see Section III), n is a prime.

Conversely, let $p = 2q + 1$ be a prime. Since $p \equiv 7 \pmod 8$, then $(2/p) = 1$, so there exists m such that $2 \equiv m^2 \pmod p$. It follows that $2^q \equiv 2^{(p-1)/2} \equiv m^{p-1} \equiv 1 \pmod p$, so p divides M_q.

If, moreover, $q > 3$, then $M_q = 2^q - 1 > 2q + 1 = p$, so M_q is composite. $\square$

Thus if $q = 11, 23, 83, 131, 179, 191, 239, 251$, then M_q has the factor 23, 47, 167, 263, 359, 383, 479, 503, respectively.

Around 1825, Sophie Germain considered, in connection with

Fermat's last theorem, the primes q such that $2q + 1$ is also a prime. These primes are now called Sophie Germain primes, and I'll return to them in Chapter 5.

It is also very easy to determine the form of the factors of Mersenne numbers:

If n divides M_q, then $n \equiv \pm1$ (mod 8) and $n \equiv 1$ (mod q).

Proof. It suffices to show that each prime factor p of M_q is of the form indicated.

If p divides $M_q = 2^q - 1$, then $2^q \equiv 1$ (mod q); so by Fermat's little theorem, q divides $p - 1$, that is, $p - 1 = 2kq$ (since $p \neq 2$). So

$$\left(\frac{2}{p}\right) \equiv 2^{(p-1)/2} \equiv 2^{qk} \equiv 1 \quad (\text{mod } p),$$

therefore $p \equiv \pm1$ (mod 8), by the property of Legendre symbol already indicated in Section II. $\square$

The best method presently known to find out whether M_q is a prime or a composite number is based on the computation of a recurring sequence, indicated by Lucas (1878), and Lehmer (1930, 1935); see also Western (1932), Hardy & Wright (1938, p. 223), and Kaplansky (1945). However, explicit factors cannot be found in this manner.

If n is odd, $n \geqslant 3$, then $M_n = 2^n - 1 \equiv 7$ (mod 12). Also, if $N \equiv 7$ (mod 12), then the Jacobi symbol

$$\left(\frac{3}{N}\right) = \left(\frac{N}{3}\right)(-1)^{(N-1)/2} = -1.$$

Primality Test for Mersenne Numbers. Let $P = 2$, $Q = -2$, and consider the associated Lucas sequences $(U_m)_{m \geqslant 0}$, $(V_m)_{m \geqslant 0}$, which have discriminant $D = 12$. Then $N = M_n$ is a prime if and only if N divides $V_{(N+1)/2}$.

Proof. Let N be a prime. By (IV.2)

$$V^2_{(N+1)/2} = V_{N+1} + 2Q^{(N+1)/2} = V_{N+1} - 4(-2)^{(N-1)/2} \equiv V_{N+1} - 4\left(\frac{-2}{N}\right)$$

$$\equiv V_{N+1} + 4 \quad (\text{mod } N),$$

because

$$\left(\frac{-2}{N}\right) = \left(\frac{-1}{N}\right)\left(\frac{2}{N}\right) = -1,$$

since $N \equiv 3 \pmod 4$ and $N \equiv 7 \pmod 8$. Thus it suffices to show that $V_{N+1} \equiv -4 \pmod 4$.

By (IV.4), $2V_{N+1} = V_N V_1 + DU_N U_1 = 2V_N + 12U_N$; hence, by (IV.14) and (IV.13):

$$V_{N+1} = V_N + 6U_N \equiv 2 + 6(12/N) \equiv 2 - 6 \equiv 4 \pmod N.$$

Conversely, assume that N divides $V_{(N+1)/2}$. Then N divides U_{N+1} [by (IV.2)]. Also, by (IV.6) $V_{(N+1)/2}^2 - 12U_{(N+1)/2}^2 = 4(-1)^{(N+1)/2}$; hence, $\gcd(N, U_{(N+1)/2}) = 1$. Since $\gcd(N,2) = 1$, then by the Test 1 (Section V)), N is a prime. $\square$

For the purpose of calculation, it is convenient to replace the Lucas sequence $(V_m)_{m \geqslant 0}$ by the following sequence $(S_k)_{k \geqslant 1}$, defined recursively as follows:

$$S_0 = 4, \quad S_{k+1} = S_k^2 - 2;$$

thus the sequence begins with 4, 14, 194, ...

Then the test is phrased as follows:

$M_n = 2^n - 1$ is *prime if and only if* M_n *divides* S_{n-2}.

Proof. $S_0 = 4 = V_2/2$. Assume that $S_{k-1} = V_{2^k}/2^{2^{k-1}}$; then

$$S_k = S_{k-1}^2 - 2 = \frac{V_{2^k}^2}{2^{2^k}} - 2 = \frac{V_{2^{k+1}} + 2^{2^k+1}}{2^{2^k}} - 2 = \frac{V_{2^{k+1}}}{2^{2^k}}.$$

By the test, M_n is prime if and only if M_n divides

$$V_{(M_n+1)/2} = V_{2^{n-1}} = 2^{2^{n-2}} S_{n-2},$$

or equivalently, M_n divides S_{n-2}. $\square$

The repetitive nature of the computations makes this test quite suitable. In this way, all examples of large Mersenne primes have been discovered. Lucas himself showed, in 1876, that M_{127} is a prime, while M_{67} is composite. Not much later, Pervushin showed that M_{61} is also a prime. Finally, in 1927 (published in 1932) Lehmer showed that M_{257} is also composite, settling one way or another, what Mersenne had asserted. Note that M_{127} has 39 digits and was the largest prime known before the age of computers.

Record

There are now 30 known Mersenne primes. Here is a complete list, with the discoverer and year of discovery (the recordman discoverer is Robinson):

q	Year	Discoverer
2	--	--
3	--	--
5	--	--
7	--	--
13	1461	Anonymous*
17	1588	P. A. Cataldi
19	1588	P. A. Cataldi
31	1750	L. Euler
61	1883	I. M. Pervushin
89	1911	R. E. Powers
107	1913	E. Fauquembergue
127	1876	E. Lucas
521	1952	R. M. Robinson
607	1952	R. M. Robinson
1279	1952	R. M. Robinson
2203	1952	R. M. Robinson
2281	1952	R. M. Robinson
3217	1957	H. Riesel
4253	1961	A. Hurwitz
4423	1961	A. Hurwitz
9689	1963	D. B. Gillies
9941	1963	D. B. Gillies
11213	1963	D. B. Gillies
19937	1971	B. Tuckerman
21701	1978	C. Noll & L. Nickel
23209	1979	C. Noll
44497	1979	H. Nelson & D. Slowinski
86243	1982	D. Slowinski
132049	1983	D. Slowinski
216091	1985	D. Slowinski

*See Dickson's *History of the Theory of Numbers*, Vol. I, p. 6.

The Mersenne primes with $q \leqslant 127$ were discovered before the computer age. Turing made, in 1951, the first attempt to find Mersenne primes using an electronic computer; however, he was unsuccessful. In 1952, Robinson carried out Lucas's test using a computer SWAC (from the National Bureau of Standards in Los Angeles), with the assistance of D. H. and E. Lehmer. He discovered the Mersenne primes M_{521}, M_{607} on January 30, 1952 — the first such discoveries with a computer. The primes M_{1279}, M_{2203}, M_{2281} were found later in the same year.

The largest known prime is M_{216091}, it has 65050 digits and was discovered on September 1, 1985. In a letter to me, Slowinski wrote that "the program ran three hours on one processor of a two CPU CRAY XMP. Several CRAY computers have been searching for a new Mersenne prime in their idle moments for the past four years. The search has not been very organized and has not concentrated on any special areas. We did check certain interesting exponents, with special runs through. For example, M_{131071} was checked in 1979 and M_{132049} was not checked until 1983." I have also asked Slowinski about certain heuristic predictions on the size of Mersenne primes, among them Eberhart's conjecture, which I'll explain in Chapter 6. Slowinski answered: "I don't know of a justification for Eberhart's conjecture. I find the conjectures amusing, but have spent my efforts on improving my program." According to last minute information (thanks to Bob Silverman!), a systematic search has now been completed up to 150,000, which assures that 132049 provides indeed the 29th Mersenne prime (in increasing order).

On the other hand, the search for Sophie Germain primes q of the form $q = k \times 2^N - 1$ (so, $2q + 1$ is also prime) yields, as already indicated, composite Mersenne numbers M_q.

Record

The largest known composite Mersenne number is M_q with $q = 16695 \times 2^{3002} - 1$ (Keller, 1985, communicated by letter).

Riesel's book (1985) has a table of complete factorization of all numbers $M_n = 2^n - 1$, with n odd, $n \leqslant 257$. A more extensive table is in the book of Brillhart et al. (1983).

Just as for Fermat numbers, there are many open problems about Mersenne numbers:

(1) Are there infinitely many Mersenne primes?

(2) Are there infinitely many composite Mersenne numbers?

The answer to both questions ought to be "yes", as I will try to justify. For example, I will indicate in Chapter 6, Section A, after (D5), that some sequences, similar to the sequence of Mersenne numbers, contain infinitely many composite numbers.

(3) Is every Mersenne number square-free?

Rotkiewicz showed in 1965 that if p is a prime and p^2 divides some Mersenne number, then $2^{p-1} \equiv 1 \pmod{p^2}$, the same congruence which already appeared in connection with Fermat numbers having a square factor; for the easy proof see Chapter 5, III.

I wish to mention two other problems involving Mersenne numbers, one of which has been solved, while the other one is still open.

Is it true that if M_q is a Mersenne prime, then M_{M_q} is also a prime number?

The answer is negative, since despite M_{13} being prime, $M_{M_{13}} = 2^{8191} - 1$ is composite [shown by Wheeler, see Robinson (1954)]. Note that $M_{M_{13}}$ has more than 2400 digits. In 1976, Keller discovered the prime factor

$$p = 2 \times 20644229 \times M_{13} + 1 = 338193759479$$

of the Mersenne number $M_{M_{13}}$, thus providing an easier verification that it is composite; only 13 squarings modulo p were needed to show that $2^{2^{13}} \equiv 2 \pmod{p}$. This has been kindly communicated to me by Keller in a recent letter, which contained also a wealth of other useful information. This factor is included in G. Haworth's *Mersenne Numbers*, quoted as reference 203 in Shanks' book (you guess which), now in its third edition.

The second problem, by Catalan (1876), reported in Dickson's *History of the Theory Numbers*, Volume I, page 22, is the following. Consider the sequence of numbers

$$C_1 = 2^2 - 1 = 3 = M_2, \quad C_2 = 2^{C_1} - 1 = 7 = M_3,$$

$$C_3 = 2^{C_2} - 1 = 2^7 - 1 = 127 = M_7,$$

$$C_4 = 2^{C_3} - 1 = 2^{127} - 1 = M_{127}, \ ..., \ C_{n+1} = 2^{C_n} - 1, \ ....$$

Are all numbers C_n primes? Are there infinitely many which are prime?

At present, it is impossible to test C_5, which has more than 10^{38} digits!

Addendum on Perfect Numbers

I shall consider now the perfect numbers, which are closely related to the Mersenne numbers.

A natural number $n > 1$ is said to be perfect if it is equal to the sum of all its aliquot parts, that is, its divisors d, with $d < n$. For example, $n = 6, 28, 496, 8128$ are the perfect numbers smaller than 10,000.

Perfect numbers were already known in ancient times. The first perfect number 6 was connected, by mystic and religious writers, to perfection, thus explaining that the Creation required 6 days, so PERFECT it is.

Euclid showed, in his *Elements*, Book IX, Proposition 36, that if q is a prime and $M_q = 2^q - 1$ is a prime, then $n = 2^{q-1}(2^q - 1)$ is a perfect number.

In a posthumous paper, Euler proved the converse: any even perfect number is of the form indicated by Euclid. Thus, the knowledge of even perfect numbers is equivalent to the knowledge of Mersenne primes.

And what about odd perfect numbers? Do they exist? Not even one has ever been found! This is a question which has been extensively searched, but its answer is still unknown.

Quick information on the progress made toward the solution of the problem may be found in Guy's book (1981) (quoted in General References), Section B1.

The methods to tackle the problem have been legion. I believe it is useful to describe them so the reader will get a feeling of what to do when nothing seems reasonable. The idea is to assume that there exists an odd perfect number N and to derive various consequences, concerning the number $\omega(N)$ of its distinct prime factors, the size of N, the multiplicative form, and the additive form of N, etc. I'll review what has been proved in each count.

(a) Number of distinct prime factors $\omega(N)$:

1980 (announced in 1975), Hagis: $\omega(N) \geqslant 8$.

1983 Hagis, and independently, Kishore: if $3 \nmid N$, then $\omega(N) \geqslant 11$.

Previous work was by Kühnel (1949), Robbins (1972), Pomerance (1974, 1975), and Kishore (1977). Chein proved also in his thesis, independently, that $\omega(N) \geqslant 8$ (this was in 1979).

Another result in this line was given by Dickson in 1913: for every $k \geqslant 1$ there are at most finitely many odd perfect numbers N, such that $\omega(N) = k$. In 1949, Shapiro gave a simpler proof.

Dickson's theorem was generalized in 1956 by Kanold, for numbers N satisfying the condition $\sigma(N)/N = \alpha$ (α is a given rational number); the proof involved the fact that the equation $aX^3 - bY^3 = c$ has at most finitely many solutions in integers x, y. Since an effective estimate for the number of solutions was given by Baker, with his celebrated method of linear forms in logarithms, it became possible for Pomerance to show in 1977 (taking $\alpha = 2$), for every $k \geqslant 1$: If the odd perfect number N has k distinct prime factors, then

$$N < (4k)^{(4k)^{2^{k^2}}}.$$

(b) Lower bound for N:

I have been informed by Silverman that Brent has just established that $N > 10^{150}$. A previously published bound was by Hagis: $N > 10^{50}$ (in 1973).

In 1976, Buxton & Elmore have claimed that $N > 10^{200}$, but this statement has not been substantiated in detail, so it should not be acceptable.

Previous work was by Bernhard (1949), Kühnel (1949), Kanold (1957), Tuckerman (1967), Kishore (1977).

(c) Multiplicative structure of N:

The first result is by Euler. $N = p^e k^2$, where p is a prime not dividing k, and $p \equiv e \equiv 1 \pmod 4$.

There have been numerous results on the kind of number k, which are more or less technical, but useful anyway. This work has been done by Brauer (1943), Kanold (1941, 1942, 1953), Levit (1947), Webber (1951), McCarthy (1957), and Hagis & McDaniel (1972). For example, the latter authors showed that $N \neq p^e k^6$, with k square-free, p a prime not dividing k, $p \equiv e \equiv 1 \pmod 4$.

(d) Largest prime factor of N:

In 1975, Hagis & McDaniel showed that the largest prime factor of N should be greater than 100110.

Already in 1955, Muskat had even shown better: there must be a prime factor larger than 10^8. However, his result, which was based on calculations of Mills, remained unpublished (but it is quoted at the end of Ore's book, 1956).

For prime-power factors, Muskat showed in 1966 that N must have one which is greater than 10^{12}.

(e) Other prime factors of N:

In 1975, Pomerance showed that the second largest prime factor of N should be at least 139.

Previous work was by Kanold (1944).

In 1952, Grün showed that the smallest prime factor p_1 of N should satisfy the relation $p_1 < \frac{2}{3}\,\omega(N) + 2$. Technical improvements were obtained by Norton in 1961.

In his thesis (1977), Kishore showed that if $i = 1, 3, 4, 5, 6$, the ith smallest prime factor of N is less than $2^{2^{i-1}}(\omega(N) - i + 1)$.

In 1958, Perisastri proved that

$$\frac{1}{2} < \sum_{p|N} \frac{1}{p} < 2 \log \frac{\pi}{2}.$$

This has been sharpened by Suryanarayana (1963), Suryanarayana & Hagis (1970), and Cohen (1978).

(f) Additive structure of N:

In 1953, Touchard proved that $N \equiv 1 \pmod{12}$ or $N \equiv 9 \pmod{36}$. An easier proof was later given by Satyanarayana (1959).

(g) Ore's conjecture:

In 1948, Ore considered the harmonic mean of the divisors of N, namely,

$$H(N) = \frac{\tau(N)}{\sum\limits_{d|n} \frac{1}{d}} = \frac{n\tau(n)}{\sigma(n)} \; ,$$

where $\tau(N)$ denotes the number of divisors of N, and $\sigma(N)$ the sum of divisors.

If N is a perfect number, then $H(N)$ is an integer; indeed, whether N is even or odd, this follows from Euler's results.

Actually, Laborde noted in 1955, that N is an even perfect number if and only if $N = 2^{H(N)-1}(2^{H(N)} - 1)$, hence $H(N)$ is an integer, and in fact a prime.

Ore conjectured that if N is odd, then $H(N)$ is not an integer. The truth of this conjecture would imply, therefore, that there do not exist odd perfect numbers.

Ore verified that the conjecture is true if N is a prime-power or if $N < 10^4$. Since 1954 (published only in 1972), Mills checked its truth for $N < 10^7$, as well as for numbers of special form, in particular, if all prime-power factors of N are smaller than 65551^2.

Pomerance (unpublished) verified Ore's conjecture when $\omega(N) \leqslant 2$, by showing that if $\omega(N) \leqslant 2$ and $H(N)$ is an integer, then N is an even perfect number (kindly communicated to me by letter).

The next results do not distinguish between even or odd perfect numbers. They concern the distribution of perfect numbers. The idea is to define, for every $x \geqslant 1$, the function $V(x)$, which counts the perfect numbers less or equal to x:

$$V(x) = \#\{N \text{ perfect} \mid N \leqslant x\}.$$

The limit $\lim_{x \to \infty} V(x)/x$ represents a natural density for the set of perfect numbers. In 1954, Kanold showed the $\lim_{x \to \infty} V(x)/x = 0$. Thus, $V(x)$ grows to infinity slower than x does.

The following more precise result of Wirsing (1959) tells how slow $V(x)$ grows: there exist x_0 and $C > 0$ such that if $x \geqslant x_0$ then

$$V(x) \leqslant e^{\frac{C \log x}{\log \log x}} \; .$$

Earlier work was done by Hornfeck (1955, 1956), Kanold (1957) and Hornfeck & Wirsing (1957), who had established that for every

$\epsilon > 0$ there exists a positive constant C such that $V(x) < Cx^{\epsilon}$.

All the results that I have indicated about the problem of the existence of odd perfect numbers represent a considerable amount of work, sometimes difficult and delicate. Yet, I believe the problem stands like an unconquerable fortress. For all that is known, it would be almost by luck that an odd perfect number would be found. On the other hand, nothing that has been proved is promising to show that odd perfect numbers do not exist. New ideas are required.

I wish to conclude this overview of perfect numbers with the following results of Sinha (1974) (the proof is elementary and should be an amusing exercise — just get your pencil ready!):

28 is the only even perfect number that is of the form $a^n + b^n$, with $n \geqslant 2$, and $\gcd(a,b) = 1$. It is also the only even perfect number of the form $a^n + 1$, with $n \geqslant 2$. And finally, there is no even perfect number of the form

$$a^{n^{n^{.^{.^{n}}}}} + 1$$

with $n \geqslant 2$ and at least two exponents n.

Looking back, perfect numbers are defined by comparing N with $\sigma(N)$, the sum of its divisors. Demanding just that N divides $\sigma(N)$ leads to the multiple perfect numbers. Numbers with $2N < \sigma(N)$ are called abundant, while those with $2N > \sigma(N)$ are called deficient.

Let $s(N) = \sigma(N) - N$, the sum of aliquot pairs of N, that is, the sum of proper divisors of N. Since some numbers are abundant and others are deficient, it is natural to iterate the process of getting $s(N)$, namely to build the sequence: $s(N)$, $s^2(N) = s(s(N))$, $s^3(N) = s(s^2(N))$, etc.

This leads to many fascinating questions, as they are described in Guy's book. I am forced to abstain from discussing this matter.

VIII. Pseudoprimes

A problem, commonly attributed to the ancient Chinese, was to ascertain whether a natural number n must be a prime if it satisfies the congruence

$$2^n \equiv 2 \pmod{n}.$$

However, it is incorrect to ascribe this question to the ancient Chinese, who could not even state it, since they never formulated the concept of prime number. In fact, thanks to Erdős, I got into direct contact with Man-Keung Siu, from Hong Kong, self-confessed "mathematician, who is deeply interested in the history of mathematics and believes in its value." In a letter to me, Siu wrote that this myth originated in a paper of J. H. Jeans, in the *Messenger of Mathematics*, **27**, 1897/8, who wrote that "a paper found among those of the late Sir Thomas Wade and dating from the time of Confucius" contained the theorem that $2^n \equiv 2 \pmod{n}$ holds if and only if n is a prime number. However, in a footnote of his monumental work *Science and Civilization in China*, Volume 3, Chapter 19 (Mathematics), J. Needham dispels Jeans' assertion, which is due to an erroneous translation of a passage of the famous book *The Nine Chapters of Mathematical Art*. This mistake has been perpetuated by several Western scholars.   In Dickson's *History of the Theory of Numbers*, Volume I, page 91, it is quoted that Leibniz believed to have proved that the so-called Chinese congruence indicated above implies that n is a prime.

The story is also repeated, for example in Honsberger's very nicely written chapter "An Old Chinese Theorem and Pierre de Fermat" in his book *Mathematical Gems*, Volume I, 1973.

A. Pseudoprimes in Base 2 (psp)

Now I return to the problem concerning the congruence $2^n \equiv 2 \pmod{n}$, which might be appropriately called, if not as a joke, the "pseudo-Chinese congruence on pseudoprimes."

The question remained open until 1819, when Sarrus showed that $2^{341} \equiv 2 \pmod{341}$, yet $341 = 11 \times 31$ is a composite number. In particular, a crude converse of Fermat's little theorem is false.

Other composite numbers with this property are, for example: 561, 645, 1105, 1387, 1729, 1905.

A composite number n satisfying the congruence $2^{n-1} \equiv 1 \pmod{n}$ is called a pseudoprime, or also a Poulet number since they were the focus of his attention. In particular, he computed, as early as 1926, a table of pseudoprimes up to 5×10^7, and in 1938 up to 10^8; see reference in Chapter 4.

Every pseudoprime n is odd and satisfies the congruence $2^n \equiv 2 \pmod{n}$; conversely, every odd composite number satisfying this congruence is a pseudoprime.

Clearly, every odd prime number satisfies the above congruence, so if $2^{n-1} \not\equiv 1 \pmod{n}$, then n must be composite. This is useful as a first step in testing primality.

In fact, Lehmer suggested the following practical method to test the primality of numbers below a given limit L: first, to make a list of all pseudoprimes less than L (this he did in 1936, for $L = 2 \times 10^8$); if $n < L$ and $2^{n-1} \not\equiv 1 \pmod{n}$, then n is composite; if $2^{n-1} \equiv 1 \pmod{n}$ and n is in the list of pseudoprimes, then it is composite; otherwise, n is prime.

In order to know more about primes, it is natural to study the pseudoprimes.

Suppose I would like to write a chapter about pseudoprimes for the *Guinness Book of Records*. How would I organize it?

The natural questions should be basically the same as those for prime numbers. For example: How many pseudoprimes are there? Can one tell whether a number is a pseudoprime? Are there ways of generating pseudoprimes? How are the pseudoprimes distributed?

As it turns out, not surprisingly, there are infinitely many pseudo-primes, and there are many ways to generate infinite sequences of pseudoprimes.

The simplest proof was given in 1903 by Malo, who showed that if n is a pseudoprime, and if $n' = 2^n - 1$, then n' is also a pseudoprime. Indeed, n' is obviously composite, because if $n = ab$ with $1 < a, b < n$, then $2^n - 1 = (2^a - 1)(2^{a(b-1)} + 2^{a(b-2)} + \cdots + 2^a + 1)$. Also n divides $2^{n-1} - 1$, hence n divides $2^n - 2 = n' - 1$; so $n' = 2^n - 1$ divides $2^{n'-1} - 1$.

In 1904, Cipolla gave another proof, using the Fermat numbers:

If $m > n > \cdots > s > 1$ are integers and N is the product of the Fermat numbers $N = F_m F_n \cdots F_s$, then N is a pseudoprime if and only if $2^s > m$.

Indeed, the order of 2 modulo N is $2^{2^{m+1}}$, which is equal to the least common multiple of the orders $2^{2^{m+1}}$, $2^{2^{n+1}}$, ..., $2^{2^{s+1}}$ of 2 modulo each factor factor $F_m, F_{n'}$ ..., F_s of N. Thus $2^{N-1} \equiv 1 \pmod{N}$ if and only if $N - 1$ is divisible by 2^{m+1}. But $N - 1 = F_m F_n \cdots F_s - 1 = 2^{2^s} Q$, where Q is an odd integer. Thus, the required condition is $2^s > m$.   □

As it was indicated in Chapter 1, the Fermat numbers are pairwise relatively prime, so the above method leads to pairwise relatively prime pseudoprimes. See below, for another method also indicated by Cipolla.

There also exist even composite integers satisfying the congruence $2^n \equiv 2 \pmod{n}$ − they may be called even pseudoprimes. The smallest one is $m = 2 \times 73 \times 1103 = 161038$, discovered by Lehmer in 1950. In 1951, Beeger showed the existence of infinitely many even pseudoprimes; each one must have at least two odd prime factors.

How "far" are pseudoprimes from being primes? From Cipolla's result, there are pseudoprimes with arbitrarily many prime factors. This is not an accident. In fact, in 1949, Erdös proved that for every $k \geqslant 2$ there exist infinitely many pseudoprimes, which are the product of exactly k distinct primes.

In 1936, Lehmer gave criteria for the product of two or three distinct odd primes to be a pseudoprime:

$p_1 p_2$ is a pseudoprime if and only if ord($2 \bmod p_2$) divides $p_1 - 1$ and ord($2 \bmod p_1$) divides $p_2 - 1$.

If $p_1 p_2 p_3$ is a pseudoprime, then the least common multiple of ord($2 \bmod p_1$) and ord($2 \bmod p_2$) divides $p_3(p_1 + p_2 - 1) - 1$.

Here is an open question: Are there infinitely many integers $n > 1$ such that

$$2^{n-1} \equiv 1 \pmod{n^2} ?$$

This is equivalent to each of the following problems (see Rotkiewicz, 1965):

Are there infinitely many pseudoprimes that are squares?

Are there infinitely many primes p such that $2^{p-1} \equiv 1 \pmod{p^2}$?

This congruence was already encountered in the question of

square factors of Fermat numbers and Mersenne numbers. I shall return to primes of this kind in Chapter 5, Section III.

On the other hand, a pseudoprime need not be square-free. The smallest such examples are $1,194,649 = 1093^2$, $12,327,121 = 3511^2$, $3,914,864,773 = 29 \times 113 \times 1093^2$.

B. Pseudoprimes in Base a (psp(a))

It is also useful to consider the congruence $a^{n-1} \equiv 1 \pmod{n}$, for $a \neq 1$. If n is a prime and $1 < a < n$, then the above congruence holds necessarily. So, if, for example, $2^{n-1} \equiv 1 \pmod{n}$, but, say, $3^{n-1} \not\equiv 1 \pmod{n}$, then n is not a prime.

This leads to the more general study of the a-pseudoprimes n, namely, the composite integers $n > a$ such that $a^{n-1} \equiv 1 \pmod{n}$.

In 1904, Cipolla indicated also how to obtain a-pseudoprimes. Let $a \geq 2$, let p be any odd prime such that p does not divide $a(a^2 - 1)$. Let

$$n_1 = \frac{a^p - 1}{a - 1}, \quad n_2 = \frac{a^p + 1}{a + 1}, \quad n = n_1 n_2;$$

then n_1 and n_2 are odd and n is composite. Since $n_1 \equiv 1 \pmod{2p}$ and $n_2 \equiv 1 \pmod{2p}$, then $n \equiv 1 \pmod{2p}$. From $a^{2p} \equiv 1 \pmod{n}$ it follows that $a^{n-1} \equiv 1 \pmod{n}$, so n is an a-pseudoprime.

Since there exist infinitely many primes, then there also exist infinitely many a-pseudoprimes (also when $a > 2$).

There are other methods in the literature to produce very quickly increasing sequences of a-pseudoprimes.

Steuerwald proceeded as follows in 1948 (this was brought to my attention by Schinzel — thanks Andrzej!). Let n be an a-pseudoprime, which is prime to $a - 1$. For example, if $a - 1 = q$ is a prime and $p > a^2 - 1$, in the Cipolla construction

$$n_1 = \frac{a^p - 1}{a - 1} = a^{p-1} + a^{p-2} + \cdots + a + 1 \equiv p \pmod{q},$$

$$n_2 = \frac{a^p + 1}{a + 1} = a^{p-1} - a^{p-2} + \cdots + a^2 - a + 1 \equiv 1 \pmod{q},$$

so $n = n_1 n_2 \equiv p \pmod{q}$. Let $f(n) = (a^n-1)/(a-1) > n$. Then $f(n)$ is also an a-pseudoprime. Indeed, $f(n)$ is composite and n divides

$$\frac{a^n - a}{a - 1} = f(n) - 1,$$ so $f(n)$ divides $a^n - 1$, which divided $a^{(a^{n-1})/(a-1)} - 1$

$= a^{f(n)-1} - 1$. Also

$$f(n) = \frac{[(a-1)+1]^n - 1}{a - 1} = (a-1)^{n-1} + \binom{n}{1}(a-1)^{n-2} + \cdots + n \equiv n \pmod{a-1},$$

so $f(n)$ is an a-pseudoprime, which is prime to $a - 1$. This process may be iterated, and leads to an infinite increasing sequence of a-pseudoprimes $n < f(n) < f(f(n)) < f(f(f(n))) < \cdots$, which grows as

$$n, a^n, a^{a^n}, a^{a^{a^n}}, \ldots .$$

Crocker showed in 1962 how to generate other infinite sequences of a-pseudoprimes. Let a be even, but not of the form 2^{2^r}, with $r \geqslant 0$. Then, for every $n \geqslant 1$, the number $a^{a^n} + 1$ is an a-pseudoprime.

It is therefore futile to wish to write the largest a-pseudoprime.

In 1959, Schinzel showed that for every $a \geqslant 1$, there exist infinitely many pseudoprimes in base a that are products of two distinct primes.

In 1971, in his thesis, Lieuwens extended simultaneously this result of Schinzel and Erdös' result about pseudoprimes in base 2: for every $k \geqslant 2$ and $a \geqslant 1$, there exist infinitely many pseudoprimes in base a, which are products of exactly k distinct primes.

In 1972, Rotkiewicz showed that if $p \geqslant 2$ is a prime not dividing $a \geqslant 2$, then there exist infinitely many pseudoprimes in base a that are multiples of p; the special case when $p = 2$ dates back to 1959, also by Rotkiewicz.

Here is a table, from the paper of Pomerance, Selfridge & Wagstaff (1980), which gives the smallest pseudoprimes for various bases, or simultaneous bases.

Bases	Smallest psp
2	$341 = 11 \times 31$
3	$91 = 7 \times 13$
5	$217 = 7 \times 31$
7	$25 = 5 \times 5$
2,3	$1105 = 5 \times 13 \times 17$
2,5	$561 = 3 \times 11 \times 17$
2,7	$561 = 3 \times 11 \times 17$
3,5	$1541 = 23 \times 67$
3,7	$703 = 19 \times 37$
5,7	$561 = 3 \times 11 \times 17$
2,3,5	$1729 = 7 \times 13 \times 19$
2,3,7	$1105 = 5 \times 13 \times 17$
2,5,7	$561 = 3 \times 11 \times 17$
3,5,7	$29341 = 13 \times 37 \times 61$
2,3,5,7	$29341 = 13 \times 37 \times 61$

To be a pseudoprime for different bases, like 561 (for the bases 2, 5, 7) is not abnormal. Indeed, Baillie & Wagstaff and Monier showed independently (1980):

If n is a composite number, then the numbers $B_{psp}(n)$ of bases a, $1 < a \leqslant n - 1$, $\gcd(a,n) = 1$, for which n is a psp(a), is equal to $\{\Pi_{p|n}\gcd(n-1, p-1)\} - 1$.

It follows that if n is an odd composite number, which is not a power of 3, then n is a pseudoprime for at least two bases a, $1 < a < n - 1$, $\gcd(a,n) = 1$.

It will be seen in Section IX that there exist composite numbers n, which are pseudoprimes for all bases a, $1 < a < n$.

By experimentation, it is observed that, usually, $B_{psp}(n)$ is very small compared to n, and, in fact, a statement of this kind was shown by Pomerance (it is in Baillie & Wagstaff's paper).

This matter was given more attention in the recent paper of Erdös & Pomerance (1986); for convenience, they considered the function $F(n) = B_{psp}(n) + 1$ and gave the following estimates for the arithmetic and geometric means of $F(n)$, where n is composite, $n \leqslant x$, x is large:

$$x^{15/23} < \frac{1}{x} \sum_n F(n) \leqslant \frac{x}{\ell(x)^{1+o(1)}}$$

where

$$\ell(x) = \exp^{(\log x \, \log \log \log x)/(\log \log x)} ,$$

and

$$\left[\prod_{n \leqslant x} F(n) \right]^{1/x} = C(\log x)^{C'} + o(1).$$

[For those who need, the symbol $o(1)$ will be explained in Chapter 4, Section I.]

On the other hand, the following problem has never been given an answer: to find whether there exist integers $a, a' \geqslant 2$ (which are not simultaneously of the form $a = b^h$, $a' = b^k$ with $b, h, k \geqslant 1$), such that there exist infinitely many integers which are pseudoprimes simultaneously for the bases a, a'. This problem was first proposed by Duparc, in 1952.

As I have said, if there exists a such that $1 < a < n$ and $a^{n-1} \not\equiv 1$ (mod n), then n is composite, but not conversely. This gives therefore a very practical way to ascertain that many numbers are composite. There are other congruence properties, similar to the above, which give also easy methods to discover that certain numbers are composite.

I shall describe several of these properties; their study has been justified by the problem of primality testing. As a matter of fact, without saying it explicitly, I have already considered these properties in Sections III and V. First, there are properties about the congruence $a^m \equiv 1$ (mod n), which lead to the Euler a-pseudoprimes and strong a-pseudoprimes. In another section, I will examine the Lucas pseudoprimes, which concern congruence properties satisfied by terms of Lucas sequences.

C. Euler Pseudoprimes in Base a (epsp(a))

According to Euler's congruence for the Legendre symbol, if $a \geqslant 2$, p is a prime and p does not divide a, then

$$\left(\frac{a}{p} \right) \equiv a^{(p-1)/2} \pmod{p}.$$

This leads to the notion of an Euler pseudoprime n in base a (epsp(a)), proposed by Shanks in 1962. These are odd composite numbers n, such that $\gcd(a,n) = 1$ and the Jacobi symbol satisfies the congruence

$$\left[\frac{a}{n}\right] \equiv a^{(n-1)/2} \pmod{n}.$$

Clearly, every epsp(a) is an a-pseudoprime.

There are many natural questions about epsp(a) which I enumerate now:

(e1) Are there infinitely many epsp(a), for each a?

(e2) Are there epsp(a) with arbitrary large number of distinct prime factors, for each a?

(e3) For every $k \geqslant 2$ and base a, are there infinitely many epsp(a), which are equal to the product of exactly k distinct prime factors?

(e4) Can an odd composite number be an epsp(a) for every possible a, $1 < a < n$, $\gcd(a,n) = 1$?

(e5) For how many bases a, $1 < a < n$, $\gcd(a,n) = 1$, can the number n be an epsp(a)?

In 1986, Kiss, Phong & Lieuwens showed that given $a \geqslant 2$, $k \geqslant 2$, and $d \geqslant 2$, there exist infinitely many epsp(a), which are the product of k distinct primes and are congruent to 1 modulo d.

This gives a strong affirmative answer to (e3), and therefore also to (e2) and (e1).

In 1976, Lehmer showed that if n is odd composite, then it cannot be an epsp(a), for every a, $1 < a < n$, $\gcd(a,n) = 1$. So the answer to (e4) is negative.

In fact, more is true, as shown by Solovay & Strassen in 1977: a composite integer n can be an Euler pseudoprime for at most $\frac{1}{2} \phi(n)$ bases a, $1 < a < n$, $\gcd(a,n) = 1$. This gives an answer to question (e5). The proof is immediate, noting that the residue classes a mod n, for which $(a/n) \equiv a^{(n-1)/2} \pmod{n}$ form a subgroup of $(\mathbb{Z}/n)^{\times}$ (group of invertible residue classes modulo n), which is a proper subgroup (by Lehmer's result); hence it has at most $\frac{1}{2} \phi(n)$ elements — by dear old Lagrange's theorem.

Let n be an odd composite integer. Denote by $B_{\text{epsp}}(n)$ the number of bases a, $1 < a < n$, $\gcd(a,n) = 1$, such that n is an epsp(a). Monier showed in 1980 that

$$B_{epsp}(n) = \delta(n) \prod_{p|n} \gcd\left[\frac{n-1}{2}, p-1\right] - 1.$$

Here

$$\delta(n) = \begin{cases} 2 \text{ if } v_2(n) - 1 = \min_{p|n}\{v_2(p-1)\}, \\ \frac{1}{2} \text{ if there exists a prime } p \text{ dividing } n \text{ such that } v_{p(n)} \text{ is} \\ \quad \text{odd and } v_2(p-1) < v_2(n-1), \\ 1 \text{ otherwise,} \end{cases}$$

and for any integer m and prime p, $v_p(m)$ denotes the exponent of p in the factorization of m.

D. Strong Pseudoprimes in Base a (spsp(a))

A related property is the following: Let n be an odd composite integer, let $n - 1 = 2^s d$, with d odd and $s \geqslant 0$; let a be such that $1 < a < n$, $\gcd(a,n) = 1$.

n is a strong pseudoprime in base a (spsp(a)) if $a^d \equiv 1 \pmod{n}$ or $2^{2^r d} \equiv -1 \pmod{n}$, for some r, $0 \leqslant r < s$.

Note that if n is a prime, then it satisfies the above condition, for every a, $1 < a < n$, $\gcd(a,n) = 1$.

Selfridge showed (see the proof in Williams' paper, 1978) that every spsp(a) is an epsp(a).

There are partial converses.

By Malm (1977): if $n \equiv 3 \pmod{4}$ and n is an epsp(a), then n is a spsp(a).

By Pomerance, Selfridge & Wagstaff (1980): if n is odd, $(a/n) = -1$ and n is an epsp(a), then n is also a spsp(a). In particular, if $n \equiv 5 \pmod{8}$ and n is an epsp(2), then it is a spsp(2).

Concerning the strong pseudoprimes, I may ask questions (s1) – (s5), analogous to the questions about Euler pseudoprimes given in Section VIIIC.

In 1980, Pomerance, Selfridge & Wagstaff proved that for every base $a > 1$, there exist infinitely many spsp(a), and this answers in the affirmative, questions (s1), as well as (e1). I shall say more about this in the study of the distribution of pseudoprimes (Chapter 5, Section VIII).

For base 2, it is possible to give infinitely many spsp(2) explicitly, as I indicate now.

If n is a psp(2), then $2^n - 1$ is a spsp(2). Since there are infinitely many psp(2), this gives explicitly infinitely many spsp(2); among these are all composite Mersenne numbers. It is also easy to see that if a Fermat number is composite, then it is a spsp(2).

Similarly, since there exist pseudoprimes with arbitrarily large numbers of distinct prime factors, then (s2), as well as (e2), have a positive answer; just note that if $p_1, p_8, ..., p_k$ divide the pseudoprime n, then 2^{p_i-1} ($i = 1, ..., k$) divides the spsp(2) $2^n - 1$.

In virtue of Lehmer's negative answer to (e4) and Selfridge's result, then clearly (s4) has also a negative answer. Very important — as I shall indicate later, in connection with the Monte Carlo primality testing methods — is the next theorem by Rabin, corresponding to Solovay & Strassen's result for Euler pseudoprimes. And it is tricky to prove:

If $n > 4$ is composite, there are at least $3(n-1)/4$ integers a, $1 < a < n$, for which n is not a spsp(a). So, the number of bases a, $1 < a < n$, gcd(a,n) = 1, for which an odd composite integer is spsp(a), is at most $(n-1)/4$. This answers question (s5).

Monier (1980) has also determined a formula for the number $B_{spsp}(n)$, of bases a, $1 < a < n$, gcd(a,n) = 1, for which the odd composite integer n is spsp(a). Namely:

$$B_{spsp}(n) = \left[1 + \frac{2^{\omega(n)\nu(n)} - 1}{2^{\omega(n)} - 1}\right]\left[\prod_{p|n} \gcd(n^*, p^*)\right] - 1,$$

where $\omega(n)$ = number of distinct prime factors of n,

$\nu_p(m)$ = exponent of p in the factorization of m (any natural number),

m^* = largest odd divisor of $m - 1$,

$\nu(n) = \min_{p|n} \{v_2(p - 1)\}$.

Just for the record, the smallest spsp(2) is 2407 = 29 × 83.

The smallest number which is a strong pseudoprime relative to the bases 2 and 3 is 1,373,653 = 829 × 1657; and relative to the bases 2,

3, and 5 is 25,326,001 = 2251 × 11251. There are only 13 such numbers less than 25 × 10⁹.

The following table is taken from the paper by the 3 Knights of Numerology:

Numbers Less Than 25 × 10⁹, which are spsp in Bases 2, 3, 5

Number	psp to base 7	11	13	Factorization
25,326,001	no	no	no	2251 × 11251
161,304,001	no	spsp	no	7333 × 21997
960,946,321	no	no	no	11717 × 82013
1,157,839,381	no	no	no	24061 × 48121
3,215,031,751	spsp	psp	psp	151 × 751 × 28351
3,697,278,427	no	no	no	30403 × 121609
5,764,643,587	no	no	spsp	37963 × 151849
6,770,862,367	no	no	no	41143 × 164569
14,386,156,093	psp	psp	psp	397 × 4357 × 8317
15,579,919,981	psp	spsp	no	88261 × 176521
18,459,366,157	no	no	no	67933 × 271729
19,887,974,881	psp	no	no	81421 × 244261
21,276,028,621	no	psp	psp	103141 × 206281

To this table, I add the list of pseudoprimes up to 25 × 10⁹ which are not square-free and their factorizations:

$1194649 = 1093^2$,
$12327121 = 3511^2$,
$3914864773 = 29 \times 113 \times 1093^2$,
$5654273717 = 1093^2 \times 4733$,
$6523978189 = 43 \times 127 \times 1093^2$,
$22178658685 = 5 \times 47 \times 79 \times 1093^2$.

With the exception of the last two, the numbers in the above list are strong pseudoprimes.

Note that the only prime factors to the square are 1093, 3511. The occurrence of these numbers will be explained in Chapter 5, Section III.

Addendum on The Congruence $a^{n-k} \equiv b^{n-k} \pmod{n}$

Let $a > b \geqslant 1$, $\gcd(a,b) = 1$ and $k \geqslant 1$.

Let $R_{(a,b,k)}$ denote the set of all composite integers $n > k$ such that $a^{n-k} \equiv b^{n-k} \pmod{n}$.

Note that if $k \geqslant 2$, then at most finitely many primes n satisfy the congruence.

The question is: for which triples (a,b,k) is $R_{(a,b,k)}$ an infinite set?

When $b = k = 1$, then $R_{(a,1,1)}$ is the set of pseudoprimes in base a: it is infinite, as was shown by Cipolla.

The determination of the sets $R_{(a,b,k)}$ that are infinite is due to McDaniel (1987). This paper has been enlarged and improved in a preprint still to appear (1988?).

Previously, the following partial results had been obtained by Rotkiewicz:

In 1959: $R_{(a,b,1)}$ is infinite.
In 1972: $R_{(a,b,3)}$ is infinite.
In 1984: $R_{(2,1,2)}$ is infinite.

As an illustration, here is a sketch of the proof of the last mentioned result: it is a question of proving that there exist infinitely many integers n such that $2^{n-2} \equiv 1 \pmod{n}$. Note that such integers n must be composite.

All begins with Lehmer & Lehmer who found the number $m = 4700063497 = 19 \times 47 \times 5263229$, which, at present, is the only known integer such that $2^m \equiv 3 \pmod{m}$. With this starting point, Rotkiewicz showed that if $n = 2^m - 1$, then $2^{n-2} \equiv 1 \pmod{n}$. Next, if p is a primitive prime factor of $2^{n-2} - 1$, then $n_1 = np$ satisfies also $2^{n_1-2} \equiv 1 \pmod{n_1}$. This result was extended by Shen in 1986.

There exists an infinite sequence $k_1 < k_2 < k_3 < \cdots$ such that each of the congruences $2^{n-k_i} \equiv 1 \pmod{n}$ holds for infinitely many integers. More explicitly, take $k_1 = 2$, $k_2 = 2^{k_1} - 1 = 3$, ..., $k_{i+1} = 2^{k_i} - 1$, ...

The theorem of McDaniel is the following: $R_{(a,b,k)}$ is an infinite set, but for the possible exceptional triples: $(2^v \times 5 + 1, 2^v \times 5 - 1, 3)$ for $v \geqslant 1$, $(2^u + 1, 2^u - 1, 3)$ for $u \geqslant 2$.

In particular, the congruence $a^{n-k} \equiv 1 \pmod{n}$ is satisfied by infinitely many composite integers n. This special case was proved by Kiss & Phong in 1987.

The congruence $a^{f(n)} \equiv b^{f(n)} \pmod{n}$, where $f(x)$ is a polynomial with integer coefficients, has been studied by McDaniel and by Phong, in preprints still to appear (1988?).

IX. Carmichael Numbers

In 1912, Carmichael considered a more rare kind of numbers. They are the composite numbers n such that $a^{n-1} \equiv 1 \pmod{n}$ for every integer a, $1 < a < n$, such that a is relatively prime to n.

The smallest Carmichael number is $561 = 3 \times 11 \times 17$.

I shall now indicate a characterization of Carmichael numbers.

Recall that I have introduced, in Section II, Carmichael's function $\lambda(n)$, which is the maximum of the orders of a mod n, for every a, $1 \leqslant a < n$, $\gcd(a,n) = 1$; in particular, $\lambda(n)$ divides $\phi(n)$.

Carmichael showed that n is a Carmichael number if and only if n is composite and $\lambda(n)$ divides $n - 1$. (It is the same as saying that if p is any prime dividing n then $p - 1$ divides $n - 1$.)

It follows that every Carmichael number is odd and the product of three or more distinct prime numbers.

Explicitly, if $n = p_1 p_2 \cdots p_r$ (product of distinct primes), then n is a Carmichael number if and only if $p_i - 1$ divides $(n/p_i) - 1$ (for $i = 1, 2, ..., r$).

Therefore, if n is a Carmichael number, then also $a^n \equiv a \pmod{n}$, for every integer $a \geqslant 1$.

Schinzel noted in 1958 that for every $a \geqslant 0$ the smallest pseudo-prime m_a in base a, satisfies necessarily $m_a \leqslant 561$. Moreover, there exists a such that $m_a = 561$. Explicitly, let p_i ($i = 1, ..., s$) be the primes such that $2 < p_i < 561$; for each p_i let e_i be such that $p_i^{e_i} < 561 < p_i^{e_i+1}$; let g_i be a primitive root modulo $p_i^{e_i}$, and by the Chinese remainder theorem, let a be such that $a \equiv 3 \pmod{4}$ and $a \equiv g_i \pmod{p_i^{e_i}}$ for $i = 1, ..., s$. Then $m_a = 561$.

Carmichael and Lehmer determined the smallest Carmichael numbers:

$561 = 3 \times 11 \times 17$	$15841 = 7 \times 31 \times 73$	$101101 = 7 \times 11 \times 13 \times 101$
$1105 = 5 \times 13 \times 17$	$29341 = 13 \times 37 \times 61$	$115921 = 13 \times 37 \times 241$
$1729 = 7 \times 13 \times 19$	$41041 = 7 \times 11 \times 13 \times 41$	$126217 = 7 \times 13 \times 19 \times 73$
$2465 = 5 \times 17 \times 29$	$46657 = 13 \times 37 \times 97$	$162401 = 17 \times 41 \times 233$
$2821 = 7 \times 13 \times 31$	$52633 = 7 \times 73 \times 103$	$172081 = 7 \times 13 \times 31 \times 61$
$6601 = 7 \times 23 \times 41$	$62745 = 3 \times 5 \times 47 \times 89$	$188461 = 7 \times 13 \times 19 \times 109$
$8911 = 7 \times 19 \times 67$	$63973 = 7 \times 13 \times 19 \times 37$	$252601 = 41 \times 61 \times 101$
$10585 = 5 \times 29 \times 73$	$75361 = 11 \times 17 \times 31$	

Open problem: Are there infinitely many Carmichael numbers?

As I have mentioned before, even the more restricted problem (Duparc, 1952) is open: Are there infinitely many composite integers n such that $2^{n-1} \equiv 1 \pmod{n}$ and $3^{n-1} \equiv 1 \pmod{n}$?

I have also considered in Section II the question of existence of composite n such that $\phi(n)$ divides $n - 1$. Since $\lambda(n)$ divides $\phi(n)$, any such n must be a Carmichael number.

Since it is not known whether there exist infinitely many Carmichael numbers, a fortiori, it is not known whether there exist Carmichael numbers with arbitrarily large number of prime factors. In this respect, there is a result of Duparc (1952) (see also Beeger, 1950):

For every $r \geqslant 3$, there exists only finitely many Carmichael numbers with r prime factors, of which the smallest $r - 2$ factors are given in advance. I shall return to these questions in Chapter 4.

In 1939, Chernick gave the following method to obtain Carmichael numbers. Let $m \geqslant 1$ and $M_3(m) = (6m + 1)(12m + 1)(18m + 1)$. If m is such that all three factors above are prime, then $M_3(m)$ is a Carmichael number. This yields Carmichael numbers with three prime factors.

Similarly, if $k \geqslant 4$, $m \geqslant 1$, let

$$M_k(m) = (6m + 1)(12m + 1) \prod_{i=1}^{k-2} (9 \times 2^i m + 1).$$

If m is such that all k factors are prime numbers and, moreover, 2^{k-4} divides m, then $M_k(m)$ is a Carmichael number with k prime factors. This method has been used by Wagstaff and Yorinaga to obtain large Carmichael numbers or Carmichael numbers with many factors.

Record

The largest known Carmichael number has been discovered by Woods & Huenemann in 1982, and it has 432 digits.

Previously, Wagstaff printed in his 1980 paper, a Carmichael number with 321 digits, equal to $M_3(m)$, where m has 106 digits.

The largest published Carmichael number with more than 3 factors was found by Wagstaff in 1980. It is $M_6(m)$, with $m = 1810081824371200$. It has 101 digits, 6 prime factors and it is also printed in Wagstaff's paper.

In 1978, Yorinaga found eight Carmichael numbers with 13 prime factors — these are the ones with the largest number of prime factors known up to now.

In 1985, Dubner found much larger Carmichael numbers, but he hasn't yet published his discovery. Among them, I mention:

$$M_3(m), \quad \text{with } m = 5{\times}7{\times}11{\times}13{\times}\cdots{\times}397{\times}882603{\times}10^{185},$$

which has 1057 digits and is the largest known Carmichael number;

$M_4(m)$, with $m = \frac{1}{6}\,323323{\times}655899{\times}10^{40}$, with 207 digits;
$M_5(m)$, with $m = \frac{1}{6}\,323323{\times}426135{\times}10^{16}$, with 139 digits;
$M_6(m)$, with $m = \frac{1}{6}\,323323{\times}239556{\times}10^7$, with 112 digits;
$M_7(m)$, with $m = 323323{\times}160{\times}8033$ with 93 digits.

I shall discuss the distribution of Carmichael numbers in Chapter 4, Section VIII.

Addendum on Knödel Numbers

For every $k \geqslant 1$, let C_k be the set of all composite integers $n > k$ such that if $1 < a < n$, $\gcd(a,n) = 1$, then $a^{n-k} \equiv 1 \pmod{n}$.

Thus C_1 is the set of Carmichael numbers. For $k \geqslant 2$, the numbers of C_k were considered by Knödel in 1953. Even though it is not known that there exist infinitely many Carmichael numbers, nevertheless, Makowski proved in 1962:

For each $k \geqslant 2$, the set C_k is infinite.

Proof. For every a, $1 \leqslant a < k$, $\gcd(a,k) = 1$, let r_a be the order of a modulo k. Let $r = \Pi r_a$ (product for all a as above). So $a^r \equiv 1 \pmod{k}$.

There exist infinitely many primes p such that $p \equiv 1 \pmod{r}$; see Chapter 4, Section IV for a proof of this very useful theorem. For each such $p > k$, write $p - 1 = hr$, and let $n = kp$. Then $n \in C_k$.

Indeed, let $1 < a < n$, $\gcd(a,n) = 1$, so $\gcd(a,k) = 1$, hence

$$a^{n-k} = a^{k(p-1)} = a^{khr} \equiv 1 \pmod{k},$$
$$a^{n-k} = a^{k(p-1)} \equiv 1 \pmod{p}.$$

Since $p \nmid k$, then $a^{n-k} \equiv 1 \pmod{n}$, showing that $n = kp$ is in C_k. $\square$

It follows from the above proof that if $k = 2$, then $2p \in C_2$ for every prime $p > 2$. If $k = 3$, then $3p \in C_3$ for every prime $p > 3$; this last fact was proved by Morrow in 1951.

X. Lucas Pseudoprimes

In view of the analogy between sequences of binomials $a^n - 1$ ($n \geqslant 1$) and Lucas sequences, it is no surprise that pseudoprimes should have a counterpart involving Lucas sequences. For each parameter $a \geqslant 2$, there were the a-pseudoprimes and their cohort of Euler pseudoprimes and strong pseudoprimes in base a. In this section, to all pairs (P,Q) of nonzero integers, will be associated the corresponding Lucas pseudoprimes, the Euler–Lucas pseudoprimes, and the strong Lucas pseudoprimes. Their use will parallel that of pseudoprimes.

Let P, Q be nonzero integers, $D = P^2 - 4Q$ and consider the associated Lucas sequences $(U_n)_{n \geqslant 0}$, $(V_n)_{n \geqslant 0}$.

Recall (from Section IV) that if n is an odd prime, then:

(X.1)   If $\gcd(n,D) = 1$, then $U_{n-(D/n)} \equiv 0 \pmod{n}$.

(X.2) $U_n \equiv (D/n) \pmod{n}$.

(X.3) $V_n \equiv P \pmod{n}$.

(X.4)   If $\gcd(n,D) = 1$, then $V_{n-(D/n)} \equiv 2Q^{(1-(D/n))/2} \pmod{n}$.

If n is an odd composite number and the congruence (X.1) holds, then n is called a Lucas pseudoprime (with the parameters (P,Q)), abbreviated ℓpsp(P,Q).

It is alright to make such a definition, but do these numbers exist? If so, are they worthwhile to study?

A. Fibonacci Pseudoprimes

To begin, it is interesting to look at the special case of Fibonacci numbers, where $P = 1$, $Q = -1$, $D = 5$. In this situation, it is more appropriate to call Fibonacci pseudoprimes the ℓpsp$(1,-1)$.

The smallest Fibonacci pseudoprimes are $323 = 17 \times 10$ and $377 = 13 \times 29$; indeed, $(5/323) = (5/377) = -1$ and it may be calculated that $U_{324} \equiv 0 \pmod{323}$, $U_{378} \equiv 0 \pmod{377}$.

E. Lehmer showed in 1964 that there exist infinitely many Fibonacci pseudoprimes; more precisely, if p is any prime greater than 5, then U_{2p} is a Fibonacci pseudoprime.

Property (X.2) was investigated by Parberry (in 1970) and later by Yorinaga (1976).

Among his several results, Parberry showed that if $\gcd(h,30) = 1$ and condition (X.2) is satisfied by h, then it is also satisfied by $k = U_h$; moreover, $\gcd(k,30) = 1$ and, if h is composite clearly, U_h is also composite. This shows that if there exists one composite Fibonacci number U_n such that $U_n \equiv (5/n) \pmod{n}$, then there exist infinitely many such numbers. As I shall say (in a short while) there do exist such Fibonacci numbers.

Actually, this follows also from another result of Parberry:

If p is prime and $p \equiv 1$ or $4 \pmod{15}$, then $n = U_{2p}$ is odd composite and it satisfies both properties (X.1) and (X.2). In particular, there are infinitely many Fibonacci pseudoprimes which, moreover, satisfy (X.2). (Here I use the fact — to be indicated later in Chapter

4, Section IV – that there exist infinitely many primes p such that
$p \equiv 1 \pmod{15}$, resp. $p \equiv 4 \pmod{15}$.)

If $p \not\equiv 1$ or $4 \pmod{15}$, then (X.2) is not satisfied, as follows from
various divisibility properties and congruences indicated in Section
IV.

Yorinaga considered the primitive part of the Fibonacci number
U_n. If you remember, I have indicated in Section IV that every
Fibonacci number U_n (with $n \neq 1, 2, 6, 12$) admits a primitive prime
factor p – these are the primes that divide U_n, but do not divide U_d,
for every d, $1 < d < n$, d dividing n. Thus $U_n = U_n^* \times U_n'$, where
$\gcd(U_n^*, U_n') = 1$ and p divides U_n^* if and only if p is a primitive
prime factor of U_n.

Yorinaga showed that if m divides U_n^* (with $n > 5$) then $U_m \equiv (5/m)$
$\pmod{m}$.

According to Schinzel's result (1963), discussed in Section IV, there
exist infinitely many integers n such that U_n^* is not a prime. So,
Yorinaga's result implies that there exist infinitely many odd
composite n such that the congruence (X.2) is satisfied.

Yorinaga published a table of all 109 composite numbers n up to
707000, such that $U_n \equiv (5/n) \pmod{n}$. Some of these numbers give also
Fibonacci pseudoprimes, like $n = 4181 = 37 \times 113$, $n = 5777 = 53 \times 109$,
and many more. Four of the numbers in the table give pseudoprimes in
base 2:

$$219781 = 271 \times 811$$
$$252601 = 41 \times 61 \times 101$$
$$399001 = 31 \times 61 \times 211$$
$$512461 = 31 \times 61 \times 271.$$

Another result of Parberry, later generalized by Baillie & Wagstaff,
is the following:

If n is an odd composite number, not a multiple of 5, if congruences
(X.1) and (X.2) are satisfied, then

$$\begin{cases} U_{(n-(5/n))/2} \equiv 0 \pmod{n} & \text{if } n \equiv 1 \pmod{4}, \\ V_{(n-(5/n))/2} \equiv 0 \pmod{n} & \text{if } n \equiv 3 \pmod{4}. \end{cases}$$

In particular, since there are infinitely many composite integers n such that $n \equiv 1 \pmod 4$, then there are infinitely many odd composite integers n satisfying the congruence $U_{(n-(5/n))/2} \equiv 0 \pmod n$.

The composite integers n such that $V_n \equiv 1 \pmod n$ [where $(V_k)_{k \geqslant 0}$ is the sequence of Lucas numbers] have also been studied. They have been called Lucas pseudoprimes, but this name is used here with a different meaning.

In 1983, Singmaster found the following 25 composite numbers $n < 10^5$ with the above property:

$$705, \ 2465, \ 2737, \ 3745, \ 4181, \ 5777, \ 6721,$$
$$10877, \ 13201, \ 15251, \ 24465, \ 29281, \ 34561,$$
$$35785, \ 51841, \ 54705, \ 64079, \ 64681, \ 68251,$$
$$75077, \ 80189, \ 90061, \ 96049, \ 97921.$$

B. Lucas Pseudoprimes (ℓpsp(P,Q))

I shall now consider ℓpsp(P,Q) associated to arbitrary pairs of parameters (P,Q). To stress the analogy with the pseudoprimes in base a, the discussion should follow the same lines. But, it will be clear that much less is known about these numbers. For example, there is no explicit mention of any algorithm to generate infinitely many ℓpsp(P,Q), when P, Q are given — except the results mentioned for Fibonacci pseudoprimes.

However, in his thesis in 1971, Lieuwens stated that for every $k \geqslant 2$, there exist infinitely many Lucas pseudoprimes with given parameters (P,Q), which are the product of exactly k distinct primes.

It is quite normal for an odd integer n to be a Lucas pseudoprime with respect to many different sets of parameters. Let $D \equiv 0$ or 1 $\pmod 4$, let $B_{\ell \text{psp}}(n,D)$ denote the number of integers P, $1 \leqslant P \leqslant n$, such that there exists Q, with $P^2 - 4Q \equiv D \pmod n$ and n is a ℓpsp(P,Q). Baillie & Wagstaff showed in 1980 that

$$B_{\ell \text{psp}}(n,D) = \prod_{p|n} \left\{ \gcd\left[n - \left(\frac{D}{n} \right), p - \left(\frac{D}{p} \right) \right] - 1 \right\}.$$

In particular, if n is odd and composite, there exists D and, correspondingly, at least three pairs (P,Q), with $P^2 - 4Q = D$ and distinct values of P modulo n, such that n is a $\ell\mathrm{psp}(P,Q)$.

Another question is the following: If n is odd, for how many distinct D modulo n, do there exist (P,Q) with $P^2 - 4Q \equiv D \pmod{n}$, $P \not\equiv 0 \pmod{n}$, and n is a $\ell\mathrm{psp}(P,Q)$?   Baillie & Wagstaff discussed this matter also, when $n = p_1 p_2$, where p_1, p_2 are distinct primes.

C. Euler–Lucas Pseudoprimes (eℓpsp(P,Q)) and Strong Lucas Pseudo-primes (sℓpsp(P,Q))

Let P, Q be given, $D = P^2 - 4Q$, as before. Let n be an odd prime number. If $\gcd(n,QD) = 1$, it was seen in Section V that

$$\begin{cases} U_{(n-(D/n))/2} \equiv 0 \pmod{n} \quad \text{when} \quad (Q/n) = 1, \\ V_{(n-(D/n))/2} \equiv D \pmod{n} \quad \text{when} \quad (Q/n) = -1. \end{cases}$$

This leads to the following definition. An odd composite integer n, such that $\gcd(n,QD) = 1$, satisfying the above condition is called a Euler–Lucas pseudoprime with parameters (P,Q), abbreviated eℓpsp(P,Q).

Let n be an odd composite integer, with $\gcd(n,D) = 1$, let $n - (D/n) = 2^s d$, with d odd, $s \geqslant 0$. If

$$\begin{cases} U_d \equiv 0 \pmod{n}, \text{ or} \\ V_{2^r d} \equiv 0 \pmod{n}, \text{ for some } r, \ 0 \leqslant r < s, \end{cases}$$

then n is called a strong Lucas pseudoprime with parameters (P,Q), abbreviated sℓpsp(P,Q). In this case, necessarily, $\gcd(n,Q) = 1$.

If n is an odd prime, and $\gcd(n,QD) = 1$, then n is an eℓpsp(P,Q) and also a sℓpsp(P,Q). It is also clear that if n is an eℓpsp(P,Q) and $\gcd(n,Q) = 1$, then n is a $\ell\mathrm{psp}(P,Q)$.

What are the relations between eℓpsp(P,Q) and sℓpsp(P,Q)? Just as in the case of Euler and strong pseudoprimes in base a, Baillie & Wagstaff showed that if n is a sℓpsp(P,Q), then n is an eℓpsp(P,Q) — this is the analogue of Selfridge's result.

Conversely, if n is an $e\ell$psp(P,Q) and either $(Q/n) = -1$ or $n - (D/n) \equiv 2$ (mod 4), then n is a $s\ell$psp(P,Q) − this is the analogue of Malm's result.

If gcd$(n,Q) = 1$, n is a ℓpsp(P,Q), $U_n \equiv (D/n)$ (mod n) and if, moreover, n is an epsp(Q), then n is also a $s\ell$psp(P,Q).

The special case for Fibonacci numbers was proved by Parberry, as already indicated.

Previously, I mentioned the result of Lehmer, saying that no odd composite number can be an epsp(a), for all possible bases.

Here is the analogous result of Williams (1977):

Given $D \equiv 0$ or 1 (mod 4), if n is an odd composite integer, and gcd$(n,D) = 1$, there exist P, Q, nonzero integers, with $P^2 - 4Q = D$, gcd$(P,Q) = 1$, gcd$(n,Q) = 1$, and such that n is not an $e\ell$psp(P,Q).

With the present terminology, I have mentioned already that Parberry had shown, for the Fibonacci sequence, that there exist infinitely many $e\ell$psp$(1,-1)$.

This has been improved by Kiss, Phong & Lieuwens (1986): Given (P,Q) such that the sequence $(U_n)_{n \geqslant 0}$ is nondegenerate (that is, $U_n \neq 0$ for every $n \neq 0$), given $k \geqslant 2$, there exist infinitely many $e\ell$psp(P,Q), each being the product of k distinct primes. Moreover, given also $d \geqslant 2$, if $D = P^2 - 4Q > 0$, then the prime factors may all be chosen to be of the form $dm + 1$ $(m \geqslant 1)$.

As for Fibonacci numbers, now I consider the congruences (X.2) and also (X.3), (X.4).

It may be shown that if gcd$(n, 2PQD) = 1$ and if n satisfies any two of the congruences (X.1) to (X.4), then it satisfies the other two.

In 1986, Kiss, Phong & Lieuwens extended a result of Rotkiewicz (1973) and proved:

Given, P, $Q = \pm 1$ [but $(P,Q) \neq (1,1)$], given $k \geqslant 2$, $d \geqslant 2$, there exist infinitely many integers n, which are Euler pseudoprimes in base 2, and which satisfy the congruences (X.1) − (X.4); moreover, each such number n is the product of exactly k distinct primes, all of the form $dm + 1$ (with $m \geqslant 1$).

D. Carmichael–Lucas Numbers

Following the same line of thought that led from pseudoprimes to Carmichael numbers, it is natural to consider the following numbers.

Given $D \equiv 0$ or 1 (mod 4), the integer n is called a Carmichael–Lucas number (associated to D), if $\gcd(n,D) = 1$ and for all nonzero relatively prime integers P, Q with $P^2 - 4Q = D$ and $\gcd(n,Q) = 1$, the number is an $\ell psp(P,Q)$.

Do such numbers exist? A priori, this is not clear. Of course, if n is a Carmichael–Lucas number associated to $D = 1$, then n is a Carmichael number.

Williams, who began the consideration of Carmichael–Lucas numbers, showed in (1977):

If n is a Carmichael–Lucas number associated to D, then n is the product of $k \geqslant 2$ distinct primes p_i such that $p_i - (D/p_i)$ divides $n - (D/n)$.

Note that $323 = 17 \times 19$ is a Carmichael–Lucas number (with $D = 5$); but it cannot be a Carmichael number, because it is the product of only two distinct primes.

Adapting the method of Chernick, it is possible to generate many Carmichael–Lucas numbers. Thus, for example, $1649339 = 67 \times 103 \times 239$ is such a number (with $D = 8$).

XI. Last Section on Primality Testing and Factorization!

Only one section — and a small one — to treat a burning topic, full of tantalizing ideas and the object of intense research, in view of immediate direct applications.

Immediate direct applications of number theory! Who would dream of it, even some 30 years ago? Von Neumann yes, not me, not many people. Poor number theory, the Queen relegated (or raised?) to be the object of a courtship inspired by necessity not by awe.

In recent years progress on the problems of primality testing and factorization has been swift. More and more deep results of number theory have been invoked. Brilliant brains devised clever procedures, not less brilliant technicians invented tricks, shortcuts to implement the methods in a reasonable time — and thus, a whole new branch is evolving.

I have attempted to develop the foundations needed to present in a lucid way the main procedures for primality testing. But this was doomed to failure. Indeed, with the latest developments, I would need, for example, to use facts about the theory of Jacobi sums, elliptic curves, etc. This is far beyond what I intend to discuss. It is more reasonable to assign supplementary reading for those who are avidly interested in the problem. Happily enough, there are now many excellent expository articles, which I will recommend at the right moment.

Despite the shortcomings just mentioned, I feel that presenting an overview of the question, even one with gaps, will still be useful. Having apologized, I may now proceed with my incomplete treatment.

First, money: how much it costs to see the magic. Then, primality, a few words about factorization and the application to public key cryptography.

Even though I have already presented several primality tests in Sections III and V, it is good to clarify the exact meaning of what is a primality test and to discuss the various types of testing procedures.

A (deterministic) primality test is a finite procedure T, which is applicable to any natural number N, assigning a certificate $T(N)$, which is either "prime" or "composite." Moreover, if the certificate is "prime" then N is indeed a prime number. A non-primality test is defined similarly, but now if the certificate is "composite" then N is composite.

An example of a non-primality test is given by Fermat's little theorem: given N, if there exists a, $1 \leqslant a < N$, such that $a^{N-1} \not\equiv 1$ (mod N), then N is composite.

A probabilistic (or Monte Carlo) primality test is defined similarly, but now if the certificate of N is "prime" then it can only be asserted that there is a high probability that N is a prime.

A corresponding definition may be formulated for a probabilistic non-primality test. It should be said that these notions can, and have been formulated in a rigorous way — but I am happy to leave the matter at an intuitive level.

A. The Cost of Testing

The cost is proportional to the time required to perform the test, thus to the number of operations involved. The operations should be counted in an appropriate way, since it is clear that addition or multiplication of very large numbers is more time consuming than if the numbers were small. So, in the last analysis, the cost is proportional to the number of operations with digits — such indivisible operations are called bit operations. Thus, for the calculation, the input is not the integer N, but the number of its digits in a base system (for example base 10, or better, base 2). If $N = a_0 b^k + a_1 b^{k-1} + \cdots + a_k$ $(0 \leqslant a_i \leqslant b - 1; a_0 \neq 0)$, then the number of digits $k + 1$ is about $(\log N)/(\log b)$, so it is proportional to $\log N$. For example, the multiplication of two numbers less than N needs at most $0((\log N)^2)$ bit operations.

[For those who need it, the symbol $0(f(n))$ will be explained in Chapter 4, Section I.]

If N is composite, what is the minimum number of bit operations required to prove it? It is enough to be able to write $N = ab$, with $1 < a, b < N$; paraphrasing Lenstra (1982), it doesn't matter whether a,b were found consulting a clairvoyant, or, after three years of Sundays, like Cole's factorization of the Mersenne number

$$2^{67} - 1 = 193707721 \times 761838257287.$$

One multiplication suffices. So the number of bit operations is $C(\log N)^2$, for some positive constant C.

If N is a prime, what is the minimum number of bit operations to prove it? This is not so easy to answer, even more so because I have not yet discussed the prime-representing polynomials (please wait until Chapter 3, Section III). At any rate, I will give explicitly a polynomial P in 26 indeterminates and degree 25, such that the set of positive values of P, at natural numbers, coincides with the set of prime numbers. So, to prove that N is prime it is necessary to consult your favorite clairvoyant (no predictable number of Sundays would suffice) to be given natural numbers $x_1, x_2, ..., x_{26}$ such that $N = P(x_1, x_2, ..., x_{26})$. The number of arithmetic operations requird is bounded by a number depending only on the polynomial P; it is the same bound for every prime N. However, something unfortunate occurs − the size of $x_1, ..., x_{26}$ cannot be controlled and the largest of the numbers $x_1, ..., x_{26}$ in fact must exceed $N^{N^{N^{N^{N^{N}}}}}...$. So, there are much too many bit operations involved.

However, another approach adopted by Pratt (1975) and appealing to Lehmer's converse of Fermat's little theorem, allowed him to conclude that $C(\log N)^4$ bit operations suffice to prove that N is a prime. And Pomerance has just now (1987) improved this record to $C(\log N)^3$ bit operations, thus deserving once more the inclusion of his name in this book!

The above considerations refer to the number of bit operations required to prove that a prime number is a prime, and that a composite number is composite.

But the problem is really to decide with a fast algorithm whether a given large number is prime or not. So, the running time depends on the type of primality test used. Each primality test being a succession of routines or algorithms of simpler kind, it is therefore necessary to estimate the running time of each of these basic algorithms. This is discussed in a very accessible way by Lehmer in his expository paper of 1969. Here are the main facts, assuming that the numbers involved are large, so multiprecision operations are used. Thus, the bounds indicated are for the number of bit operations; moreover, the various constants C reflect the computer's speed, and the quality of the program.

(a) Algorithm to find $\gcd(M,N)$ where $M < N$: running time $t \leqslant C(\log M)^3$.

This follows from the classical result of Lamé (1844), which states that the number of divisions necessary to find the $\gcd(M,N)$ is at most five times the number of decimal digits of M, so $C(\log M)$.

(b) Calculation of the Jacobi symbol (M/N), with $1 \leqslant M \leqslant N$: running time $t \leqslant C(\log M)^3$.

This is derived from the reciprocity law for the Jacobi symbol.

In fact, with a more effective method indicated by Lehmer, which avoids divisions, it is possible in (a) and (b) to have $t \leqslant C(\log M)^2$.

(c) Calculation of $b^N \bmod m$: running time $t \leqslant C(\log N)(\log m)^2$.

The method uses the binary expansion of N, as I have already indicated in Section III.

(d) Calculation of $U_N \bmod m$ or $V_N \bmod m$, where $U = (U_n)_{n \geqslant 0}$ and $V = (V_n)_{n \geqslant 0}$ are Lucas sequences: running time $t \leqslant C(\log N)(\log m)^2$.

The method uses binary expansion of N, as I have already indicated in Section V.

(e) Calculation of a primitive root modulo p:

The algorithm given by Gauss, indicated in Section II, requires a number of operations proportional to p. Only when the prime factors of $p - 1$ are known, it is possible to prove that there is an algorithm with at most $C(\log p)^3$ bit operations.

Now I consider the primality tests.

If sufficiently many prime factors of $N - 1$ or $N + 1$ are known, the tests indicated in Sections III and V are applicable. The preceding indications allow to conclude that the running time of these tests is at most $C(\log N)^3$. Thus, these tests may be performed in a running time that is a polynomial on the input (number of digits of N).

For numbers that are not of a special form, the very naive primality test is by trial division of N by all primes $p < \sqrt{N}$. It will be seen in Chapter 4 that, for any large integer n, the number of primes less than n is about $n/(\log n)$ (this statement will be made much more precise later on); thus there will be at most $C[\sqrt{N}/(\log N)]$ operations, which tells that the running time is at most $C\sqrt{N} \log N$. So this procedure does not run in polynomial time on the input.

Up to now, it has not been found any deterministic primality test, applicable to arbitrary numbers which runs in polynomial time. But it has also not been proved that no such test exists.

On the other hand, it will be seen that there exist probabilistic tests in polynomial time. There exist also tests, which run in polynomial time, but can only be justified by appealing to unproved conjectures of analytic number theory (namely a generalization of

Riemann's hypothesis; I will discuss the classical Riemann's hypothesis in Chapter 4).

If the aim is to make the list of primes up to N, there are two obvious procedures. First, to test each integer $n \leqslant N$ for primality. Even discarding the obvious composite numbers, the number of operations turns out to be $C'\Sigma_{n=2}^{N}\log n < CN \log N$ (see Chapter 4 for the evaluation of this and similar sums); so the running time is $CN(\log N)^3$. The second procedure uses the sieve of Eratosthenes, and the number of operations is $C'\Sigma_{p \leqslant \sqrt{N}}(N/p) < CN \log \log N$, so the running time is $CN(\log \log N)(\log N)^2$. In conclusion, the running time is not a polynomial on the input.

The preceding discussion of the running time to perform various algorithms is an illustration of the problems of the complexity theory of algorithms — a branch of theoretical computer science now in development. In short, one of the aims is to classify problems with respect to their complexity. There are four main classes, called

P = class of polynomial-time problems.
RP = class of random polynomial-time problems.
NP = class of nondeterministic polynomial-time problems.
EXP = class of exponential-time problems.

Each of these classes is more restricted than the following:

$$P \subset RP \subset NP \subset EXP.$$

Given a problem, for which the solution is computable, the main question is to prove that it belongs to one of the above classes.

Thus, the problem of testing a number for primality has many algorithms, but none is yet known with polynomial time. This is an outstanding open problems, as I have already mentioned.

I realize that I have just been name-dropping. It is far beyond my purpose to explain these concepts. If you are curious, a little bit is said in Wagon's article on primality testing (1986); a good introduction is in the expository paper of van Emde Boas (1982). There is also the volume edited by Traub (1976) on algorithms and complexity and the very recent preprint by Adleman & McCurley (1986).

B. Recent Primality Tests

(a) *Miller's Test.*

In 1975, Miller proposed a primality test. The proof depended on the validity of a generalized form of the Riemann hypothesis. I will not explain the exact meaning of this hypothesis or conjecture, but in Chapter 4, Section I, I shall discuss the classical Riemann hypothesis.

To formulate Miller's test, which involves the congruences used in the definition of strong pseudoprimes, it is convenient to use the terminology introduced by Rabin.

If N is an integer, if $1 < a < N$, if $n - 1 = 2^s d$, with $s \geqslant 0$, d odd, a is said to be a witness for N when $a^{N-1} \not\equiv 1 \pmod{N}$ or there exists r,

$0 \leqslant r < s$ such that $1 < \gcd(a^{2^r d} - 1, N) < N$.

If N has a witness, it is composite.

If N is composite, if $1 < a < N$, and a is not a witness, then $\gcd(a,N) = 1$ and N is a spsp(a). Conversely, if N is odd and N is a spsp(a) then a is not a witness for N.

In the new terminology, in order to certify that N is prime, it suffices to show that no integer a, $1 < a < N$, $\gcd(a,N) = 1$, is a witness. Since N is assumed to be very large, this task is overwhelming! It would be wonderful just to settle the matter by considering small integers a, and checking whether any one is a witness for N. Here is where the generalized Riemann's hypothesis is needed. It is used to establish the following fact.

Let G be a set of residue classes a mod N, with $1 < a < N$, $\gcd(a,N) = 1$, such that:

(a) 1 mod $N \in G$;

(b)   if a mod N, b mod $N \in G$, then ab mod $N \in G$;

(c)   not every a mod N (with a as above) is in G.

In other words, G is a proper subgroup of the group of invertible residue classes modulo N.

Using the generalized Riemann hypothesis, it is possible to show that there exists a, $1 < a < N$, $\gcd(a,N) = 1$, such that a mod $N \notin G$, and, moreover, a is small: $a < 2(\log N)^2$.

The result with the constant 2 is due to Bach (1985), improving a previous result of Ankeny; the constant 70, figured in the paper of Adleman, Pomerance & Rumely (1983). Moreover, for every $\epsilon > 0$ there exists N_0 such that if $N > N_0$, then 2 may be replaced by $1 + \epsilon$.

In the present situation, G is taken to be the subgroup of all $a \mod N$ such that $1 < a < N$, $\gcd(a,N) = 1$, and $(a/N) \equiv a^{(N-1)/2} \pmod N$, where (a/N) denotes the Jacobi symbol. By Lehmer's result, it is indeed a proper subgroup of the one of residue classes modulo N.

These considerations lead to the following primality test of Miller:

Miller's Test. Let N be an odd integer. If there exists a, $1 < a < 2(\log N)^2$, $\gcd(a,N) = 1$, which is a witness for N, then N is "guilty," that is, composite. Otherwise, N is "innocent," that is, a prime.

I should add here that for numbers up to 25×10^9, because of the calculations reported in Section VIII, the only composite integers $N < 3 \times 10^9$ that are strong pseudoprimes simultaneously to the bases 2, 3, 5 are the four numbers 25,326,001; 161,304,001; 960,946,321; 1,157,839,381. So, if N is not one of these numbers and 2, 3, 5 are not witnesses for N, then N is a prime. Similarly, 3,215,031,751 is the only number $N < 25 \times 10^9$ which is a strong pseudoprime for the bases 2, 3, 5, 7. So if $N < 25 \times 10^9$, N not the above number, and 2, 3, 5, 7 are not witnesses, then N is a prime.

This test may be easily implemented on a pocket calculator.

The number of operations for testing whether a number is a witness for N is $0((\log N)^3)$; thus, the number of bit operations needed is $0((\log N)^5)$. So, this test runs in polynomial time on the input (provided the generalized Riemann hypothesis is assumed true).

Miller's test is very easy to program. The only problem is that it is not justified rigorously. If ever a composite N cannot be declared by a witness a, $1 < a < 2(\log N)^2$, there is something wrong. What? The generalized Riemann's hypothesis!!

How unsettling ... I better move to primality tests which are fully justified.

Miller's test is discussed in Lenstra's paper (1982) as well as in the nice recent expository paper of Wagon (1986).

(b)   *The APR Test.*

The primality test devised by Adleman, Pomerance & Rumely (1983), usually called the APR test, represents a breakthrough. To wit:

(i)   It is applicable to arbitrary natural numbers N, without requiring the knowledge of factors of $N - 1$ or $N + 1$.

(ii)   The running time $t(N)$ is almost polynomial: more precisely, there exist effectively computable constants $0 < C' < C$, such that

$$(\log N)^{C' \log \log \log N} \leqslant t(N) \leqslant (\log N)^{C \log \log \log N}.$$

(iii) The test is justified rigorously, and for the first time ever in this domain, it is necessary to appeal to deep results in the theory of algebraic numbers; it involves calculations with roots of unity and the general reciprocity law for the power residue symbol. (Did you notice that I have not explained these concepts? It is far beyond what I plan to treat.)

The running time of the APR is at the present the world record for a deterministic primality test.

The APR test was given in a probabilistic and also in a deterministic version. The first implementation of the algorithm was by W. Dubuque.

Soon afterwards, Cohen & Lenstra (1984) modified the APR test, making it more flexible, using Gauss sums in the proof (instead of the reciprocity law), and having the new test programmed for practical applications. As stated in the paper: "it is the first primality test in existence that can routinely handle numbers of up to 100 decimal digits, and it does so in approximately 45 seconds."

A presentation of the APR test was made by Lenstra in the Séminaire Bourbaki, exposé 576 (1981). It was also discussed in papers of Lenstra (1982) and Nicolas (1984).

It is likely that at the time of writing, this performance has been surpassed.[*]

---

[*]It has! "Primality of prime numbers up to 213 decimal digits can now routinely be proved within approximately ten minutes," as stated in the abstract of the newest paper by Cohen & Lenstra (not Jr., but Br. = brother), 1987.

(c) *Tests with Elliptic Curves.*

The present developments in primality testing derive from the following idea: to replace the consideration of the multiplicative group of invertible residue classes modulo the number N (and results like Fermat's little theorem) by groups of points in elliptic curves, and even methods of abelian varieties. There is quite an intense activity and a great ferment of ideas − a Pandora's box is open.

Unless one is an insider, it is difficult to be kept informed of the latest developments. My best course is therefore to advise reading the preprints of Lenstra (1986), Bosma (1985), and Adleman & Huang (1986).

C. Monte Carlo Methods

Early in this century, the casino in Monte Carlo attracted aristocracy and adventurers, who were addicted to gambling. Tragedy and fortune were determined by the spinning wheel.

I read with particular pleasure the novel by Luigi Pirandello, telling how the life of Mattia Pascal was changed when luck favored him, both at Monte Carlo and in his own Sicilian village. When a corpse was identified as being his, he could escape the inferno of his married life, and became RICH and FREE! But Monte Carlo is not always so good. More often, total ruin, followed by suicide, is the price paid.

As you will enter into the Monte Carlo primality game, if your Monte Carlo testing will be unsuccessful, I sincerely hope that you will not be driven to suicide.

I shall describe three Monte Carlo methods, the first one advocated by Baillie & Wagstaff in 1980, which combines the use of pseudoprimes in various bases and of Lucas pseudoprimes for different Lucas sequences.

(a) *Baillie & Wagstaff's Method.*

To test a large odd number N for primality, proceed as follows:

Step 1. By trial division, be sure that every small prime p (say up to 1000) does not divide N.

Step 2. Verify that N satisfies the congruence $a^{N-1} \equiv 1 \pmod{N}$ for various selected bases, like 2, 3, ... If this is not so, certify that N is composite.

Numerical evidence, analyzed by Baillie & Wagstaff, suggests that the properties of satisfying Fermat's congruence for different bases are not mutually independent.

It is therefore preferable to do:

Alternate Step 2. Verify that N satisfies the condition in the definition of a strong pseudoprime in base 2. If not, certify that N is composite.

The next step of this method will be to submit N to tests in view of establishing that, for appropriate parameters (P,Q) and the corresponding Lucas sequences, congruences $U_{N-(D/N)} \equiv 0 \pmod{N}$, or stronger congruences, are satisfied. I will be more precise, but first I discuss the choice of the parameters (P,Q).

Method A, of Selfridge: consider the sequence 5, −7, 9, −11, 13, ..., let D be the first integer in the sequence, such that $(D/N) = -1$; define $P = 1$, $Q = \frac{1}{4}$.

Method B, of Baillie & Wagstaff: consider the sequence 5, 9, 13, 17, 21, ... , let D be the first integer in the sequence, such that $(D/N) = -1$; define P to be the least odd integer such that $P > \sqrt{D}$, and $Q = \frac{1}{4}(P^2 - D)$.

Step 3. Verify that N satisfies the conditions in the definition of a strong Lucas pseudoprime with parameters P, Q. That is, let $N + 1 = 2^s d$, with d odd, $s \geqslant 0$, and check that $U_d \equiv 0 \pmod{N}$ or $V_{2^r d} \equiv 0 \pmod{N}$ for some r, $0 < r < s$. If not, certify that N is composite. If yes, certify that N is prime!

Note that certifying that N is a prime implies only that the probability that N is a prime is high.   This is supported only by numerical evidence, but the probability has not been estimated.

One point of importance in the above method is the need to estimate the size of the first D with the required property. It should not be too time consuming. In this respect, what can be done is an evaluation of the average size of D.

Let N be odd and define

$$f(N) = \begin{cases} 0, & \text{if } N \text{ is a square,} \\[2ex] \text{smallest positive integer } d \text{ such that } (d/N) \neq 1, \\ \text{if } N \text{ is not a square.} \end{cases}$$

Using sophisticated analytical methods of Burgess (1962), Baillie & Wagstaff estimated that for every $\epsilon > 0$ and for every sufficiently large N, $f(N) < N^{\frac{1}{4}+\epsilon}$.

On average

$$\lim_{x \to \infty} \frac{1}{\pi(x)} \sum_{p \leq x} f(p) = \sum_{n=1}^{\infty} \frac{p_n}{2^n} = 3.674... \, ,$$

where $\pi(x)$ denotes the number of primes $p \leq x$, and $p_1 = 2 < p_2 = 3 < p_3 < \cdots$ is the increasing sequence of primes. This was proved by Erdös in 1961. Using ideas of Elliott (1970), Baillie & Wagstaff extended the above result, by showing that

$$\lim_{x \to \infty} \frac{1}{x/2} \sum_{\substack{N \leq x \\ N \text{ odd}}} f(N) = 1 + \sum_{n=1}^{\infty} \frac{p_n + 1}{2^{n-1}} \prod_{i=1}^{n-1} \left(1 - \frac{1}{p_i}\right) = 3.147... \, .$$

When these considerations are applied to the determination of D, in method A or B, it is seen that the average number of trials to find D is indeed really small. So, in practice, this is not a weakness of the method.

(b) *Method of Solovay & Strassen.*

Recall that Solovay & Strassen showed in 1977 that if N is a composite integer, then there exists at most $\frac{1}{2}\phi(n)$ bases a, $1 < a < N$, $\gcd(a,N) = 1$, such that

$$\left(\frac{a}{N}\right) \equiv a^{(N-1)/2} \pmod{N}.$$

This seemingly unimportant result is the justification of a rather accurate Monte Carlo primality method, which I describe now.

Step 1. Choose $k > 1$ numbers a, $1 < a < N$, $\gcd(a,N) = 1$. Usually a is taken not too large, so it is reasonably easy by trial division to confirm that $\gcd(a,N) = 1$.

Step 2. Test in succession whether N satisfies

$$\left[\frac{a}{N}\right] \equiv a^{(N-1)/2} \pmod{N},$$

for the chosen values of a (that is, N satisfies the condition of an Euler pseudoprime in base a). If an a is found for which the above congruence does not hold, declare that N is composite. Otherwise, declare that N is prime.

The probability, in each try, that the congruence indicated is satisfied, is at most $\frac{1}{2}$, according to Solovay & Strassen's result.

If the trials with different bases are assumed to be mutually independent (there is no reason or numerical evidence to believe the contrary for the property in the definition of Euler pseudoprimes in different bases), then the probability that N is indeed a prime, when declared prime by the test, is at least $1 - 1/2^k$. For example, if $k = 30$, then the likely error is at most one in one billion tests.

Quite accurate and practical, but not a true primality test.

(c) *Method of Rabin.*

You may wish to sell prime numbers − yes, I say sell − to be used in public key cryptography (be patient, I will soon come to this application of primality and factorization). And you wish to be sure, or sure with only a negligible margin of error, that you are really selling a prime number, so that you may advertise: "Satisfaction guaranteed or money back."

Then it would be safer to have an even more accurate test than Solovay & Strassen's.

Of course, there are tests which are deterministic, as seen in Sections III and V, but they apply to numbers of special forms and it would be relatively easy to crack codes built on such numbers. So, what you need is a practical test for arbitrary numbers, with absolutely high accuracy.

Such a test was discovered by Rabin (1976, 1980). As I have already stated in Section VIII, when studying strong pseudoprimes, if N is an odd composite integer, then there exist at most $(n-1)/4$ bases a, $1 < a < N$, $\gcd(a,N) = 1$, such that N is a strong pseudoprime in base a.

Based on the same idea of Solovay & Strassen, Rabin proposed the following test:

Step 1. Choose, at random, $k > 1$ numbers a, $1 < a < N$, $\gcd(a,N) = 1$ (same remark about checking this last condition, as in Solovay & Strassen's test).

Step 2. Test, in succession, for each chosen basis a, whether N satisfies the condition in the definition of a strong pseudoprime in base a; writing $N - 1 = 2^s d$, with d odd, $s \geqslant 0$, either $a^d \equiv 1 \pmod{N}$

or $a^{2^r d} \equiv -1 \pmod{N}$ for some r, $0 \leqslant r < s$.

If an a is found for which the above condition does not hold, then declare N to be composite. Otherwise, declare N to be prime.

Now, the probability that N is a prime, when certified prime, is at least $1 - 1/4^k$. So, for $k = 30$, the likely error is at most one in 10^{18} tests.

I think one can safely develop a business on this basis, and honestly back the product sold.

D. Titanic Primes

In an article of 1983/4, Yates coined the expression "titanic prime" to name any prime with more than 1000 digits. In the paper, with the suggestive title "*Sinkers of the Titanics*" (1984/5), Yates compiled a list of the 110 largest known titanic primes. By January 1, 1985, he knew 581 titanic primes, of which 170 have more than 2000 digits.

It is not surprising that these primes have special forms, the largest ones being Mersenne primes, others being of the form $k \times 2^n \pm 1$, $k^2 \times 2^n + 1$, $k^4 \times 2^n + 1$, $k \times 10^n + 1$, $(10^n - 1)/9$.

Of course, it is for numbers of these forms that the more efficient primality testing algorithms are available.

A presentation of 10 extraordinary primes (with various properties) is in Ondrejka's paper (1985/6).

E. Factorization

The factorization of large integers is a hard problem: there is no known algorithm that runs in polynomial time. It is also an important problem, because it has found a notorious application to public key cryptography.

Nevertheless, I shall not discuss here the methods of factorization — this would once again lead me too far from the subject of records on prime numbers.

Here are a few indications in the literature.

Among the expository papers, the following deserve attention: Guy (1975) discusses the methods, now considered classical; Williams (1984) covers about the same ground, being naturally more up to date — it is pleasant reading; Dixon (1984) writes about factorization as well as primality; Lenstra (1986) has a preprint of a lecture describing methods of factorization using elliptic curves, while the theory itself is in another of his preprints of the same year. The lecture notes of a short course by Pomerance (1984) contain an annotated bibliography. The analysis of running times of various methods is in papers by Vorhoeve (1982) and Pomerance (1982).

The recent volume by Brillhart, Lehmer, Selfridge, Tuckerman & Wagstaff (1983) contains tables of known factors of $b^n \pm 1$ ($b = 2, 3, 5, 6, 7, 10, 11, 12$) for various ranges of n. For example, the table of factors of $2^n - 1$ extends for $n < 1200$; for larger bases b, the range is smaller. A second edition of the book is scheduled to appear in 1987 and it will contain 2045 new factorizations, reflecting the important progress accomplished in the last few years, both in the methods and in the technology. All ten "most wanted" factorizations of the first edition have now been performed. To wit, it was known that

$$2^{211} - 1 = 15193 \times C\ 60$$

where $C\ 60$ indicates a number with 60 digits — the first "most wanted" number to be factored. This has now been done.

At the origin, this collective work, also dubbed "the Cunningham project," was undertaken to extend the tables published by

Cunningham & Woodall in 1925.

The book of Riesel (1985), often quoted, discusses factorization (and primality) at length. It contains also tables of factors of Fermat numbers, of Mersenne numbers, of numbers of the forms $2^n + 1$, $10^n + 1$, or repunits $(10^n - 1)/9$, and many more. It is a good place to study the techniques of factorization, which are exposed in a coherent and unified way.

Just as an illustration, and for the delight of lovers of large numbers, I will give now explicit factorizations of some Mersenne, Fermat, and other numbers [for the older references, see Dickson's *History of the Theory of Numbers*, vol. I, pages 22, 29, 377, and Archibald (1935)].

$M_{59} = 2^{59} - 1 = (179951) \times (3203431780337)$, by Landry in 1869;

$M_{67} = 2^{67} - 1 = (193707721) \times (761838257287)$, by Cole in 1903, already indicated;

$M_{73} = 2^{73} - 1 = (439) \times (2298041) \times (9361973132609)$, the factor 439 by Euler, the other factors by Poulet in 1923;

$F_6 = 2^{2^6} + 1 = (1071 \times 2^8 + 1) \times (262814145745 \times 2^8 + 1) =$
$(274177) \times (67280421310721)$, by Landry in 1880.

The above factorizations were obtained before the advent of computers!

More recently, the following factorizations were obtained:

$M_{113} = 2^{113} - 1 = (3391) \times (23279) \times (65993) \times (1868569) \times (1066818132868207)$, the smallest factor by Reuschle (1856) and the remaining factors by Lehmer in 1947;

$M_{193} = 2^{193} - 1 = (13821503) \times (61654440233248340616559) \times (14732265321145317331353282383)$, by Pomerance & Wagstaff in 1983;

$M_{257} = 2^{257} - 1 = (535006138814359) \times$
$(1155685395246619182673033)$
$\times (374550598501810936581776630096313181393)$, by Penk & Baillie (1979, 1980);

$F_7 = 2^{2^7} + 1 = (59649589127497217) \times (5704689200685129054721)$,
by Morrison & Brillhart (in 1970, published in 1971);

$F_8 = 2^{2^8} + 1 = (1238926361552897) \times$
(93461639715357977769163558199606896584051237541638188558028032
1) by Brent & Pollard in 1980;

$(10^{103}+1)/11 = (1237) \times (44092859) \times (102860539) \times (984385009) \times$
$(612053256358933) \times (18272511486652115564716I) \times$
(14718654539938553026608876141375219 79), factorization completed
by Atkin & Rickert, in 1984.

In a paper, dedicated to Dov Jarden, and still to appear.
Brillhart, Montgomery & Silverman, give the known factors of
Fibonacci numbers U_n (for $n \leqslant 999$) and of Lucas numbers V_n (for
$n \leqslant 500$). This pushes much further the work which had been done
by many other numerologists, among whom Jarden (see the 3rd
edition of his book, 1958), and more recently Lehmer, Morrison,
Naur, Selfridge, Williams, Wunderlich.   For details, see the above
mentioned paper by Brillhart et al.

Anyone interested in primality testing, factorization, or similar
calculations with very large numbers needs, of course, access to high-
speed sophisticated computers of the latest generation. However, it
should be noted that there is still pioneering work to be done in the
development of gadgets adaptable to personal computers. These already allow the reaching of substantial results in the comfort of
home (see Dubner, 1985/6). So, if it is snowing outside − as is often
the case in Canada − you may test your prime, keeping warm feet.

F. Public Key Cryptography

Owing to the proliferation of means of communication and the need
of sending messages − like bank transfers, love letters, instructions
for buying stocks, secret diplomatic informations, as, for example,
reports of spying activities − it has become very desirable to develop
a safe method of coding messages. In the past, codes have been kept
secret, known only to the parties sending and receiving the messages.

But, it has been often possible to study the intercepted messages and crack the code. In war situations, this had disastrous consequences.

Great progress in cryptography came with the advent of public key crypto-systems.

The main characteristics of the system are its simplicity, the public key, and the extreme difficulty to crack it. The idea was proposed in 1976 by Diffie & Hellman, and the effective implementation was proposed in 1978 by Rivest, Shamir, & Adleman. This crypto-system is therefore called the RSA-system. I shall describe it now.

Each message corresponds to a number, in the following way. Each letter or sign corresponds to a two-digit number, a blank to the number 00:

For example,

$$a = 01$$
$$b = 02$$
$$c = 03$$
$$\text{etc.}$$

Thus a message M may be viewed as a positive integer, namely, writing the string of double digit numbers corresponding to the letters and signs of the message.

Each user A of the system lists in a public directory his key, which is a pair of positive integers: (n_A, s_A). The first integer n_A is a product of two primes $n_A = p_A q_A$, which are chosen to be large and are kept secret. Moreover, s_A is chosen to be relatively prime with both $p_A - 1$, $q_A - 1$.

To send a message M to another user B, A encrypts M — the way to encode M depends on who will receive it. Upon receiving the encoded message from A, the user B decodes it using his own secret decoding method.

In detail, the process goes as follows. If the message $M \geqslant n_B$, it suffices to break M into smaller blocks; so it may be assumed that $M < n_B$. If $\gcd(M, n_B) \neq 1$, the letter "a", that is 01, or some other dummy letter is added to the end of M, and for the new message, $\gcd(M, n_B) = 1$.

A sends to B the encoded message $E_B(M) = M'$, $1 \leqslant M' < n_B$,

$$M' \equiv M^{s_B} \pmod{n_B}.$$

In order to decode M', the user B calculates t_B, $1 \leqslant t_B <$ $(p_B-1)(q_B-1) = \phi(n_B)$, such that $t_B s_B \equiv 1 \pmod{\phi(n_B)}$; this is done once and for all. Then

$$D_B(M') = M'^{t_B} \equiv M^{s_B t_B} \equiv M \pmod{n_B},$$

so B may read the message M.

How simple!

It is also possible for A to send a "signed message" to B, so that B is able to know that unmistakably the message does come from A and could not have been forged.

The process is the following. If $n_A < n_B$, A breaks his message into blocks $M < n_A$.

A scrambles the message M, using his secret decoding method D_A, namely: if

$$1 \leqslant t_A < (p_A - 1)(q_A - 1) = \phi(n_A), \text{ with } t_A s_A \equiv 1 \pmod{\phi(n_A)},$$

let $D_A(M) \equiv M^{t_A} \pmod{n_A}$, with $1 \leqslant D_A(M) < n_A < n_B$. Then A sends

to B the encoded message $L = E_B(D_A(M)) \equiv M^{t_A s_B} \pmod{n_B}$. This is

decoded by B, to produce the message $D_B(L) = L^{t_B} \equiv M^{t_B t_A s_B} \equiv M^{t_A}$

$\pmod{n_B}$. This message $D_B(L)$ is then unscrambled with the public

encoding method of A: $E_A(D_B(L)) \equiv M^{s_A t_A} \equiv M \pmod{n_A}$. So, the message read by M must indeed come from A.

If $n_B < n_A$, A breaks his messages into blocks $M < n_B$. He sends

$$L = D_A(E_B(M)) \equiv M^{t_A s_B} \pmod{n_B}$$

and B decodes as before with

$$E_A(D_B(L)) \equiv E_A(M^{t_A}) \equiv M \pmod{n_A}.$$

To crack the crypto-system it is necessary to discover $\phi(n_A)$ for each user A. This is equivalent to the factorization of n_A. Indeed, if p_A, q_A are known, then $\phi(n_A) = (p_A - 1)(q_A - 1)$. Conversely, putting $p = p_A$, $q = q_A$, $n = n_A$, from $\phi(n) = (p - 1)(q - 1) = n + 1 - (p + q)$, $(p + q)^2 - 4n = (p - q)^2$ (if $p > q$), then

$$p + q = n + 1 - \phi(n),$$

$$p - q = \sqrt{[n + 1 - \phi(n)]^2 - 4n} \, ,$$

and from this, p, q are expressed in terms of n, $\phi(n)$.

If p, q are large primes, say with 100 or more digits, and chosen at random, the factorization of n, with 200 or more digits, is, in general, unfeasible with the methods known today.

Just after the RSA method was proposed, some criticisms were raised. If the message M is sent to B, as $E_B(M) \equiv M^{s_B} \pmod{n_B}$, any user may decode it as follows. For simplicity, write $s = s_B$, $n = n_B$ and let m be the order of s modulo $\phi(N)$, so m is the smallest positive integer such that $s^m \equiv 1 \pmod{\phi(n)}$. Then applying E_B successively m times, leads to $M^{s^m} \equiv M \pmod{n}$, so the message was decoded using only information which is public, namely, $s = s_B$, $n = n_B$. The value of m need not, and cannot be computed [note that $\phi(N)$ is secret]. But this method is only efficient when m, the iteration index of the message, is small.

It is therefore a good precaution to try to guarantee that, with high probability, the iteration index of any message received by the user B is not small. Rivest suggested (1978) that $p = p_B$, $q = q_B$ be chosen as follows: $p = ap' + 1$, $p' = bp'' + 1$, and $q = cq' + 1$, $q' = dq'' + 1$, where p', p'', q', q'' are distinct primes and a, b, c, d are small integers.

The iteration exponent m of M is the following integer: if k is the order of M mod n, then m is the order of s mod k.

Let G be the multiplicative group of invertible residue classes modulo $n = pq$; it has order $\phi(n) = (p - 1)(q - 1) = acp'q'$. The group G is the direct product of cyclic groups, among those are the cyclic groups $C_{p'}$, $C_{q'}$ of orders p', q', respectively. If $\alpha \in G$, denote by $o(\alpha)$, its order in the multiplicative group. Let H be the subset of G consisting of all α such that $o(\alpha)$ is not divisible by $p'q'$.

G is the direct product of a group D of order ac and the cyclic groups $C_{p'}$, $C_{q'}$ (note that a, c being small, then ac, $p'q'$ are relatively prime). The elements $\alpha \in G$ correspond to triples $(\alpha_1, \alpha_2, \alpha_3)$ with $\alpha_1 \in D$, $\alpha_2 \in C_{p'}$, $\alpha_3 \in C_{q'}$. The elements of H correspond to those with $\alpha_2 = 1$, or $\alpha_3 = 1$. Hence H has $acq' + acp' - ac = ac(p' + q' - 1)$ elements.

If M is a message, $1 \leqslant M \leqslant n$, the probability that $\gcd(M,n) \neq 1$ is

$$\frac{1}{n} \left[\frac{n}{p} + \frac{n}{q} - 1 \right] < \frac{1}{p} + \frac{1}{q},$$

so it is very small.

If $\gcd(M,n) = 1$ then $M \bmod n$ is in the group G; the probability that $k = \mathrm{ord}(M \bmod n)$ will not be a multiple of $p'q'$ is

$$\frac{ac(p' + q' - 1)}{\phi(n)} = \frac{p' + q' - 1}{p'q'} < \frac{1}{q'} + \frac{1}{p'},$$

so again, it is very small.

Now, let $\gcd(M,n) = 1$ and assume that $p'q'$ divides k. Let G' be the multiplicative group of invertible residue classes modulo $p'q'$; it has $\phi(p'q') = bdp''q''$ elements; since $\gcd(s,p'q') = 1$, then $s \bmod p'q'$ belongs to G'. The preceding argument shows that the probability for the order of $s \bmod p'q'$ to be not divisible by $p''q''$ is

$$\frac{bd(p'' + q'' - 1)}{\phi(p'q')} = \frac{p'' + q'' - 1}{p''q''} < \frac{1}{q''} + \frac{1}{p''},$$

and, once more, it is very small.

So, it is possible to assume, with very high probability, that $\gcd(M,n) = 1$, $p'q'$ divides k, and $p''q''$ divides the order of $s \bmod p'q'$.

Finally, m is the order of $s \bmod k$ and $p'q'$ divides k, hence the order of $s \bmod p'q'$ divides m; therefore, $p''q''$ divides m. Since p'', q'' were chosen to be large primes, then the iteration index m is large, with very large probability.

It still remains the question of how to find, for each user, the pair of primes p, q that will be his most cherished possession. How to produce a prime with, say, about 100 digits!

Suppose I pick a number N, at random, of that size. What is the probability that it will be a prime? For this I must know the approximate number $\pi(N)$ of primes $p < N$. By the result of Tschebycheff to be discussed in Chapter 4, Section I, the number $\pi(N)$ is approximately $N/(\log N)$; thus, the probability that N is a prime is about $1/(\log N)$, which in this case turns out to $1/230$. So, one would expect to try about 230 numbers N, each time testing for primality, until a prime is found.

It is better to proceed in a more intelligent way. Pick any number N_0 of the required size, about 10^{100}. Then begin deleting all the numbers between N_0 and $N_0 + 5000$, which are multiples of 2, 3, 5, 7, 11, ..., p, ... (for all primes $p < 1000$, for example). This is the method first discovered by Eratosthenes (discussed in Section I). The numbers which remain are those having no prime factor less than 1000. Each one of these numbers should then be tested for primality, until a prime is found.

Proceeding in the above manner is indeed more economical. This can be shown by calculating the function $\phi(n,k)$, equal to the number of integers $m \leqslant n$, such that m is not divisible by any one of the first k primes; see Chapter 4, Section I, where I give some indications in connection with Meissel's formula.

In this presentation of the RSA − crypto-system, I have benefited from the article of Hoogendoorn (1982).

More detailed information may be obtained in the original paper of Rivest, Shamir & Adleman, as well as in the book by Riesel (1985), where a numerical example is worked out in full detail. For a more complete treatment of cryptography in general, see the book by Konheim (1981).

Chapter 3

ARE THERE FUNCTIONS DEFINING PRIME NUMBERS?

To determine prime numbers, it is natural to ask for functions $f(n)$ defined for all natural numbers $n \geqslant 1$, which are computable in practice and produce some or all prime numbers.

For example, one of the following conditions should be satisfied:

(a) $f(n) = p_n$;

(b) $f(n)$ is always a prime number, and if $n \neq m$, then $f(n) \neq f(m)$;

(c) the set of prime numbers is equal to the set of positive values assumed by the function.

Clearly, condition (a) is more demanding than (b) and than (c).

The efforts in this direction have been rather disappointing, except for results of theoretical importance, related to condition (c).

I. Functions Satisfying Condition (a)

Sierpiński showed in 1952:

If $\alpha = \sum_{m=1}^{\infty} (p_m / 10^{2^m})$ (this series is convergent), then

$$p_n = [10^{2^n} \alpha] - 10^{2^{n-1}} [10^{2^{n-1}} \alpha].$$

As already indicated, here $[x]$ denotes the largest integer in x, that is, $[x] \leqslant x < [x] + 1$.

This formula is useless, because to calculate p_n one needs to know not only $p_1, p_2, \ldots, p_{n-1}$, but also $p_n \ldots$.

The result of Willans (1964) is more interesting, even though still of no practical use. It provides definite answers to Hardy & Wright's questions in their famous book:

(1) Is there a formula for the nth prime number?

(2) Is there a formula for a prime, in terms of the preceding prime?

What is intended here is to find a closed expression for the nth prime p_n, in terms of n, by means of functions that are computable and, if possible, classical. Intimately related with this problem is to find reasonable expressions for the function counting primes.

For every real number $x > 0$ let $\pi(x)$ denote the number of primes p such that $p \leqslant x$.

This is a traditional notation for one of the most important functions in the theory of prime numbers. I shall return to it in Chapter 4. Even though the number $\pi = 3.14\ldots$ and the function $\pi(x)$ do occur below in the same formula, this does not lead to any ambiguity.

First I indicate a formula for $\pi(m)$, given by Willans. It is based on the classical Wilson's theorem which I proved in Chapter 2.

For every integer $j \geqslant 1$ let

$$F(j) = \left[\cos^2\pi\,\frac{(j-1)! + 1}{j}\right].$$

So for any integer $j > 1$, $F(j) = 1$ when j is a prime, while $F(j) = 0$ otherwise. Also $F(1) = 1$.

Thus

$$\pi(m) = -1 + \sum_{j=1}^{m} F(j).$$

Willans also expressed:

$$\pi(m) = \sum_{j=2}^{m} H(j) \quad \text{for} \quad m = 2,3, \ldots$$

where

$$H(j) = \frac{\sin^2\pi\,\dfrac{\{(j-1)!\}^2}{j}}{\sin^2(\pi/j)}\,.$$

Mináč gave an alternate (unpublished) expression, which involves neither the cosine nor the sine:

$$\pi(m) = \sum_{j=2}^{m} \left[\frac{(j-1)! + 1}{j} - \left[\frac{(j-1)!}{j} \right] \right].$$

Proof. The proof of Mináč's formula is quite simple, and since it is not published anywhere, I'll give it here.

First a remark: if $n \neq 4$ is not a prime, then n divides $(n-1)!$. Indeed, either n is equal to a product $n = ab$, with $2 \leqslant a, b \leqslant n - 1$ and $a \neq b$, or $n = p^2 \neq 4$. In the first alternative, n divides $(n-1)!$; in the second case, $2 < p \leqslant n - 1 = p^2 - 1$, so $2p \leqslant p^2 - 1$ and n divides $2p^2 = p \times 2p$, which divides $(n-1)!$.

For any prime j, by Wilson's theorem, $(j-1)! + 1 = kj$ (where k is an integer), so

$$\left[\frac{(j-1)! + 1}{j} - \left[\frac{(j-1)!}{j} \right] \right] = \left[k - \left[k - \frac{1}{j} \right] \right] = 1.$$

If j is not a prime and $j \geqslant 6$, then $(j-1)! = kj$ (where k is an integer), by the above remark. Hence

$$\left[\frac{(j-1)! + 1}{j} - \left[\frac{(j-1)!}{j} \right] \right] = \left[k + \frac{1}{j} - k \right] = 0.$$

Finally, if $j = 4$, then

$$\left[\frac{3! + 1}{4} - \left[\frac{3!}{4} \right] \right] = 0.$$

This is enough to prove the formula indicated for $\pi(m)$. $\square$

With the above notations, Willans gave the following formula for the nth prime:

$$p_n = 1 + \sum_{m=1}^{2^n} \left[\left[\left(\frac{n}{\sum_{j=1}^{m} F(j)} \right)^{1/n} \right] \right]$$

or, in terms of the prime counting function:

$$p_n = 1 + \sum_{m=1}^{2^n} \left[\left[\left(\frac{n}{1 + \pi(m)} \right)^{1/n} \right] \right].$$

For the related problem of expressing a prime q in terms of the prime p immediately preceding it, Willans gave the formula:

$$q = 1 + p + F(p+1) + F(p+1)F(p+2) + \cdots + \prod_{j=1}^{p} F(p+j),$$

where $F(j)$ was defined above.

Another formula for the smallest prime greater than $m \geqslant 2$ was given by Ernvall, while still a student, and published in 1975: Let

$$d = \gcd((m!)^{m!} - 1, (2m)!),$$

let

$$t = \frac{d^d}{\gcd(d^d, d!)}$$

and let a be the unique integer such that d^a divides t, but d^{a+1} does not divide t. Then the smallest prime larger than m is

$$p = \frac{d}{\gcd(t/d^a, d)} .$$

Taking $m = p_{n-1}$, this yields a formula for p_n.

Despite the nil practical value, I tend to believe that such formulas may be of some relevance to logicians wishing to understand clearly how various parts of arithmetic may be deduced from different axiomatizations or fragments of Peano axioms.

In 1971, Gandhi gave a formula for the nth prime p_n. To explain it, I require the Möbius function, which is one of the most important arithmetic functions. The Möbius function is defined as follows:

$$\begin{cases} \mu(1) = 1, \\[6pt] \text{if } n \text{ is the product of } r \text{ distinct primes, then } \mu(n) = (-1)^r, \\[6pt] \text{if the square of a prime divides } n, \text{ then } \mu(n) = 0. \end{cases}$$

Let $P_{n-1} = p_1 p_2 \cdots p_{n-1}$. Then Gandhi showed:

$$p_n = \left[1 - \frac{1}{\log 2} \log\left(-\frac{1}{2} + \sum_{d | P_{n-1}} \frac{\mu(d)}{2^d - 1} \right) \right]$$

or equivalently, p_n is the only integer such that

$$1 < 2^{P_n} \left[-\frac{1}{2} + \sum_{d|P_{n-1}} \frac{\mu(d)}{2^d - 1} \right] < 2.$$

The following simple proof was given by Vanden Eynden in 1972.

Proof. For simplicity of notation, let $Q = P_{n-1}$, $p_n = p$, and

$$S = \sum_{d|Q} \frac{\mu(d)}{2^d - 1}.$$

So

$$(2^Q - 1)S = \sum_{d|Q} \mu(d) \frac{2^Q - 1}{2^d - 1} = \sum_{d|Q} \mu(d)(1 + 2^d + 2^{2d} + \cdots + 2^{Q-d}).$$

If $0 \leqslant t < Q$, the term $\mu(d)2^t$ occurs exactly when d divides $\gcd(t,Q)$. So the coefficient of 2^t in the last sum is $\sum_{d|\gcd(t,Q)}\mu(d)$; in particular, for $t = 0$ it is equal to $\sum_{d|Q}\mu(d)$.

But, for any integer $m \geqslant 1$ it is well known and easy to show, that

$$\sum_{d|m} \mu(d) = \begin{cases} 1 & \text{if } m = 1, \\ 0 & \text{if } m > 1. \end{cases}$$

Writing $\sum'_{0<t<Q}$ for the sum extended over all t, such that $0 < t < Q$ and $\gcd(t,Q) = 1$, then $(2^Q - 1)S = \sum'_{0<t<Q}2^t$; the largest index t in this summation is $t = Q - 1$. It follows that

$$2(2^Q - 1)\left[-\frac{1}{2} + S\right] = -(2^Q - 1) + \sum'_{0<t<Q} 2^{t+1} = 1 + \sum'_{0<t<Q-1} 2^{t+1}.$$

If $2 \leqslant j < p_n = p$, there exist some prime q such that $q < p_n = p$ (so $q|Q$) and $q|Q - j$. Hence every index t in the above sum satisfies $0 < t \leqslant Q - p$. Thus

$$\frac{2^{Q-p+1}}{2 \times 2^Q} < -\frac{1}{2} + S = \frac{1 + \sum'_{0<t\leqslant Q-p}2^{t+1}}{2(2^Q - 1)} < \frac{2^{Q-p+2}}{2 \times 2^Q},$$

where the inequalities are easy to establish.

Hence multiplying with 2^p, it follows that

$$1 < 2^p\left[-\frac{1}{2} + S\right] < 2. \qquad \square$$

In 1974, Golomb gave another proof, which I find illuminating.

He described it as being the sieve of Eratosthenes (see Chapter 4, Section I) performed on the binary expansion of 1.

To each positive integer n, assign the probability or weight $W(n) = 2^{-n}$; note that $\sum_{n=1}^{\infty} W(n) = 1$.

In this distribution, the probability that a random integer be a multiple of a fixed integer $d \geqslant 1$ is

$$M(d) = \sum_{n=1}^{\infty} W(nd) = \sum_{n=1}^{\infty} 2^{-nd} = \frac{1}{2^d - 1}.$$

The probability that a random integer be relatively prime to a fixed integer $m \geqslant 1$ is easily seen to be

$$R(m) = 1 - \sum_{p|m} M(p) + \sum_{pp'|m} M(pp') - \sum_{pp'p''|m} M(pp'p'') + \cdots$$

$$= \sum_{d|m} \mu(d)M(d) = \sum_{d|m} \frac{\mu(d)}{2^d - 1}.$$

As before, let $Q = p_1 p_2 \cdots p_{n-1}$. So

$$R(Q) = \sum_{d|Q} \frac{\mu(d)}{2^d - 1},$$

but, on the other hand, in this distribution, $R(Q)$ is given directly as

$$R(Q) = \sum_{\gcd(m,Q)=1} W(m) = \frac{1}{2} + \frac{1}{2^{P_n}} + \frac{1}{2^{P_{n+1}}} + \alpha$$

(where α is a sum of reciprocals of some higher powers of 2). Thus

$$R(Q) - \frac{1}{2} = \sum_{d|Q} \frac{\mu(d)}{2^d - 1} - \frac{1}{2} = \frac{1}{2^{P_n}} + \frac{1}{2^{P_{n+1}}} + \alpha.$$

Hence

$$2^{P_n} \left[\sum_{d|Q} \frac{\mu(d)}{2^d - 1} - \frac{1}{2} \right] = 1 + \theta_n,$$

where $0 < \theta_n < 1$. So p_n is the only integer m such that

$$1 < 2^m \left[\sum_{d|Q} \frac{\mu(d)}{2^d - 1} - \frac{1}{2} \right] < 2$$

and this is just another way of writing Gandhi's formula. Note that $0 < \theta_n < 1/2$, because $p_{n+1} \geqslant p_n + 2$.

In binary notation, all this becomes even more transparent. Now $W(n) = .000...1$ (with digit 1 at nth place), so $\sum_{n=1}^{\infty} W(n) = .1111... = 1$.

For the even integers

$$\sum_{n=1}^{\infty} W(2n) = .010101... = \frac{1}{2^2 - 1} = \frac{1}{3}.$$

Subtracting, with $P_1 = p_1 = 2$: $R(P_1) = .101010... = 1 - 1/3$. By subtracting the multiples of 3 and adding back the twice-subtracted multiples of 6, obtain

$$Q(3) = .001001001... = \frac{1}{2^3 - 1} = \frac{1}{7},$$

$$Q(6) = .000001000001... = \frac{1}{2^6 - 1} = \frac{1}{63},$$

and with $P_2 = p_1 p_2 = 6$

$$R(P_2) = R(P_1) - Q(3) + Q(6) = .1000101000101000...$$
$$= 1 - \frac{1}{3} - \frac{1}{7} + \frac{1}{63}.$$

Continuing in this way,

$$R(P_{n-1}) = .100...010...10... = \frac{1}{2} + \frac{1}{2^{p_n}} + \frac{1}{2^{p_{n+1}}} + \alpha$$

and

$$R(P_{n-1}) - \frac{1}{2} = .000...010...$$

where the first digit 1 appears in position p_n.

II. Functions Satisfying Condition (b)

The number $f(n) = [\theta^{3^n}]$ is a prime for every $n \geq 1$; here θ is a number which is roughly equal to 1.3064... (see Mills, 1947). Similarly,

$$g(n) = [2^{2^{2^{\cdot^{\cdot^{2^\omega}}}}}]$$

(a string of n exponents) is a prime for every $n \geq 1$; here ω is a number which is roughly equal to 1.9287800... (see Wright, 1951).

The fact that θ, ω are known only approximately and the numbers grow very fast, make these formulas no more than curiosities. For example, $g(1) = 3$, $g(2) = 13$, $g(3) = 16381$, $g(4)$ has more than 5000 digits.

There are many other formulas of a similar kind in the literature, but they are just as useless; see Dudley (1969).

At this point one might wonder: Why not try some polynomial with integral coefficients instead of these awkward functions involving exponentials and the largest integer function?

The reason is simply given by the following negative result:

If f is a polynomial with integral coefficients, in one indeterminate, which is not constant, then there exist infinitely many integers n such that $|f(n)|$ is not a prime number.

Proof. I may assume that there exists some integer $n_0 \geqslant 0$ such that $|f(n_0)| = p$ is a prime number. Since the polynomial is not constant, $\lim_{x \to \infty} |f(x)| = \infty$, so there exists $n_1 > n_0$ such that if $n \geqslant n_1$, then $|f(n)| > p$. For any h such that $n_0 + ph \geqslant n_1$, $f(n_0 + ph) = f(n_0) + $ multiple of $p = $ multiple of p. Since $|f(n_0 + ph)| > p$, then $|f(n_0 + ph)|$ is a composite integer. $\square$

Now that no polynomial with integral coefficients in one indeterminate is fit for the purpose, could a polynomial with several indeterminates be suitable?

Once more, this is excluded by the following strong negative result:

If f is a polynomial with complex coefficients in m indeterminates such that its values at natural numbers are, in absolute value, prime numbers, then f must be a constant.

More generally, Reiner showed in 1943:

If f_i, g_i ($i = 1, ..., n$) are polynomials with integral coefficients and positive leading coefficients, then the values of $\sum_{i=1}^{n} f_i(x)^{g_i(x)}$ at the positive integers cannot always be prime numbers.

Similarly, Buck showed in 1946 that if a rational function f is not a constant, then there exists a natural number n such that $|f(n)|$

is not a prime number.

Even though nonconstant polynomials $f(X)$ with integral coefficients have composite values (in absolute value) at infinitely many natural numbers, Euler discovered in 1772 one such polynomial $f(X)$ with a "long string" of prime values. For these polynomials, there exist natural numbers m, n with $0 \leqslant m < n$ (and $n - m$ not too small) such that $f(k)$ is a prime, for every k, $m \leqslant k \leqslant n$.

Here is Euler's famous example: $f(X) = X^2 + X + 41$. For $k = 0, 1,$ 2, 3, ..., 39, all its values are prime numbers, namely, 41, 43, 47, 53, 61, 71, 83, 97, 113, 131, 151, 173, 197, 223, 251, 281, 313, 347, 383, 421, 461, 503, 547, 593, 641, 691, 743, 797, 853, 911, 971, 1033, 1097, 1163, 1231, 1301, 1373, 1447, 1523, 1601. For $k = 40$ the value is $1681 = 41^2$.

The same is true for $X^2 - X + 41$ and $1 \leqslant k \leqslant 40$ (just by changing X into $X - 1$).

Changing X into $X - 40$ gives the polynomial indicated by Escott (1899): $X^2 - 79X + 1601$, which has prime values for $k = 0, 1, ..., 79$ — but these are the same primes listed above, each produced twice. Other polynomials have been found which provide the same string of primes repeated more times, or in reverse order. See Dickson's *History of the Theory of Numbers*, vol. I, page 420, for references.

For polynomials of the form $X^2 + X + q$, where q is a prime number, I note the interesting equivalent properties:

(1) $q = 2, 3, 5, 11, 17,$ or 41.

(2) $X^2 + X + q$ assumes prime values for $k = 0, 1, ..., q - 2$.

(3)   The field $\mathbf{Q}(\sqrt{1 - 4q})$ of all algebraic numbers of the form $r + s\sqrt{1 - 4q}$ (where r, s are rational numbers) has "class number" 1 (which I will explain henceforth).

The implication (1) → (2) was noted by Euler (in a letter to Bernoulli, dated 1772). Frobenius (1912) was the first to show that (3) implies (2), and of course, it was known since Gauss that (1) implies (3). In 1912, Rabinovitch proved the equivalence of (2) and (3). I note that Lehmer proved in 1936 once more that (2) → (3), and more recently, Szekeres (1974) and Ayoub & Chowla (1981) established again the implication (3) → (2). A presentation of this proof is on page 155 of Cohn's book (1962) and, for those who like to practice their Spanish, I have a forthcoming article (1987) on this question, where all the calculations and proofs are given in full detail, almost from first principles.

And now, I elaborate on the meaning of condition (3). It is equivalent to say that each "algebraic integer" may be decomposed into "prime algebraic integers," in a unique way, except for "units." Thus, the arithmetic properties of such fields are like those of ordinary integers, at least in respect to the fundamental theorem of factorization into primes. The main difference lies in the fact that the only integers that divide 1 are 1, −1, while in a field of algebraic numbers, there may be many more units, that is, algebraic integers dividing 1.

It remained to determine for which primes q the field $\mathbf{Q}(\sqrt{1-4q})$ has class number 1.

In the present situation of imaginary quadratic fields, I like to tell an interesting story. In *Disquisitiones Arithmeticae* (1801), already quoted in Chapter 2, Gauss showed that if $d = 1, 2, 3, 7, 11, 19, 43, 67, 163$, then $\mathbf{Q}(\sqrt{-d})$ has class number 1. He also conjectured, in article 303, that these are the only possible values of d for which $\mathbf{Q}(\sqrt{-d})$ has class number 1.

[To say the truth, Gauss did not refer to imaginary quadratic fields, but rather to negative discriminants having only one class of binary quadratic forms. I will soon explain this concept in more detail and its relationship with quadratic fields.]

In a classical paper, Heilbronn & Linfoot showed, in 1934, that there exists at most one other value of $d > 0$ for which $\mathbf{Q}(\sqrt{-d})$ has class number 1, and if such d exists, Lehmer showed that it must be quite large: $d > 5 \times 10^9$. In 1952, Heegner proved that no other such d could exist, but his proof contained some steps which were unclear, perhaps even a gap.

Baker reached the same conclusion in 1966, with his method involving effective lower bounds on linear forms of three logarithms; this is also reported in his article of 1971. At about the same time, unaware of Heegner's results but with similar ideas concerning elliptic modular functions, Stark proved that no further possible value for d exists. So were determined all the imaginary quadratic fields with class number 1. It was somewhat an anticlimax when, in

1968, Deuring was able to straighten out Heegner's proof, and when Stark also showed (1969) that, by 1949, Gelfond and Linnik could have used the theorem on linear independence of just two logarithms (known at that time) to reach the same conclusion. The technical details involved in these proofs are far beyond the scope of this book.

I have only given the unfolding of this story to show how an apparently simple problem of determination of quadratic polynomials assuming prime values at consecutive integers has not been so innocent as it seemed; its complete solution has indeed required quite sophisticated theories.

Summarizing, the nine values of d yield the six values of q mentioned above.

Record

For polynomials of the type $X^2 + X + q$, the best possible result is already Euler's, as follows from the preceding discussion.

What happens for general quadratic polynomials with integral coefficients?

Euler considered also the polynomials $2X^2 + p$, with $p = 3, 5, 11, 29$, and showed that they assume prime values for all $k = 0, 1, ..., p-1$. Are there other polynomials of this form with the same property?

In the paper already quoted, Frobenius studied many quadratic polynomials with similar properties and he noted for the first time a relation with the class number of a certain quadratic field. I will be more precise shortly.

For this purpose, I need to return to the concept of class number, which I have only mentioned much too briefly. I need even to go back to Lagrange, who first developed the theory of binary quadratic forms in order to study the representation of integers m as values of $aX^2 + bXY + cY^2$ (where a, b, c are integers, $ac \neq 0$).

If a form $a'X'^2 + b'X'Y' + c'Y'^2$ is obtained from the above form by a linear change of variables

$$\begin{cases} X = \alpha X' + \beta Y', \\ Y = \gamma X' + \delta Y', \end{cases}$$

where α, β, γ, δ are integers and the determinant is $\alpha\delta - \beta\gamma = 1$, then the two forms represent the same integers. In this sense, it is reasonable to consider such forms as being equivalent. Moreover, these forms have also the same discriminant $D = b^2 - 4ac = b'^2 - 4a'c'$. Each possible discriminant D must be either a multiple of 4 or else congruent to 1, modulo 4. Let $d = D/4$, respectively $d = D$.

In *Disquisitiones Arithmeticae*, Gauss classified the binary quadratic forms with given discriminant D. He developed a very beautiful theory, which is the fundamental stone on which the theory of quadratic fields was subsequently built.

Among the many interesting facts established by Gauss, I mention first the discovery that there is an operation of composition between equivalence classes of forms of a given discriminant. As a consequence, the classes constitute a group with this operation. But what concerns me more directly here is the important fact that for every discriminant there are only finitely many equivalence classes of forms with the given discriminant.

The theory was later reinterpreted, associating to each form $aX^2 + bXY + cY^2$ of discriminant D, the ideal I of all numbers

$$\gamma = ra + s\,\frac{-b + \sqrt{D}}{2} \qquad (r,s \text{ arbitrary integers})$$

in the quadratic field $Q(\sqrt{D}) = Q(\sqrt{d})$. This establishes a one-to-one correspondence. It is easy to see that equivalent forms correspond to equivalent ideals, in the following sense: I, I' are equivalent ideals if there exist some nonzero numbers $\alpha \in Q(\sqrt{d})$ such that $I' = \{\alpha\gamma \mid \gamma \in I\}$. Also, the composition of forms corresponds to the multiplication of ideals. So the number of classes of forms with the given discriminant D is the same as the number of classes of ideals of $Q(\sqrt{d})$, which is usually denoted by $h(d)$, or simply h — if there is no ambiguity.

It should be added that $h(d) = 1$ exactly when the fundamental theorem of factorization into prime elements holds in $Q(\sqrt{d})$ — but I have said it already.

Gauss conjectured that for each integer $h \geqslant 1$ there are only finitely many negative discriminants for which the number of equivalence classes of binary forms is equal to h. In the language of quadratic fields, for each $h \geqslant 1$ there are only finitely many square-free integers $d > 0$ such that $Q(\sqrt{-d})$ has class number h.

As I have indicated above, this has been shown true for $h = 1$. Soon I will say more about this question. Now it is time to examine the imaginary quadratic fields with class number $h = 2$.

A theory of genera, developed also by Gauss, implies that if the class number of $Q(\sqrt{-d})$ is a multiple of 2, then d has at most two distinct prime factors. This gives three possible types of fields.

(I) $d = 2p$, where p is an odd prime.

(II) $d = p$, where p is a prime, $p \equiv 1 \pmod 4$.

(III) $d = pq \equiv 3 \pmod 4$, where $p < q$ are odd primes.

So, what are the possible fields with class number 2?

More generally, how to prove Gauss' conjecture and how to determine the imaginary quadratic fields with a given class number?

The solution of this problem took a long time and has required deep methods.

The first progress was made by Dirichlet (1839/40), who gave an explicit formula to compute the class number of imaginary quadratic fields. This is well explained in the book by Borevich & Shafarevich (1964). So, in principle, it was not difficult to determine the class number of several fields $Q(\sqrt{-d})$.

More recently, thanks to the successful work of Hecke (result published by Landau in 1918), Deuring (1933), Mordell (1934), and Heilbronn (1934), it was established that if $|d|$ tends to infinity, then so does $h(-d)$. This says that, for every integer $h \geqslant 1$, there are only finitely many fields $Q(\sqrt{-d})$ having class number equal to h; in particular, with class number 2.

Siegel has even shown in 1935 that $h(-d) > C|d|^{\frac{1}{2} - \epsilon}$ for every $\epsilon > 0$; nevertheless, the constant C was not explicitly determined.

But how many and which are these fields? For this purpose, it was imperative to compute explicitly, for each $h \geqslant 1$, a number $C(h) > 0$ such that if $h(-d) = h$, then $|d| < C(h)$. And then, it would remain to compute for every d, with $|d| < C(h)$, the actual value of $h(-d)$, using Dirichlet's formula, or any other speedier process.

Easier said than done. As it turned out — and had already happened in the search for those $Q(\sqrt{-d})$ with class number 1 — it was necessary to use high powered transcendental methods. Baker (1969, 1971), Montgomery & Weinberger (1974), and Stark (1975) were finally able to write the complete list of the 18 fields $Q(\sqrt{-d})$ with $h(-d) = 2$, namely d = 5, 6, 10, 13, 15, 22, 35, 37, 51, 58, 91, 115, 123, 187, 235, 267, 403, 427.

Before continuing with the discussion of Gauss' problem, I return briefly to prime values of polynomials.

For each one of the three types of fields seen above, consider the polynomials

$$f_I(X) = 2X^2 + p$$

$$f_{II}(X) = 2X^2 + 2X + \frac{p + 1}{2} \qquad \text{where } p \equiv 1 \ (\text{mod } 4)$$

$$f_{III}(X) = pX^2 + pX + \frac{p + q}{2} \qquad \text{where } p < q \text{ and } pq \equiv 3 \ (\text{mod } 4).$$

Frobenius observed that if the class number of $Q(\sqrt{-2p})$ is 2, then $f_I(k)$ is prime for every k = 0, 1, ..., $p - 1$, and also that if the class number of $Q(\sqrt{-p})$ is 2 then $f_{II}(k)$ is prime for k = 0, 1, ..., $(p - 3)/2$.

Hendy proved in 1974 the converse, as well as the fact that $f_{III}(k)$ is prime for k = 0, 1, ..., $(p+q-1)/2$ if and only if the class number of $Q(\sqrt{-p})$ is 2.

From these calculations, it follows:

(I)   The only polynomials of type I are $2X^2 + p$, with p = 3, 5, 11, 29, and they assume prime values for k = 0, 1, ..., $p - 1$.

(II)   The only polynomials of type II are $2X^2 + 2X + n$, with n = 3, 7, 19 and they assume prime values for k = 0, 1, ..., $n - 2$.

(III)   The only polynomials of type III are $pX^2 + pX + n$, with (p,n) equal to the following pairs: (3,2), (3,5), (3,11), (3,23), (5,3), (5,7), (5,13), (7,5), (7,17), (11,7), (13,11) and they assume prime values for k = 0, 1, ..., $n - 2$.

Returning to arbitrary values of h, the problem of Gauss was finally solved through the combined efforts of Goldfeld (1976), and Gross & Zagier (1983). In short, the result is the following (it is weaker than Siegel's estimate given above, but it is effective):

For every $\epsilon > 0$, there is an effectively computable constant $C > 0$ such that $h(-d) > C(\log|d|)^{1-\epsilon}$.

Even more refined considerations allowed Oesterlé (1984) to show that

$$h(-d) > \frac{1}{55} \log|d| \prod_{\substack{p|d \\ p \neq d}} \left(1 - \frac{[2\sqrt{p}]}{p + 1}\right).$$

I realize that I'm saying very little about a great achievement, but to enter into details would require ideas and techniques far beyond my aims.   Instead, I suggest reading the beautifully written paper of Goldfeld (1985), who knows the problem firsthand and has made a decisive contribution towards its solution.

[The analogous question for real quadratic fields will be considered in Chapter 6, Section II.   If you are impatient, go directly to that section and see how this is also related to prime values of quadratic polynomials.]

For polynomials of higher degree nothing in this respect seems to be known.   I shall return to this point in Chapter 6.

For polynomials of degree one, $f(X) = AX + q$, there are at most q successive values $|f(0)|, |f(1)|, ..., |f(q - 1)|$ which are primes.   This leads, therefore, to the open problem:

Is it true that for every prime q there exists an integer $A \geqslant 1$ such that $q, A + q, 2A + q, ..., (q-1)A + q$ are primes?   For example,

$q = 3$, $A = 2$ yield the primes 3, 5, 7;
$q = 5$, $A = 6$ yield the primes 5, 11, 17, 23, 29;
$q = 7$, $A = 150$ yield the primes 7, 157, 307, 457, 607, 757, 907.

I shall examine this and other questions about primes in arithmetic progressions in Chapter 4, Section IV.

Another investigation that has been undertaken is the search of quadratic polynomials $f(X)$ assuming prime values often.   If $N > 1$, let $v(f,N) = \#\{x \mid x = 0,1, ..., N$ such that $|f(x)|$ is equal to 1 or to a prime\}.   (This function is closely related to another counting function to be introduced in Chapter 4, Interlude.)   The problem is to determine, for given N, the polynomial $f(X)$ for which $v(f,N)$ is maximum.

Record

For $N = 1000$ the polynomial $f(X) = 2X^2 - 199$, indicated by Karst in 1973, provides the maximum value of $v(f,1000)$ up-to-date. Namely,

$$v(2X^2 - 199, 1000) = 598.$$

It should be noted that for Euler's polynomial

$$v(X^2 + X + 41, 1000) = 581.$$

Later, in Chapter 6, Section III, I shall also consider the contrary phenomenon of polynomials assuming composite values for all integers $n = 0, 1, 2, \ldots$ up to some large N.

III. Functions Satisfying Condition (c)

Surprisingly, if one requires only condition (c), the situation is totally different. This was discovered as a by-product in the investigation of Hilbert's tenth problem. The ideas come from logic and the results are quite extraordinary — even if at the present they have not yet found immediate practical application.

In my presentation, I will not enter into technical details, which would take me too far from the prime numbers. Thus, I'll have to trade rigor for intuition, and I'm counting on the good will of the reader. Please do not interpret what I'm going to write in any undesirable way!

Hilbert's tenth problem asked about the solution in integers $(x_1, \ldots, x_n)$ of diophantine equations $P(X_1, \ldots, X_n) = 0$, where P is any polynomial with integral coefficients, and any number of indeterminates. More exactly: To give an algorithm that may be applied to any diophantine equation and will tell if it has solution in integers.

An algorithm should be understood as a definite procedure, which could be implemented as a computer program consisting of finitely many successive steps and leading to an answer "yes" or "no" — the kind of manipulations that mathematicians agree as legitimate. This is all I will say, since I do not want to introduce notions belonging to the theory of recursive sets or functions, nor do I wish to mention Turing machines. These concepts are very important, however rather distant from my main subject, the prime numbers.

Here is a simple and useful comment. Lagrange showed, as everyone knows, that every non-negative integer is the sum of four squares. If $P(X_1, ..., X_n)$ is a polynomial with integral coefficients, and if Q is the polynomial in $4n$ indeterminates defined by

$$Q = P(1 + X_1^2 + Y_1^2 + Z_1^2 + T_1^2, ..., 1 + X_n^2 + Y_n^2 + Z_n^2 + T_n^2),$$

then if there is an algorithm to tell whether the equation $Q = 0$ has a solution in integers (positive or not) then this algorithm tells whether $P = 0$ has a solution in positive integers. Conversely, if R is the polynomial in $2n$ indeterminates defined by

$$R = Q(y_1 - z_1, ..., y_n - z_n)$$

if there exists an algorithm to tell that the equation $R = 0$ has a solution in positive integers, then this algorithm tells that $Q = 0$ has a solution in integers. Thus, in Hilbert's problem, it is equivalent to ask for an algorithm that will decide whether any polynomial diophantine equation has solution in positive integers. This shifts the focus to the study of sets S of n-tuples $(x_1, ..., x_n)$ of positive integers.

The central concept is the following: S is called a diophantine set if there exists a polynomial P with integral coefficients, in indeterminates $X_1, ..., X_n, Y_1, ..., Y_m$ ($m \geqslant 0$), such that $(x_1, ..., x_n) \in S$ if and only if there exist positive integers $y_1, ..., y_m$ satisfying

$$P(x_1, ..., x_n, y_1, ..., y_m) = 0.$$

First, the trivial examples. Every finite set S of n-tuples of positive integers is diophantine. Indeed, if S consists of the n-tuples $(a_1, a_2, ..., a_n), (b_1, b_2, ..., b_n), ..., (k_1 k_2, ..., k_n)$, let

$$P(X_1, X_2, ..., X_n) = [(X_1 - a_1)^2 + (X_2 - a_2)^2 + \cdots + (X_n - a_n)^2]$$
$$\times [(X_1 - b_1)^2 + (X_2 - b_2)^2 + \cdots + (X_n - b_n)^2]$$

$$\cdots$$

$$\times [(X_1 - k_1)^2 + (X_2 - k_2)^2 + \cdots + (X_n - k_n)^2].$$

By its very definition, P has integral coefficients, so the set of all n-tuples $(x_1, ..., x_n)$ of positive integers such that $P(x_1, ..., x_n) = 0$, is a diophantine set.

Here is another example: the set S of all composite positive integers. Indeed, x is composite if and only if there exist positive integers y, z such that $x = (y + 1)(z + 1)$. Thus, x is composite whenever there exist y, z such that (x,y,z) is a solution of $X - (Y + 1)(Z + 1) = 0$.

The following fact, noted by Putnam in 1960, is not difficult to show:

A set S of positive integers is diophantine if and only if there exists a polynomial Q with integral coefficients (in $m \geqslant 1$ indeterminates) such that $S = \{Q(x_1, ..., x_m) \geqslant 1 \mid x_1 \geqslant 1, ..., x_m \geqslant 1\}$.

Proof. If S is the set of positive values of Q at positive integers, let $P = Q(X_1, ..., X_m) - X$; then $x \in S$ if and only if there exist positive integers $x_1, ..., x_m$ such that $P(x,x_1, ..., x_m) = 0$; this means that S is a diophantine set.

Conversely, let S be a diophantine set, let P be a polynomial with integral coefficients, in $m + 1$ indeterminates, such that $x \in S$ if and only if there exist $x_1 \geqslant 1, ..., x_m \geqslant 1$ such that $P(x,x_1, ..., x_m) = 0$. Let $Q = X(1 - P^2)$. If $x \in S$, and $x_1 \geqslant 1, ..., x_m \geqslant 1$ are such that $P(x_0, x_1, ..., x_m) = 0$, then $Q(x,x_1, ..., x_m) = x$; thus, x is a positive value of Q at positive integers. On the other hand, if z is a positive value of Q at positive integers, $z = Q(x,x_1, ..., x_m) = x[1 - \{P(x,x_1, ..., x_m)\}^2]$, then necessarily $P(x,x_1, ..., x_m) = 0$, otherwise $1 - \{P(x,x_1, ..., x_m)\}^2 \leqslant 0$; hence, $z \leqslant 0$. So $z = x$ and therefore $z \in S$. □

The next step in this theory consists in establishing that the set of prime numbers is diophantine.

For this purpose, it is necessary to examine the definition of prime numbers from the vantage of the theory of diophantine sets.

A positive integer x is a prime if and only if $x > 1$ and for any integers y, z such that $y \leqslant x$ and $z \leqslant x$, either $yz < x$, or $yz > x$, or $y = 1$, or $z = 1$. This definition of prime numbers contains bounded universally quantified occurrences of y, z, namely, $y \leqslant x$, $z \leqslant x$.

Another possible definition of prime numbers is the following. The positive integer x is a prime if and only if $x > 1$ and $\gcd((x - 1)!, x) = 1$. The latter condition is rephrased as follows: There exist positive integers a, b such that $a(x - 1)! - bx = 1$; note that if a or b is negative, taking a sufficiently large integer k, then $a' = a + kx > 0$, $b' = b + k(x - 1)! > 0$ and $a'(x - 1)! - b'x = 1$.

Using one or the other characterization of prime numbers, it was shown with the theory developed by Putnam, Davis, J. Robinson and Matijasevič the important theorem:

The set of prime numbers is diophantine.

With the first characterization of prime numbers, it is necessary to appeal to a general theorem concerning predicates with universally quantified variables; the second proof depends on showing that the set of all pairs $(n,n!)$ is a diophantine set.

A combination of these results leads to the following astonishing result:

There exists a polynomial, with integral coefficients, such that the set of prime numbers coincides with the set of positive values taken by this polynomial, as the variables range in the set of nonnegative integers.

It should be noted that this polynomial also takes on negative values, and that a prime number may appear repeatedly as a value of the polynomial.

In 1971, Matijasevič indicated a system of algebraic relations leading to such a polynomial (without writing it explicitly) with degree 37, in 24 indeterminates; in the English translation of his paper this was improved to degree 21, and 21 indeterminants.

An explicit polynomial with this property, of degree 25, in the 26 indeterminates $a,b,c, ..., z$, was given by Jones, Sato, Wada & Wiens in 1976:

$$(k+2)\{1 - [wz+h+j-q]^2 - [(gk+2g+k+1)(h+j)+h-z]^2$$
$$- [2n+p+q+z-e]^2 - [16(k+1)^3(k+2)(n-1)^2+1-f^2]^2$$
$$- [e^3(e+2)(a+1)^2+1-o^2]^2 - [(a^2-1)y^2+1-x^2]^2$$
$$- [16r^2y^4(a^2-1)+1-u^2]^2 - [((a+u^2(u^2-a))^2-1)(n+4dy)^2$$
$$+ 1 - (x-cu)^2]^2 - [n+\ell+v-y]^2$$
$$- [(a^2-1)\ell^2+1-m^2]^2 - [ai+k+1-\ell-i]^2$$
$$- [p+\ell(a-n-1)+b(2an+2a-n^2-2n-2)-m]^2$$
$$- [q+y(a-p-1)+s(2ap+2a-p^2-2p-2)-x]^2$$
$$- [z+p\ell(a-p)+t(2ap-p^2-1)-pm]^2\}.$$

Here is a remark, attributed to Dixon, in Kaplansky's preliminary edition of *Hilbert's Problems* (University of Chicago):

If P is a polynomial as above, whose positive range is the set of prime numbers, then $R = 2 + \frac{1}{2}(P - 2 + |P - 2|)$ is a function (but not a polynomial) whose range is exactly the set of primes. However, this is not a great advantage; now the value 2 is assumed by R, whenever P assumes a negative value.

One is obviously tempted to reduce the number of indeterminates, or the degree, or both. But there is a price to pay. If the number n of indeterminates is reduced, then the degree d increases, and vice versa, if the degree d is forced to be smaller, the n must increase.

This is illustrated in the table concerning prime representing polynomials.

n = number of indeterminate	d = degree	Author	Year	Remarks
24	37	Matijasević	1971	Not written explicitly
21	21	Same author	1971	
26	25	Jones, Sato, Wada & Wiens	1976	First explicit polynomial
42	5	Same authors	1976	Record low degree, not written ex- plicitly
12	13697	Matijasević	1976	
10	about 1.6×10^{45}	Same author	1977	Record low number of indetermi- nates, not written ex- plicitly

It is not known which is the minimum possible number of variables (it cannot be 2). However, Jones showed that there is a prime
representing polynomial of degree at most 5. I'll say more about this
later.

The same methods used to treat the set of prime numbers apply
also to other diophantine sets once their defining arithmetical
properties are considered from the appropriate point of view.

This has been worked out by Jones.

In a paper of 1975, Jones showed that the set of Fibonacci numbers is identical with the set of positive values at nonnegative integers, of the polynomial in 2 indeterminates and degree 5:

$$2xy^4 + x^2y^3 - 2x^3y^2 - y^5 - x^4y + 2y.$$

In 1979, Jones showed that each one of the sets of Mersenne
primes, even perfect numbers, Fermat primes, corresponds in the
same way to some polynomial in 7 indeterminates, however with

higher degree. He also wrote explicitly other polynomials with lower degree, and more indeterminates, representing the above sets.

Set	Number of indeterminates	Degree
Fibonacci numbers	2	5
Mersenne primes	13	26
	7	914
Even perfect numbers	13	27
	7	915
Fermat primes	14	25
	7	905

By a method of Skolem (see his book, 1938), for the three latter sets the degree may be reduced to 5, however the number of variables increases to about 20.

For the set of Mersenne primes the polynomial is the following one, in the 13 indeterminates $a, b, c, \ldots$:

$$n\{1-[4b+3-n]^2-b([2+hn^2-a]^2 + [n^3d^3(nd+2)(h+1)^2+1-m^2]^2$$
$$+ [db+d+chn^2+g(4a-5)-kn]^2 + [(a^2-1)c^2+1-k^2n^2]^2$$
$$+ [4(a^2-1)i^2c^4+1-f^2]^2$$
$$+ [(kn+\mathit{l}f)^2-((a+f^2(f^2-a))^2-1)(b+1+2jc)^2-1]^2)\}.$$

For the even perfect numbers, the polynomial in 13 indeterminates is

$$(2b+2)n\{1-[4b+3-n]^2-b([2+hn^2-a]^2$$
$$+ n^3d^3(nd+2)(h+1)^2+1-m^2]^2$$
$$+ [db+d+chn^2+g(4a-5)-kn]^2 + [(a^2-1)c^2+1-k^2n^2]^2$$
$$+ [4(a^2-1)i^2c^4 +1-f^2]^2$$
$$+ [(kn+\mathit{l}f)^2-((a+f^2(f^2-a))^2-1)(b+1+2jc)^2-1]^2)\}.$$

For the prime Fermat numbers, the polynomial in 14 indeterminates is

$$(6g+5)\{1-[bh+(a-12)c+n(24a-145)-d]^2$$
$$- [16b^3h^3(bh+1)(a+1)^2+1-m^2]^2$$
$$- [3g+2-b]^2 - [2be+e-bh-1]^2 - [k+b-c]^2$$
$$- [(a^2-1)c^2 + 1 - d^2]^2 - [4(a^2-1)i^2c^4 + 1 - f^2]^2$$
$$- [(d+\ell f)^2-((a+f^2(f^2-a))^2-1)(b+2jc)^2-1]^2\}.$$

In each case, the polynomials also assume negative values. As Jones showed, if a polynomial P, with $n \geqslant 1$ indeterminates, and integral coefficients, is such that $P(x_1, ..., x_n) \in S$ (set of Mersenne numbers, or even perfect numbers) for all nonnegative integers $x_1, ..., x_n$, then necessarily the polynomial has degree zero. This was already indicated in the preceding section, when S is the set of primes.

I wish now to call the attention to some very interesting developments. Every diophantine set S is a listable set, that is, there is an algorithm which allows to write, one after the other, all the elements − and only the elements − of S. This is not difficult to establish. Conversely, in order to solve Hilbert's tenth problem, Matijasevič showed that every listable set is diophantine.

The listable, hence diophantine, sets may be enumerated as S_1, S_2, S_3, ... , in such a way that the binary relation $x \in S_m$ is diophantine. This means that there exists a polynomial U, in $n + 2$ indeterminates, and integral coefficients, such that $x \in S_m$ if and only if there exist positive integers $z_1, z_2, ..., z_n$ such that $U(x, m, z_1, ..., z_n) = 0$.

Such a polynomial U is called a universal diophantine polynomial.

The existence of this polynomial is just an easy consequence of the enumeration of all listable sets; however the actual determination of a universal diophantine polynomial requires a closer analysis. In 1970, Matijasevič announced the possibility of a universal polynomial with 200 indeterminates; this number was successively reduced to $n = 35$ (J. Robinson), $n = 13$ (Matijasevič & Robinson, published in 1975), and $n = 9$, announced by Matijasevič in 1975. In his paper of 1982, Jones wrote in detail the proof of the latter result. Besides many ideas already used in the previous proofs, a basic role is played by the old result of Kummer (1852), quoted in Chapter 2, Section II, about the exact power of p dividing a binomial coefficient.

The universal diophantine polynomial with nine indeterminates has degree about 1.6×10^{45}. Jones also wrote another universal polynomial with 58 indeterminates, degree 4.

In particular, taking the set of prime numbers, which is listable, hence is labeled S_m (for a certain m), the universal diophantine polynomial of nine indeterminates (and two parameters) gives a prime representing polynomial with 10 indeterminates, as was claimed by Matijasevič in 1977.

From the explicit determination of the universal polynomial, follows the tantalizing theorem of Jones (1982):

For any axiomatizable theory T and any proposition P, if P has a proof in T, then P has another "proof" (outside the formal theory) consisting of 100 additions and multiplications of integers.

After Hatcher & Hodgson (1971) have clarified the real meaning of this theorem, it remains to see the impact it may eventually have in arithmetic.

Now, I return to Hilbert's tenth problem. It was by the consideration of diophantine sets that Matijasevič was able to show that there is no algorithm which is good enough to decide whether any diophantine equation has solution in positive integers. This resulted from his theorem that every listable set is diophantine and the already known existence of a listable set which is not recursive. Of course, I am not explaining this last concept here, since this would take me even further away from the topic of prime numbers. But, fortunately, there is an excellent and self-contained exposition by Davis (1973) — a real gem — which I recommend with no restrictions.

Even though there is no algorithm good for an arbitrary diophantine equation, wide classes of diophantine equations are decidable. Thus, Siegel showed in 1972 that every quadratic diophantine equation is decidable. It is not known if every cubic diophantine equation is decidable. On the other hand, for two indeterminates, if $P \in \mathbf{Z}[X,Y]$ is a homogeneous polynomial of degree at least 3 and c is a constant, Baker has indicated an effective bound for the solutions of the equation $P(X,Y) = c$ (see for example, his book, 1975). It is not known whether every polynomial equation in two indeterminates is decidable.

Chapter 4

HOW ARE THE PRIME NUMBERS DISTRIBUTED?

As I have already stressed, the various proofs of existence of infinitely many primes are not constructive and do not give an indication of how to determine the nth prime number. Equivalently, the proofs do not indicate how many primes are less than any given number N.

By the same token, there is no reasonable formula or function representing primes.

It will, however, be possible to predict with rather good accuracy the number of primes smaller than N (especially when N is large); on the other hand, the distribution of primes in short intervals shows a kind of built-in randomness. This combination of "randomness" and "predictability" yields at the same time an orderly arrangement and an element of surprise in the distribution of primes. According to Schroeder (1984), in his intriguing book *Number Theory in Science and Communication*, these are basic ingredients of works of art. Many mathematicians will readily agree that this topic has a great aesthetic appeal.

Recall from Chapter 3 that for every real number $x > 0$, $\pi(x)$ denotes the number of primes p such that $p \leqslant x$. $\pi(x)$ is also called the prime counting function.

The matters to consider are the following:

(I) The growth of $\pi(x)$, its order of magnitude, and comparison with other known functions.

(II)   Results about the nth prime; the difference between consecutive primes, how small, how large, how irregular it may be. This includes the discussion of large gaps between consecutive primes, but leads also to several open problems, discussed below.

(III) Twin primes, their characterization and distribution.

(IV)   Primes in arithmetic progressions.

(V)   Primes in special sequences.

(VI)   Goldbach's famous conjecture.

(VII) Waring's problem and the Waring-Goldbach problem.

(VIII) The distribution of pseudoprimes and of Carmichael numbers.

Now I elaborate on these topics.

I. The Growth of $\pi(x)$

The basic idea in the study of the function $\pi(x)$, or others related to the distribution of primes, is to compare with functions that are both classical and computable, and such that their values are as close as possible to the values of $\pi(x)$. Of course, this is not simple and, as one might expect, an error will always be present. So, for each approximating function, one should estimate the order of magnitude of the difference, that is, of the error. The following notions are therefore natural.

Let $f(x)$, $h(x)$ be positive real valued continuous functions, defined for $x \geqslant x_0 > 0$.

The notation $f(x) \sim h(x)$ means that $\lim_{x\to\infty}(f(x)/h(x)) = 1$; $f(x)$ and $h(x)$ are then said to be asymptotically equal as x tends to infinity. Note that their difference may actually tend to infinity.

If, under the above hypothesis, there exist constants $C, C', 0 < C < C'$, and x_0, x_1, with $x_1 \geqslant x_0$, such that $C \leqslant f(x)/h(x) \leqslant C'$ for all $x \geqslant x_1$, then $f(x), h(x)$ are said to have the same order of magnitude.

If $f(x), g(x), h(x)$ are real valued continuous functions defined for $x \geqslant x_0 > 0$, and $h(x) > 0$ for all $x \geqslant x_0$, the notation

$$f(x) = g(x) + O(h(x))$$

means that the functions $f(x)$ and $g(x)$ have a difference that is ultimately bounded (as x tends to infinity) by a constant multiple of $h(x)$; that is, there exists $C > 0$ and $x_1 \geqslant x_0$ such that for every $x \geqslant x_1$ the inequality $|f(x) - g(x)| \leqslant Ch(x)$ holds. This is a useful notation to express the size of the error when $f(x)$ is replaced by $g(x)$.

Similarly, the notation

$$f(x) = g(x) + o(h(x))$$

means that $\lim_{x \to \infty}[f(x) - g(x)]/h(x) = 0$, so, intuitively, the error is negligible in comparison to $h(x)$.

A. History Unfolding

It is appropriate to describe in historical order the various discoveries about the distribution of primes, culminating with the prime number theorem. This is how Landau proceeded in his famous treatise *Handbuch der Lehre von der Verteilung der Primzahlen*, which is the classical work on the subject. Another interesting presentation, perhaps a little bit unconventional and outdated, is due to Torelli (1902).

In such a succinct text, I shall only mention the main contributors: Euler, Legendre, Gauss, Tschebycheff, Riemann, de la Vallee Poussin, Hadamard, in the last century; and more recently, Erdös, Selberg. Other names, like Mertens, von Mangoldt, and Meissel will also appear. In their obituary of Landau, Hardy & Heilbronn wrote:

> The *Handbuch* was probably the most **important** book he wrote. In it the analytic theory of numbers is presented for the first time, not as a collection of a few beautiful scattered theorems, but as a systematic science. The book transformed the subject, hitherto the hunting ground of a few adventurous heroes, into one of the most fruitful fields of research in the last thirty years. Almost everything in it has been superseded, and that is the greatest tribute to the book.

Euler

First, I give a result of Euler which tells, not only that there are infinitely many primes, but also that "the primes are not so sparse as the squares." (This statement will be made clear shortly.)

Euler noted that for every real number $\sigma > 1$ the series $\sum_{n=1}^{\infty}(1/n^{\sigma})$ is convergent, and in fact, for every $\sigma_0 > 1$ it is uniformly convergent on the half-line $\sigma_0 \leqslant x < \infty$. Thus, it defines a function $\zeta(\sigma)$ (for $1 < \sigma < \infty$), which is continuous and differentiable. Moreover, $\lim_{\sigma \to \infty} \zeta(\sigma) = 1$ and $\lim_{\sigma \to 1+0}(\sigma - 1)\zeta(\sigma) = 1$. The function $\zeta(\sigma)$ is called the zeta function.

The link between the zeta function and the prime numbers is the following Eulerian product, which expresses the unique factorization of integers as product of primes:

$$\sum_{n=1}^{\infty} \frac{1}{n^{\sigma}} = \prod_{p} \frac{1}{1 - \dfrac{1}{p^{\sigma}}} \qquad (\text{for } \sigma > 1).$$

In particular, this implies that $\zeta(\sigma) \neq 0$ for $\sigma > 1$.

With the same idea used in his proof of the existence of infinitely many primes (see Chapter 1), Euler proved in 1737:

The sum of the inverses of the prime numbers is divergent: $\sum_{p}(1/p) = \infty$.

Proof. Let N be an arbitrary natural number. Each integer $n \leqslant N$ is a product, in a unique way, of powers of primes p, $p \leqslant n \leqslant N$. Also for every prime p,

$$\sum_{k=1}^{\infty} \frac{1}{p^{k}} = \frac{1}{1 - \dfrac{1}{p}}.$$

Hence

$$\sum_{n=1}^{N} \frac{1}{n} \leqslant \prod_{p \leqslant N} \left(\sum_{k=1}^{\infty} \frac{1}{p^{k}} \right) = \prod_{p \leqslant N} \frac{1}{1 - \dfrac{1}{p}}.$$

But

$$\log \prod_{p \leqslant N} \frac{1}{1 - \dfrac{1}{p}} = - \sum_{p \leqslant N} \log\left(1 - \frac{1}{p}\right),$$

and for each prime p,

$$-\log\left[1 - \frac{1}{p}\right] = \sum_{m=1}^{\infty} \frac{1}{mp^m} \leqslant \frac{1}{p} + \frac{1}{p^2}\left[\sum_{h=0}^{\infty} \frac{1}{p^h}\right]$$

$$= \frac{1}{p} + \frac{1}{p^2} \times \frac{1}{1 - \frac{1}{p}} = \frac{1}{p} + \frac{1}{p(p-1)} < \frac{1}{p} + \frac{1}{p^2}.$$

Hence

$$\log \sum_{n=1}^{N} \frac{1}{n} \leqslant \log \prod_{p \leqslant N} \frac{1}{1 - \frac{1}{p}} \leqslant \sum_{p \leqslant N} \frac{1}{p} + \sum_{p \leqslant N} \frac{1}{p^2} \leqslant \sum_{p} \frac{1}{p} + \sum_{n=1}^{\infty} \frac{1}{n^2}.$$

But the series $\sum_{n=1}^{\infty}(1/n^2)$ is convergent. Since N is arbitrary and the harmonic series is divergent, then $\log \sum_{n=1}^{\infty}(1/n) = \infty$, and therefore the series $\sum_p(1/p)$ is divergent. □

There have been many proofs of the divergence of $\sum_p(1/p)$, for example, by Erdös, Bellman, Moser, Dux, and Clarkson. They are reviewed in the article of Vanden Eynden (1980), who also gave a very simple proof.

As I have already mentioned, the series $\sum_{n=1}^{\infty}(1/n^2)$ is convergent. Thus, it may be said, somewhat vaguely, that the primes are not so sparsely distributed as the squares.

One of the beautiful discoveries of Euler was the sum of this series:

$$\sum_{n=1}^{\infty} \frac{1}{n^2} = \frac{\pi^2}{6}.$$

Euler also evaluated the sums $\sum_{n=1}^{\infty}(1/n^{2k})$ for every $k \geqslant 1$, thereby solving a rather elusive problem.

For this purpose, he made use of the Bernoulli numbers, which are defined as follows:

$$B_0 = 1, \quad B_1 = -\frac{1}{2}, \quad B_2 = \frac{1}{6}, \quad \ldots,$$

B_k being recursively defined by the relation

$$\begin{bmatrix} k+1 \\ 1 \end{bmatrix} B_k + \begin{bmatrix} k+1 \\ 2 \end{bmatrix} B_{k-1} + \cdots + \begin{bmatrix} k+1 \\ k \end{bmatrix} B_1 + B_0 = 0.$$

These numbers are clearly rational, and it is easy to see that $B_{2k+1} = 0$

for every $k \geq 1$. They appear also as coefficients in the Taylor expansion:

$$\frac{x}{e^x - 1} = \sum_{k=0}^{\infty} \frac{B_k}{k!} \, x^k.$$

Using Stirling's formula,

$$n! \sim \frac{\sqrt{2\pi} \; n^{n+\frac{1}{2}}}{e^n} \qquad (\text{as } n \to \infty),$$

it may also be shown that

$$|B_{2n}| \sim 4 \, \sqrt{\pi n} \left(\frac{n}{\pi e} \right)^{2n},$$

hence the above series is convergent in the interval $|x| < 2\pi$.

Euler had already used the Bernoulli numbers to express the sums of equal powers of consecutive numbers:

$$\sum_{j=1}^{n} j^k = S_k(n) \qquad (k \geq 1),$$

where

$$S_k(X) = \frac{1}{k+1} \left[X^{k+1} - \binom{k+1}{1} B_1 X^k + \binom{k+1}{2} B_2 X^{k-1} + \cdots + \binom{k+1}{k} B_k X \right].$$

Euler's formula giving the value of $\zeta(2k)$ is:

$$\zeta(2k) = \sum_{n=1}^{\infty} \frac{1}{n^{2k}} = (-1)^{k+1} \frac{(2\pi)^{2k} B_{2k}}{2(2k)!} .$$

In particular,

$$\zeta(2) = \sum_{n=1}^{\infty} \frac{1}{n^2} = \frac{\pi^2}{6} \qquad (\text{already mentioned}),$$

$$\zeta(4) = \sum_{n=1}^{\infty} \frac{1}{n^4} = \frac{\pi^4}{90}, \quad \text{etc.}$$

An elementary proof of the formula for $\zeta(2k)$ was given, for example, by Apostol in 1973.

Euler also considered the Bernoulli polynomials, defined by

$$B_k(X) = \sum_{i=0}^{k} \binom{k}{i} B_i X^{k-i} \qquad (k \geq 0).$$

They may be used to rewrite the expression for $S_k(X)$, but more important is their application to a far reaching generalization of Abel's summation formula, namely, the well-known Euler-MacLaurin summation formulas:

If $f(x)$ is a continuous function, continuously differentiable as many times as required, if $a < b$ are integers, then

$$\sum_{n=a+1}^{b} f(n) = \int_{a}^{b} f(t)dt + \sum_{r=1}^{k} (-1)^r \frac{B_r}{r!} \{f^{(r-1)}(b) - f^{(r-1)}(a)\}$$
$$+ \frac{(-1)^{k-1}}{k!} \int_{a}^{b} B_k(t - [t])f^{(k)}(t)dt.$$

The reader is urged to consult the paper by Ayoub, *Euler and the zeta function* (1974), where there is a description of the many imaginative relations and findings of Euler concerning $\zeta(s)$ — some fully justified, others only made plausible, but anticipating later works by Riemann.

Legendre

The first serious attempt to study the function $\pi(x)$ is due to Legendre (1808), who used the Eratosthenes sieve and proved that

$$\pi(N) = \pi(\sqrt{N}) - 1 + \Sigma \mu(d) \left[\frac{N}{d}\right].$$

The notation $[t]$ has already been explained, the summation is over all divisors d of the product of all primes $p \leqslant \sqrt{N}$, and $\mu(n)$ denotes the Möbius function, which was already defined in Chapter 3, Section I.

As a consequence, Legendre showed that $\lim_{x\to\infty}(\pi(x)/x) = 0$, but this is a rather weak result.

Experimentally, Legendre conjectured in 1798 and again in 1808 the hypothesis that

$$\pi(x) = \frac{x}{\log x - A(x)},$$

where $\lim_{x\to\infty}A(x) = 1.08366...$. That Legendre's statement cannot be true, was shown forty years later by Tschebycheff (see below). An easy proof was given by Pintz (1980).

Gauss

At age 15, in 1792, Gauss conjectured that $\pi(x)$ and the function logarithmic integral of x, defined by

$$\text{Li}(x) = \int_2^x \frac{dt}{\log t}$$

are asymptotically equal. Since $\text{Li}(x) \sim x/(\log x)$, this implies that

$$\pi(x) \sim \frac{x}{\log x} \ .$$

This conjecture was to be confirmed later, and is now known as the prime number theorem; I shall soon return to this matter.

The approximation of $\pi(x)$ by $x/(\log x)$ is only reasonably good, while it is much better using the logarithmic integral, as it will be illustrated in a table.

Tschebycheff

Important progress for the determination of the order of magnitude of $\pi(x)$ was made by Tschebycheff, around 1850. He proved, using elementary methods, that there exist constants $C, C', 0 < C' < 1 < C$, such that

$$C' \frac{x}{\log x} < \pi(x) < C \frac{x}{\log x} \qquad \text{(for } x \geqslant 2).$$

He actually computed values for C, C' very close to 1. For example, taking $x \geqslant 30$,

$$C' = \log \frac{2^{1/2}3^{1/3}5^{1/5}}{30^{1/30}} = 0.92129 \ldots \ ,$$

$C = \frac{6}{5} C' = 1.10555\ldots$. Moreover, if the limit of

$$\frac{\pi(x)}{x/(\log x)}$$

exists (as $x \to \infty$), it must be equal to 1. He deduced also that Legendre's approximation of $\pi(x)$ cannot be true, unless 1.08366 is replaced by 1 (see Landau's book, page 17).

Tschebycheff also proved Bertrand's postulate that between any natural number $n \geqslant 2$ and its double there exists at least one prime. I shall discuss this proposition in more detail, when I present the main properties of the function $\pi(x)$.

Tschebycheff worked with the function $\theta(x) = \Sigma_{p \leqslant x} \log p$, now called Tschebycheff's function, which yields basically the same information as $\pi(x)$, but is somewhat easier to handle.

Even though Tschebycheff came rather close, the proof of the fundamental prime number theorem, conjectured by Gauss, had to wait for about 50 more years, until the end of the century. During this time, important new ideas were contributed by Riemann.

Riemann

Riemann had the idea of defining the zeta function for complex numbers s having real parts greater than 1, namely, $\zeta(s) = \Sigma_{n=1}^{\infty}(1/n^s)$. The Euler product formula still holds, for every complex s with $\mathrm{Re}(s) > 1$.

Using the Euler-MacLaurin summation formula, $\zeta(s)$ is expressible as follows:

$$\zeta(s) = \frac{1}{s-1} + \frac{1}{2} + \sum_{r=2}^{k} \frac{B_r}{r!} s(s+1) \cdots (s+r-2)$$

$$- \frac{1}{k!} s(s+1) \cdots (s+k-1) \int_1^{\infty} B_k(x-[x]) \frac{dx}{x^{s+k}} .$$

Here k is any integer, $k \geqslant 1$, the numbers B_r are the Bernoulli numbers, which the reader should not confuse with $B_k(x-[x])$, the value of the kth Bernoulli polynomial $B_k(X)$ at $x-[x]$.

The integral converges to $\mathrm{Re}(s) > 1-k$, and since k is an arbitrary natural number, this formula provides the analytic continuation of $\zeta(s)$ to the whole plane.

$\zeta(s)$ is everywhere holomorphic, except at $s = 1$, where it has a simple pole with residue 1, that is, $\lim_{s \to 1}(s-1)\zeta(s) = 1$.

In 1859, Riemann established the functional equation for the zeta function. Since this equation involves the gamma function $\Gamma(s)$, I must first define $\Gamma(s)$. For $\mathrm{Re}(s) > 0$, a convenient definition is by means of the Eulerian integral

$$\Gamma(s) = \int_0^\infty e^{-u} u^{s-1} du.$$

For arbitrary complex numbers s, it may be defined by

$$\Gamma(s) = \frac{1}{s e^{\gamma s}} \prod_{n=1}^\infty \frac{e^{s/n}}{1 + \dfrac{s}{n}},$$

where γ is Euler's constant, equal to

$$\gamma = \lim_{n \to \infty} \left[1 + \frac{1}{2} + \cdots + \frac{1}{n} - \log n \right] = 0.577215665\ldots.$$

Euler's constant, also known with good reason as Mascheroni's constant, by the Italians, is related to Euler's product by the following formula of Mertens:

$$e^\gamma = \lim_{n \to \infty} \frac{1}{\log n} \prod_{i=1}^n \frac{1}{1 - \dfrac{1}{p_i}}.$$

$\Gamma(s)$ is never equal to 0; it is holomorphic everywhere except at the points 0, −1, −2, −3, ..., where it has simple poles. For every positive integer n, $\Gamma(n) = (n - 1)!$, so the gamma function is an extension of the factorial function. The gamma function satisfies many interesting relations, among which are the functional equations

$$\Gamma(s)\Gamma(1 - s) = \frac{\pi}{\sin \pi s} \quad \text{and} \quad \Gamma(s)\Gamma\left(s + \frac{1}{2}\right) = \frac{\sqrt{\pi}}{2^{2s-1}} \Gamma(2s),$$

$$\Gamma(s + 1) = s\Gamma(s).$$

Here is the functional equation for the Riemann zeta function:

$$\pi^{-s/2} \Gamma\left(\frac{s}{2}\right) \zeta(s) = \pi^{-(1-s)/2} \Gamma\left(\frac{1-s}{2}\right) \zeta(1 - s).$$

For example, it follows from the functional equation that $\zeta(0) = -\frac{1}{2}$. The zeros of the zeta function are:

(a) Simple zeroes at the points −2, −4, −6, ..., which are called the trivial zeroes.

(b) Zeroes in the critical strip, consisting of the nonreal complex numbers s with $0 \leqslant \text{Re}(s) \leqslant 1$.

Indeed, if $\text{Re}(s) > 1$, then by the Euler product, $\zeta(s) \neq 0$. If $\text{Re}(s) < 0$, then $\text{Re}(1 - s) > 1$, the right-hand side in the functional equation is not zero, so the zeroes must be exactly at $s = -2, -4, -6, \ldots$, which are the poles of $\Gamma(s/2)$.

The knowledge of the zeroes in the critical strip has a profound influence on the understanding of the distribution of primes. A first thing to note is that the zeroes in the critical strip are not real and they are symmetric about the real axis and the vertical line $\text{Re}(x) = \frac{1}{2}$.

Riemann conjectured that all nontrivial zeroes ρ of $\zeta(s)$ are on the critical line $\text{Re}(s) = \frac{1}{2}$, that is, $\rho = \frac{1}{2} + i\gamma$. This is the famous Riemann's hypothesis, which has never been proved. It is undoubtedly a very difficult and important problem, and I shall return soon to it and narrate some modern developments.

Riemann also had the idea of considering all the powers of primes $p^n \leqslant x$, with each such p weighted as $1/n$. For this purpose, he defined the function

$$J(x) = \begin{cases} \pi(x) + \frac{1}{2}\pi(x^{1/2}) + \frac{1}{3}\pi(x^{1/3}) + \frac{1}{4}\pi(x^{1/4}) + \cdots - \dfrac{1}{2m}, \\[4pt] \quad \text{if } x = p^m, \text{ where } p \text{ is a prime number, } m \geqslant 1. \\[8pt] \pi(x) + \frac{1}{2}\pi(x^{1/2}) + \frac{1}{3}\pi(x^{1/3}) + \frac{1}{4}\pi(x^{1/4}) + \cdots, \text{ if } x > 0 \text{ is a real} \\[4pt] \quad \text{number which is neither a prime nor a prime-power.} \end{cases}$$

One of the principal formulas conjectured by Riemann was an expression of $J(x)$ in terms of the logarithmic integral; this formula involves the zeroes of $\zeta(s)$.

First, define $\text{Li}(e^w)$ for any complex number $w = u + iv$, as follows.

$$\text{Li}(e^w) = \int \frac{e^t}{t}\, dt + z,$$

where the integral is over the horizontal line, from $-\infty$ to $u + iv$, and $z = \pi i, -\pi i, 0$, according to $v > 0$, $v < 0$, or $v = 0$. Riemann's formula, which was proved by von Mangoldt, is

$$J(x) = \text{Li}(x) - \Sigma \, \text{Li}(x^\rho) + \int_x^\infty \frac{dt}{t(t^2 - 1)\log t} - \log 2$$

[the sum is extended over all the nontrivial zeroes ρ of $\zeta(s)$, each with its own multiplicity].

Let

$$R(x) = \sum_{m=1}^\infty \frac{\mu(m)}{m} \text{Li}(x^{1/m})$$

be the so-called Riemann function.

Riemann gave the following exact formula for $\pi(x)$, in terms of the Riemann function:

$$\pi(x) = R(x) - \sum_\rho R(x^\rho)$$

[the sum being extended over all the nontrivial zeroes of $\zeta(x)$, each counted with its own multiplicity].

The Riemann function $R(x)$ provides a very good approximation for $\pi(x)$, as will be seen in the following table. The size of the error is expressed in terms of $R(x^\rho)$, for all the roots of $\zeta(s)$ in the critical strip.

The Riemann function is computable by this quickly converging power series, given by Gram in 1893:

$$R(x) = 1 + \sum_{n=1}^\infty \frac{1}{n\zeta(n+1)} \times \frac{(\log x)^n}{n!} \,.$$

The work of Riemann on the distribution of primes is thoroughly studied in Edwards' book (1974), which I recommend without reservations. Other books on the Riemann zeta function are the classical treatise by Titchmarsh (1951) and the recent volume of Ivić (1985).

de la Vallée Poussin and Hadamard

Riemann provided many of the tools for the proof of the *fundamental prime number theorem*:

$$\pi(x) \sim \frac{x}{\log x} \,.$$

Other tools came from the theory of complex analytic functions, which was experiencing a period of rapid growth.

The prime number theorem was raised to the status "a most wanted theorem," and it was folklore to consider that he who would prove it would become immortal.

The theorem was established, not by one, but by two eminent analysts, independently, and in the same year (1896). No, they did not become immortals, as in some old Greek legend, but ... almost! Hadamard lived to the age of 98, de la Vallée Poussin just slightly less, to the age of 96.

de la Vallée Poussin established the following fact: there exists $c > 0$ and $t_0 = t_0(c) > e^{2c}$, such that $\zeta(s) \neq 0$ for every $s = \sigma + it$ in the region:

$$\begin{cases} 1 - \dfrac{c}{\log t_0} \leqslant \sigma \leqslant 1, & \text{when } |t| \leqslant t_0 \\[2ex] 1 - \dfrac{c}{\log |t|} \leqslant \sigma \leqslant 1, & \text{when } t_0 \leqslant |t|. \end{cases}$$

Thus, in particular, $\zeta(1 + it) \neq 0$ for every t, as shown by Hadamard.

The determination of a large zero-free region for $\zeta(s)$ was an important feature in the proof of the prime number theorem.

Not only did Hadamard and de la Vallée Poussin prove the prime number theorem. They have also estimated the error as being:

$$\pi(x) = \text{Li}(x) + O(xe^{-A\sqrt{\log x}}),$$

for some positive constant A.

I shall soon tell how the error term was subsequently reduced by determining larger zero-free regions for the zeta function.

In 1903, Landau gave a proof of the prime number theorem (and a similar theorem for prime ideals in algebraic number fields), using basically only that $\zeta(s)$ can be continued a little way over the line $\sigma = 1$, but without appealing to the functional equation of the zeta function. In 1932, Landau gave another proof of the theorem based, this time, on Wiener's ideas.

As a matter of fact, there have been many variants of proofs for the prime number theorem with analytical methods, and they appear in various books and papers. One which is particularly simple was proposed recently by Newman (1980).

There are other equivalent ways of formulating the prime number theorem. Using the Tschebycheff function, the theorem may be rephrased as follows:

$$\theta(x) \sim x.$$

Another formulation involves the summatory function of the von Mangoldt function. Let

$$\Lambda(n) = \begin{cases} \log p & \text{if } n = p^{\nu} \ (\nu \geqslant 1) \text{ and } p \text{ is a prime} \\ 0 & \text{otherwise.} \end{cases}$$

This function, introduced by von Mangoldt, has the following interesting property relating to the logarithmic derivative of the zeta function:

$$-\frac{\zeta'(s)}{\zeta(s)} = \sum_{n=1}^{\infty} \frac{\Lambda(n)}{n^s} \qquad [\text{for } \text{Re}(s) > 1].$$

It is also related to the function $J(x)$ already encountered:

$$J(x) = \sum_{n \leqslant x} \frac{\Lambda(n)}{\log n}.$$

The summatory function of $\Lambda(n)$ is defined to be

$$\psi(x) = \sum_{n \leqslant x} \Lambda(n).$$

It is easily expressible in terms of Tschebycheff's function

$$\psi(x) = \theta(x) + \theta(x^{1/2}) + \theta(x^{1/3}) + \cdots.$$

The prime number theorem may also be formulated as:

$$\psi(x) \sim x.$$

In 1976, Gerig gave a proof of the prime number theorem in the above form, using only simple properties of $\zeta(s)$ for $\text{Re}(s) > 1$, and facts of harmonic analysis.

Erdös and Selberg

It was believed for a long time that analytical methods could not be avoided in the proof of the prime number theorem. Thus, the mathematical community was surprised when both Erdös and Selberg showed, in 1949, how to prove the prime number theorem using essentially only elementary estimates of arithmetical functions.

Many such estimates of sums were already known, as, for example,

$$\sum_{n \leq x} \frac{1}{n} = \log x + \gamma + O\left(\frac{1}{x}\right), \quad \text{where } \gamma \text{ is Euler's constant;}$$

$$\sum_{n \leq x} \frac{1}{n^\sigma} = \frac{x^{1-\sigma}}{1-\sigma} + \zeta(\sigma) + O\left(\frac{1}{x^\sigma}\right), \quad \text{where } \sigma > 1;$$

$$\sum_{n \leq x} \log n = x \log x - x + O(\log x);$$

$$\sum_{n \leq x} \frac{\log n}{n} = \frac{1}{2} (\log x)^2 + C + O\left(\frac{\log x}{x}\right).$$

The above estimates are obtained using the Abel or Euler-Maclaurin summation formulas, and have really no arithmetical content. The following sums involving primes are more interesting:

$$\sum_{p \leq x} \frac{\log p}{p} = \log x + O(1);$$

$$\sum_{p \leq x} \frac{1}{p} = \log \log x + C + O\left(\frac{1}{\log x}\right); \quad \text{where } C = 0.2615... ;$$

$$\sum_{n \leq x} \frac{\Lambda(n)}{n} = \log x + O(1);$$

$$\sum_{n \leq x} \frac{\Lambda(n)\log n}{n} = \frac{1}{2} (\log x)^2 + O(\log x).$$

Selberg gave in 1949 the following estimate:

$$\sum_{p \leq x} (\log p)^2 + \sum_{pq \leq x} (\log p)(\log q) = 2x \log x + O(x)$$

(where p, q are primes). A somewhat simpler proof was indicated by Shapiro (1950).

This estimate is, in fact, equivalent to each of the following:

$$\theta(x)\log x + \sum_{p \leq x} \theta\left(\frac{x}{p}\right)\log p = 2x \log x + O(x);$$

$$\sum_{n \leq x} \Lambda(n)\log n + \sum_{mn \leq x} \Lambda(m)\Lambda(n) = 2x \log x + O(x).$$

From his estimate, Selberg was able to give an elementary proof of the prime number theorem. At the same time, also using a variant of Selberg's estimate

$$\frac{\psi(x)}{x} + \frac{1}{\log x} \sum_{n \leq x} \frac{\psi(x/n)}{x/n} \frac{\Lambda(n)}{n} = 2 + O\left(\frac{1}{\log x}\right),$$

Erdös gave, with a different elementary method, his proof of the prime number theorem.

There have been other proofs of the prime number theorem based on Selberg's estimate. For example, a neat proof published in 1954, by Breusch.

Diamond published in 1982 a detailed and authoritative article about elementary methods in the study of the distribution of prime numbers.

Now I shall examine very briefly various topics pertaining to the study of $\pi(x)$.

B. Sums Involving the Möbius Function

Even before Möbius had formally defined the function $\mu(n)$, Euler had already considered it.

I shall discuss sums involving the Möbius functions, of the form $\sum_{n=1}^{\infty}\mu(n)/n^s$, where s is a complex number. It is easy to see that, if $Re(s) > 1$, then the series is absolutely convergent. Moreover,

$$\sum_{n=1}^{\infty} \frac{\mu(n)}{n^s} = \frac{1}{\zeta(s)} \qquad \text{[when } Re(s) > 1\text{]}.$$

In particular, if $s = 2$, this gives

$$\sum_{n=1}^{\infty} \frac{\mu(n)}{n^2} = \frac{1}{\zeta(2)} = \frac{6}{\pi^2}.$$

It follows that for every $x > 1$,

$$\sum_{n \leqslant x} \frac{\mu(n)}{n^2} = \frac{6}{\pi^2} + O\left(\frac{1}{x}\right).$$

The situation is even more interesting when $s = 1$. Already in 1748, Euler conjectured, on experimental evidence, that $\sum_{n=1}^{\infty} \mu(n)/n$ converges to 0. Much later, in 1884, Gram proved that this sum is bounded.

von Mangoldt used the prime number theorem of Hadamard and de la Vallée Poussin to show that $\sum_{n=1}^{\infty} \mu(n)/n = 0$. Actually, the converse is true: this fact implies the prime number theorem.

Another proof of the prime number theorem, using the Möbius function, was given by Amitsur. He considered the algebra of arithmetic functions, as I explain now.

The complex valued functions f, defined over the natural numbers (called arithmetical functions) correspond to the series $\sum_{n=1}^{\infty} f(n)/n^s$ (s complex number). For example, the constant function equal to 1 corresponds to the Riemann zeta series $\sum_{n=1}^{\infty} 1/n^s$.

It is convenient to consider the set of arithmetical functions endowed with pointwise addition and the so-called Dirichlet convolution $*$:

$$(f*g)(n) = \sum_{d|n} f(d)g\left(\frac{n}{d}\right), \quad \text{for every } n \geqslant 1.$$

Thus, if f, g correspond to series convergent for $\text{Re}(s) > \alpha$, then the series corresponding to $f*g$ is also convergent for $\text{Re}(s) > a$, and has sum equal to

$$\left[\sum_{n=1}^{\infty} \frac{f(n)}{n^s}\right]\left[\sum_{n=1}^{\infty} \frac{g(n)}{n^s}\right].$$

The set of arithmetical functions becomes an algebra, which is interesting to study in its own right.

The function u defined by $u(1) = 1$, $u(n) = 0$ for $n > 1$, is the unit of the algebra of functions.

If $f(1) \neq 1$, there exists a unique arithmetical function g such that $f * g = u$; g is the convolution inverse of f.

It is easy to see that the convolution inverse of the constant function 1 is the Möbius function μ. Thus,

if $f(n) = \sum_{d|n} g(d)$, then $g(n) = \sum_{d|n} \mu(d) f(n/d)$.

This is the so-called Möbius inversion formula.

The operator D, defined by $(Df)(n) = f(n) \log n$ (for every $n \geqslant 1$), is a derivation of the algebra of arithmetic functions, that is, $D(f+g) = D(f) + D(g)$, $D(fg) = D(f)g + fD(g)$.

In 1956, Amitsur used the formalism of the algebra of arithmetic functions and ideas of Shapiro (1950) to obtain a new proof that $\sum_{n=1}^{\infty} \mu(n)/n = 0$, thus, giving a new elementary proof of the prime number theorem.

The summatory function of the Möbius function is the Mertens function $M(x) = \sum_{n \leqslant x} \mu(n)$.

It may be shown that the prime number theorem is also equivalent to the assertion that $\lim_{x \to \infty} M(x)/x = 0$. For details relative to the preceding statements, the reader may consult the books of Landau (1909), Ayoub (1963), or Apostol (1976).

Quite recently, in 1984, Daboussi gave an elementary proof that $\lim_{x \to \infty} M(x)/x = 0$. This new proof of the prime number theorem made no appeal to Selberg's inequality.

Concerning the order of magnitude of $M(x)$, there were two conjectures made in 1897. The stronger conjecture, by von Sterneck, asserted that

$$|M(x)| < \frac{1}{2} \sqrt{x} \quad \text{(for } x > 200\text{).}$$

It was proved wrong by Jurkat in 1961; but the first counterexample was by Neubauer (1963), who discovered that if $x = 7,760,000,000$, then $M(x) = 47465 > 44046 = \sqrt{x}/2$. In fact, as calculated recently by Cohen, $x_0 = 7,725,038,629$ is the smallest integer for which von Sterneck's conjecture is wrong, since $M(x_0) = 43947$.

Mertens conjectured that $|M(x)| < \sqrt{x}$ for every $x > 1$. This question which had been in the minds of many classical mathematicians, such as Stieltjes and Hadamard, was very important to settle.

As a matter of fact, it is easy to show that if there exists a constant $A > 0$ such that $|M(x)| < A\sqrt{x}$ for every $x > 1$, then Riemann's hypothesis would be true.

Indeed, let $s = \sigma + it$ with $\sigma > 1$. Then

$$\frac{1}{\zeta(s)} = \sum_{n=1}^{\infty} \frac{\mu(n)}{n^s} = \sum_{n=1}^{\infty} \frac{\mu(n) - \mu(n-1)}{n^s} = \sum_{n=1}^{\infty} M(n)\left\{ \frac{1}{n^s} - \frac{1}{(n+1)^s} \right\}$$

$$= \sum_{n=1}^{\infty} M(n)\int_n^{n+1} \frac{s\, dx}{x^{s+1}} = s \sum_{n=1}^{\infty} \int_n^{n+1} \frac{M(x)dx}{x^{s+1}} = s\int_1^{\infty} \frac{M(x)dx}{x^{s+1}},$$

noting that $M(x)$ is constant when $n \leqslant x < n + 1$. If $|M(x)| < A\sqrt{x}$, then the last integral defines an analytic function on the domain where $\sigma > \frac{1}{2}$. So, $1/\zeta(s)$ can be analytically continued to this domain. This shows that $\zeta(s)$ has no zeroes $\sigma + it$, with $\sigma > \frac{1}{2}$ — which is the statement of Riemann's hypothesis.

Note that the converse is false. What is true is a result that looks very similar (see Titchmarsh, 1951): the Riemann hypothesis is equivalent to the statement that for every $\epsilon > 0$, $M(x) = 0(x^{\frac{1}{2}+\epsilon})$.

In 1985, Odlyzko & te Riele showed that Mertens' conjecture is false. In a recent preprint (1985), Pintz gave an effective disproof of the conjecture, by showing that if $t = e^{3.21 \times 10^{64}}$ then

$$\max_{x \leqslant t} \frac{M(x)}{\sqrt{x}} > 1.$$

As a matter of fact, Odlyzko & te Riele showed that

$$\limsup_{x \to \infty} \frac{M(x)}{\sqrt{x}} > 1.06, \quad \liminf_{x \to \infty} \frac{M(x)}{\sqrt{x}} < -1.009$$

[without, however giving any specific integer x_0 for which $M(x_0) > \sqrt{x_0}$]. It is now conjectured that these limits should be respectively $+\infty$, $-\infty$.

On the other hand, it is possible to show that

$$M(x) = 0\left[\frac{x}{(\log x)^{2/3}} \right],$$

and even more:

$$M(x) = 0\left[xe^{-C(\log x)^{3/5}(\log \log x)^{-1/5}} \right] \quad \text{with } C > 0$$

(see Walfisz, 1963).

Based on explicit calculations of the zeroes of Riemann's zeta function, Schoenfeld gave in 1969, the explicit inequality

$$|M(x)| < \frac{0.55x}{(\log x)^{2/3}} \quad (\text{for } x > 1).$$

C. The Distribution of Values of Euler's Function

I will gather here results concerning the distribution of values of Euler's function. They supplement the properties already stated in Chapter 2, Section II.

First, some indications about the growth of Euler's function. It is easy to show that

$$\phi(n) \geqslant \log 2 \, \frac{n}{\log(2n)},$$

in particular, for every $\delta > 0$, $\phi(n)$ grows ultimately faster than $n^{1-\delta}$.

Even better, for every $\epsilon > 0$ there exists $n_0 = n_0(\epsilon)$ such that, if $n \geqslant n_0$, then

$$\phi(n) \geqslant (1 - \epsilon)e^{-\gamma} \, \frac{n}{\log \log n}.$$

On the other hand, it follows from the prime number theorem, that there exist infinitely many n such that

$$\phi(n) < (1 + \epsilon)e^{-\gamma} \, \frac{n}{\log \log n}.$$

So,

$$\lim \inf \frac{\phi(n) \log \log n}{n} = e^{-\gamma}.$$

A proof of the above results may be found, for example, in the books by Landau (1909), quoted in the General References, or Apostol (1976).

And they have been made more precise. Rosser & Schoenfeld showed in 1962 that

$$\phi(n) \geq e^{-\gamma} \frac{n}{\log \log n + \dfrac{5}{2e^{\gamma}\log \log n}} \quad \text{for all} \quad n \geq 3$$

with the only exception $n = 2 \times 3 \times 5 \times 7 \times 11 \times 13 \times 17 \times 19 \times 23$, for which $\frac{5}{2}$ is replaced by 2.50637.

For the other inequality, Nicolas proved (1983) that there exist infinitely many n, for which

$$\phi(n) < e^{-\gamma} \frac{n}{\log \log n}.$$

The method of proof is interesting, in that the inequality is shown first under the assumption that the Riemann hypothesis is true, secondly under the contrary assumption.

What is the average of $\phi(n)$?

From the relation $n = \Sigma_{d|n}\phi(d)$, using Möbius inversion formula, it follows that

$$\frac{1}{x} \sum_{n \leq x} \phi(n) = \frac{3x}{\pi^2} + O(\log x).$$

So, the mean value of $\phi(n)$ is equal to $3n/\pi^2$.

As a consequence, if two integers m, $n \geq 1$ are randomly chosen, then the probability that m, n be relatively prime is $6/\pi^2$.

All these matters are well explained in the books of Hardy & Wright and Apostol (1976).

The statistical study of the function $\phi(n)$ was started by Schoenberg, who considered the function $\phi(n)/n$. In 1928, he showed that the following limit exists:

$$\lim_{N \to \infty} \frac{1}{N} \#\{n \mid 1 \leq n \leq N, \ \phi(n)/n \leq x\} \quad \text{for every } x, 0 < x < 1.$$

This number δ_x is the density of distribution of the function $\phi(n)/n$. In 1972, Wall determined bounds for δ_x which are in support of the conjecture that there exists $C > 0$ such that, if $x > 0$ is sufficiently small, then $\delta_x < Cx$.

This is discussed in the book by Kac (1959) and in more depth in Elliott's book (1979) — it concerns a theorem of Erdös (1935, 1937, 1938) and Wintner (1928).

Now I shall quote some results about the growth of the valence function of $\phi(n)$:

$$N_\phi(m) = \#\{n \geqslant 1 \mid \phi(n) = m\}.$$

Erdös showed in 1935 that there exists $\delta > 0$ such that $N_\phi(m) > m^\delta$ for infinitely many $m \geqslant 1$. In 1979, Wooldridge proved that the above statement holds in fact for every $\delta < 3 - 2\sqrt{2}$. Result improved by Pomerance (in 1980) to $\delta < 0.55092$ and Balog to $\delta < 0.65$ (indicated in Oberwolfach, 1984). These results require sophisticated proofs.

Denote by $N_\phi^\#(m)$ the number of integers k, $1 \leqslant k \leqslant m$, such that there exists $n \geqslant 1$ with $\phi(n) = k$; so $N_\phi^\#(m)$ counts the integers k such that $N_\phi(k) \neq 0$. Erdös showed in 1935 and in 1945 respectively the following estimates:

For every $\epsilon > 0$ there exists $M(\epsilon)$ such that if $m > M(\epsilon)$ then

$$N_\phi^\#(m) < \frac{m}{(\log m)^{1-\epsilon}},$$

so

$$N_\phi^\#(m) = O\left[\frac{m}{(\log m)^{1-\epsilon}}\right];$$

and, for every $k > 0$,

$$N_\phi^\#(m) > \frac{m(\log \log m)^k}{\log m}.$$

The summatory function of N_ϕ is defined for every real number $x > 0$ as follows:

$$N_\phi^*(x) = \sum_{m \leqslant x} N_\phi(m) = \#\{n \leqslant 1 \mid \phi(n) \leqslant x\} = \sum_{\phi(n) \leqslant x} 1.$$

In 1945, Erdös calculated the limit of $N_\phi^*(x)/x$:

$$\lim_{x \to \infty} \frac{N_\phi^*(x)}{x} = \prod_p \left[1 + \frac{1}{p(p-1)}\right]$$

(the product extended over all primes p).

This may be expressed in terms of the Riemann zeta function $\zeta(s)$ as follows:

$$\lim_{x \to \infty} \frac{N^*_\phi(x)}{x} = \frac{\zeta(2)\zeta(3)}{\zeta(6)} = \alpha .$$

A simpler proof was given by Dressler in 1970.

Bateman evaluated the error term $N^*_\phi(x) - \alpha x$ in 1972, obtaining, for example

$$N^*_\phi(x) \leqslant \alpha x + O\left[\frac{x}{e^{C(\log x)^{1/3}}} \right],$$

for every positive constant C .

Nicolas gave an elementary proof of a somewhat weaker estimate (1983):

$$N^*_\phi(x) = \alpha x + O\left[\frac{x \log \log \log x}{\log x} \right].$$

D. Tables of Primes

Now, I turn my attention to tables of prime numbers, and of factors of numbers (not divisible by 2, 3, or 5). The first somewhat extended tables are by Brancker in 1668 (table of least factor of numbers up to 100,000), Krüger in 1746 (primes up to 100,000), Lambert in 1770 (table of least factor of numbers up to 102,000), Felkel in 1776 (table of least factor of numbers up to 408,000), Vega in 1797 (primes up to 400,031), Chernac in 1811 (prime factors of numbers up to 1,020,000), and Burkhardt in 1816/7 (least factor of numbers up to 3,036,000).

Legendre and Gauss based their empirical observations on such tables.

Little by little, the tables were extended. Thus in 1856 Crelle presented to the Berlin Academy a table of primes up to 6,000,000, and this work was extended by Dase, before 1861, up to 9,000,000.

In this connection, the most amazing feat is Kulik's factor table of numbers to 100,330,200 (except for multiples of 2, 3, 5), entitled *Magnus Canon Divisorum pro omnibus numeris per 2, 3 et 5 non divisibilibus, et numerorum primorum interfacentium ad millies centena millia accuratius ad* 100330201 *usque.* Kulik spent about 20 years preparing this table, and at his death in 1863, the eight manuscript volumes, with a total of 4212 pages, were deposited at the Academy of Sciences in Vienna (in February 1867).

Kulik's table is mentioned in Dickson's (entertaining?) *History of the Theory of Numbers* and also in the entertaining (!) *Recreations in the Theory of Numbers* by Beiler (1966). Since I was curious to know how this table had been prepared, I wrote to Professor Hlawka in Vienna, and soon received a letter from Ms. Binder, with copies of a few sample pages. The presentation was such that each page covered about 23000 numbers, so one million did require about 43 pages. In 1914, D. N. Lehmer (after publishing his own tables — see below), pointed out numerous mistakes in Kulik's table. There are also references to the table in Volume 2, 1946/7 of *Mathematical Tables and other Aids to Computation* (now called *Mathematics of Computation*), at pages 30 and 149–150 (review by S. A. Joffe).

One unfortunate circumstance about Kulik's table is that Volume II, running from 12,642,600 to 22,852,800, has been missing for quite some time. Beiler wrote "What careless custodian, what heedless dusting woman, what furtive student was responsible for the loss?" In a letter to me, Keller commented:

> I was nearly obsessed by such an outrageous negligence! ... I became curious myself, so I telephoned to the present custodian of the Archives of the Academy of Sciences in Vienna, a Dr. Hundsam, ingenuously hoping that the missing volume might have reappeared. This apparently old aged man at once was aware of the problem, not hiding his satisfaction that "his" precious tables were enjoying attention repeatedly. He confirmed that the second volume had been lost definitely, but he had a peculiar theory of how this occurred. He said he personally remembered that about 1948 the complete(!) 8 volume set had been sent, on request, to the American Mathematial Society for copying, and that only 7 of these were returned. Subsequently, someone had blamed the International Postal Service for the unrecoverable loss. Only afterwards I examined

the descriptions of the table accessible to me. As a result, I'm convinced that Dr. Hundsam's tale must belong to the realm of fantasy. For instance, an editorial note to Joffe's report in Volume 2 of MTAC, probably written by D. H. Lehmer, explicitly says that "D. N. Lehmer mentions that Volume 2 of Kulik's table was already missing 35 years ago," i.e., since 1911 at least.

In 1909, D. N. Lehmer published a table of factor numbers up to about 10,000,000, and in 1914 he published the list of primes up to that limit. This time, the volumes were widely distributed and easily accessible to mathematicians.

With the advent of computers, numerous tables were prepared and published; some were available on cards or tapes.

In certain cases, in view of special investigations, tables covering certain intervals were prepared. For example, in 1959, Baker & Gruenberger prepared a table of primes on cards, up to 104,395,289. They have also indicated all the primes between 999×10^6 and 10^9.

Later, in 1967, I note that Jones, Lal & Blundon gave the primes in the intervals 10^n to $10^n + 150,000$, for $n = 8, 9, ..., 14, 15$.

In 1976, Bays & Hudson computed a table of primes up to 1.2×10^{12}; the program kept track not only of $\pi(x)$ but also of the number of primes up to x, in various arithmetic progressions. "This table has not been stored and is to this day sitting in my office taking up badly needed space. However our CPU time of 600 hours could probably be reduplicated on the CRAY today, using the best sieves known" − so wrote Richard Hudson to me.

It should be noted that tables of primes on cards or tape are obsolete, because it is possible to generate all the primes up to any given bound, or in a prescribed interval, with the sieve of Eratosthenes, quicker than one can read it from card or tape.

To my readers, who have faithfully arrived up to this point, as a token of appreciation, and for their utmost convenience, I include a TABLE OF PRIMES UP TO 10000 following the Bibliography! Enjoy yourself!

E. The Exact Value of $\pi(x)$ and Comparison with $x/(\log x)$, Li(x), and $R(x)$

The exact values of $\pi(x)$ may be obtained by direct counting using tables, or by an ingenious method devised in 1871 by Meissel, a German astronomer, which allowed him to go far beyond the range of the tables. In fact, to compute $\pi(x)$ the method requires the knowledge of the prime numbers $p \leqslant x^{1/2}$ as well as the values of $\pi(y)$ for some $y \leqslant x^{2/3}$. It is based on the following formula

$$\pi(x) = \phi(x,m) + m(s + 1) + \frac{s(s - 1)}{2} - 1 - \sum_{i=1}^{s} \pi\left(\frac{x}{P_{m+i}}\right),$$

where $m = \pi(x^{1/3})$, $n = \pi(x^{1/2})$, $s = n - m$, and $\phi(x,m)$ denotes the number of integers a such that $a \leqslant x$ and a is not a multiple of 2, 3, ..., P_m.

Even though the calculation of $\phi(x,m)$ is long, when m is large, it offers no major difficulty. The methods are based on the following simple facts:

Recurrence relation:

$$\phi(x,m) = \phi(x,m - 1) - \phi\left(\left[\frac{x}{P_m}\right], m - 1\right).$$

Division property: If $P_m = p_1 p_2 \cdots p_m$, if $a \geqslant 0$, $0 \leqslant r < P_m$, then $\phi(aP_m + r,m) = a\phi(P_m) + \phi(r,m)$.

Symmetry property: If $\frac{1}{2} P_m < r < P_m$, then

$$\phi(r,m) = \phi(P_m) - \phi(P_m - r - 1, m).$$

Meissel determined, in 1885, the number $\pi(10^9)$ (however he found a value which is low by 56). A simple proof of Meissel's formula was given by Brauer in 1946. In 1959, Lehmer simplified and extended Meissel's method. In 1985, Lagarias, Miller & Odlyzko have further refined the method by incorporating new sieving techniques.

Record

The largest computed value of $\pi(x)$ is $\pi(4 \times 10^{16}) = 1,075,292,778,753,150$, by Lagarias, Miller & Odlyzko (1985).

The following table gives values of $\pi(x)$, which may be compared with the calculated values of the functions $x/(\log x)$, $Li(x)$, $R(x)$.

x	$\pi(x)$	$\left[\dfrac{x}{\log x}\right] - \pi(x)$	$[Li(x)]$ $- \pi(x)$	$[R(x)]$ $- \pi(x)$
10^8	5,761,455	-332,774	754	97
10^9	50,847,534	-2,592,592	1,701	-79
10^{10}	455,052,511	-20,758,030	3,104	-1,828
10^{11}	4,118,054,813	-169,923,160	11,588	-2,318
10^{12}	37,607,912,018	-1,416,706,193	38,263	-1,476
10^{13}	346,065,536,839	-11,992,858,452	108,971	-5,773
10^{14}	3,204,941,750,802	-102,838,308,636	314,890	-19,200
10^{15}	29,844,570,422,669	-891,604,962,453	1,052,619	73,218
10^{16}	279,238,341,033,925	-7,804,289,844,393	3,214,632	327,052
2×10^{16}	547,863,431,950,008	-15,020,437,343,198	3,776,488	-225,875
4×10^{16}	1,075,292,778,753,150	-28,929,900,579,950	5,538,861	-10,980

I have already mentioned Tschebycheff's inequalities for $\pi(x)$, obtained with elementary methods and prior to the prime number theorem. In 1892, Sylvester refined Tschebycheff's method, obtaining

$$0.95695 \frac{x}{\log x} < \pi(x) < 1.04423 \frac{x}{\log x}$$

for every x sufficiently large (see also Langevin, 1977). For teaching purposes, there is a very elegant determination by Erdös (in 1932) of the appropriate constants; thus

$$\log 2 \frac{x}{\log x} < \pi(x) < 2 \log 2 \frac{x}{\log x}$$

for every sufficiently large x. See also Nair (1982) and Costa Pereira (1986).

In 1962, using a very delicate analysis, Rosser & Schoenfeld showed that, if $x \geqslant 17$, then $x/(\log x) \leqslant \pi(x)$.

On the other hand, Riemann and Gauss believed that $\mathrm{Li}(x) > \pi(x)$ for every $x > 1$. Even though in the present range of tables this is true, it had been shown by Littlewood in 1914 that the difference $\mathrm{Li}(x) - \pi(x)$ changes sign infinitely often, say, at numbers $x_0 < x_1 < x_2 < \cdots$, where x_n tends to infinity.

Assuming Riemann's hypothesis, Skewes showed in 1933 that $x_0 < 10^{10^{10^{34}}}$. For a long time, this number was famous, as being the largest number that appeared in a somewhat natural way in mathematics. In a second paper in 1955, Skewes found, without assuming the Riemann hypothesis, that

$$x_0 < e^{e^{e^{e^{7.7}}}}.$$

With other methods, Lehman showed in 1966 that the first change in sign takes place already at a much smaller number $x_0 < 1.65 \times 10^{1165}$. In a recently reported computation (1986), te Riele has reduced this bound. More precisely, he showed that between 6.62×10^{370} and 6.69×10^{370} there are more than 10^{180} successive integers x for which $\pi(x) > \mathrm{Li}(x)$. This phenomenon has not yet been detected in the tables.

About these matters, see the article of Zagier in *The Mathematical Intelligencer*, 1977. It is nicely written and a good source for more information about prime numbers.

F. The Nontrivial Zeroes of $\zeta(s)$

First, I shall discuss the zeroes in the whole critical strip and then the zeroes on the critical line $\mathrm{Re}(s) = \frac{1}{2}$.

Since $\zeta(\bar{s}) = \overline{\zeta(s)}$ (where the bar denotes the complex conjugate), then the zeroes lie symmetrically with respect to the real axis; so, it suffices to consider the zeroes in the upper half of the critical strip.

For every $T > 0$ let $N(T)$ denote the number of zeroes $\rho = \sigma + it$ in the critical strip, with $0 < t \leqslant T$. First of all, it was conjectured by

Riemann, and proved by von Mangoldt:

$$N(T) = \frac{T}{2\pi} \left\{ \log \left[\frac{T}{2\pi} \right] - 1 \right\} + O(\log T).$$

It follows that there exist infinitely many zeroes in the critical strip. Since there can be only finitely many zeroes $\sigma + it$ for each value $t > 0$, the zeroes of $\zeta(s)$ in the upper half of the critical strip may be enumerated as $\rho_n = \sigma_n + it_n$, with $0 < t_1 \leqslant t_2 \leqslant t_3 \leqslant \cdots$.

For every sufficiently large T,

$$N(T + 1) - N(T) \leqslant 2 \log T,$$

that is, the density of zeroes of $\zeta(s)$ is less than $2 \log T$ (for large T). It follows that

$$t_n \sim \frac{2\pi n}{\log n},$$

$$\sum_{|t_n| \leqslant T} \frac{1}{|t_n|} = O((\log T)^2),$$

$$\sum_{|t_n| > T} \frac{1}{t_n^2} = O\left[\frac{\log T}{T} \right].$$

Now I consider the zeroes on the critical line. For every $T > 0$ let $N_0(T)$ denote the number of zeroes $\frac{1}{2} + it$ of the zeta function, such that $0 < t \leqslant T$.

Here are the main results. Hardy showed in 1914 that $\zeta(s)$ has infinitely many zeroes on the critical line. More precisely, there exists a constant $C > 0$ such that, for every sufficiently large T, $N_0(T) > CT$. This result was improved by Selberg, in 1942, who proved that

$$N_0(T) > C \frac{T}{2\pi} \log \left[\frac{T}{2\pi} \right],$$

for some constant C, $0 < C < 1$, and all sufficiently large T.

In 1983, Conrey showed that $C \geqslant 0.3658$. Assuming various daring conjectures, Balasubramanian, Conrey & Heath-Brown claimed (1985) that they may show that $C \geqslant 0.55$.

All the known nontrivial zeroes of $\zeta(s)$ are simple and lie on the critical line.   Montgomery showed in 1973, assuming Riemann's hypothesis, that at least two thirds of the nontrivial zeroes are simple.

In 1974, Levinson proved that at least one third of the nontrivial zeroes of Riemann's zeta function are on the critical line. More precisely, if T is sufficiently large, $L = \log(T/2\pi)$, and $U = T/L^{10}$, then $N_0(T + U) - N_0(T) > \frac{1}{3}(N(T + U) - N(T))$.

Extensive computations of the zeroes of $\zeta(s)$ have been made. These began with Gram in 1903, who computed the first 15 zeroes (that is, ρ_n for $1 \leqslant n \leqslant 15$). Backlund, Hutchinson, and Haselgrove extended Gram's work. Titchmarsh calculated in 1935 the zeroes ρ_n, for $n \leqslant 1041$. With the advent of computers, Lehmer brought this up to $n = 35,337$. By 1969, Rosser, Yohe & Schoenfeld had computed the first 3,500,000 zeroes. Since then, this has been largely extended, for example by Brent (in 1979, up to 81,000,001), van de Lune, te Riele & Winter (in 1983 up to 300,000,001).

Record

Van de Lune, te Riele & Winter have determined (in 1986) that the first 1,500,000,001 nontrivial zeroes of $\zeta(s)$ are all simple, lie on the critical line, and have imaginary part with $0 < t < 545,439,823.215$. This work has involved over 1000 hours on a supercomputer.

Just not to be shamefully absent from this book, here is a table with the smallest zeroes $\rho_n = \frac{1}{2} + it_n$, $t_n > 0$:

n	t_n	n	t_n	n	t_n
1	14.134725	11	52.970	21	79.337
2	21.022040	12	56.446	22	82.910
3	25.010856	13	59.347	23	84.734
4	30.424878	14	60.833	24	87.426
5	32.935057	15	65.113	25	88.809
6	37.586176	16	67.080	26	92.494
7	40.918720	17	69.546	27	94.651
8	43.327073	18	72.067	28	95.871
9	48.005150	19	75.705	29	98.831
10	49.773832	20	77.145		

In Edwards' book (1974), there is a detailed explanation of the method used by Gram, Backlund, Hutchinson, and Haselgrove to compute the smallest 300 zeroes of $\zeta(s)$. See also Edwards' book, as well as the recent article of Wagon (1986), on the computation of further zeroes, containing, in particular, a discussion of Gram points, the method of Turing, Lehmer's phenomenon, and Rosser's rule.

The more detailed study of the zeroes on the critical line involves estimates for the difference $t_{n+1} - t_n$ between the ordinates of consecutive zeroes. For a long time, the result of Hardy & Littlewood (1918), that $t_{n+1} - t_n = O(t_n^{1/4+\epsilon})$ was the best. But it was superseded, with the use of finer methods, by Moser, Balasubramanian, Karacuba, and finally by Ivić (1983), who has the best result to date: $t_{n+1} - t_n < t_n^{\theta+\epsilon}$, with $\epsilon > 0$ and $\theta = 0.1559458...$ (see Ivić's book, 1985, for the proof and comments).

The analysis of the distribution of spacings between the zeroes of Riemann's zeta function has been the subject of a detailed numerical study by Odlyzko (1987), who has tested the facts brought to light by the recent computations against more or less credible conjectures.

A method to approach Riemann's hypothesis is to estimate for any given σ, $\frac{1}{2} \leqslant \sigma \leqslant 1$, the number $N(\sigma,t)$ of zeroes $\rho = \beta + it$ of $\zeta(s)$, with $\sigma \leqslant \beta$, and $0 < t \leqslant T$ (where T is sufficiently large). The idea is that, for $\sigma > \frac{1}{2}$, the smaller the upper bound for $N(\sigma,T)$, the more likely is the truth of Riemann's hypothesis. It is not expected to prove Riemann's hypothesis in this way, but to obtain statements which, even though weaker, will still have many of the interesting consequences of Riemann's hypothesis.

I want only to give a flavor of the results in this direction; for the fine points, I suggest the book by Ivić already quoted. Thus, for every σ, $\frac{1}{2} \leqslant \sigma \leqslant 1$ and sufficiently large T,

$$N(\sigma,T) \leqslant CT^{(12/5)(1-\sigma)+\epsilon},$$

where $C > 0$ depends only on ϵ. This is a consequence of results of Ingham (1940) and Huxley (1972).

As σ comes closer to 1, the estimates may be made sharper. Ivić showed in 1980: if $17/18 \leqslant \sigma \leqslant 1$, then

$$N(\sigma,T) \leqslant CT^{[4/(2\sigma+1)](1-\sigma)+\epsilon} \leqslant CT^{(18/13)(1-\sigma)+\epsilon}.$$

The following estimate, which is a consequence of Riemann's hypothesis, is called the density hypothesis: $N(\sigma,T) \leqslant CT^{2(1-\sigma)+\epsilon}$, for T sufficiently large and $\frac{1}{2} \leqslant \sigma \leqslant 1$.

Many of the consequences of Riemann's hypothesis may already be deduced from the density hypothesis. Here is a method to try to ascertain whether the density hypothesis holds. If $\sigma \geqslant \frac{1}{2}$, let $A(\sigma) > 0$ be the smallest real number such that $N(\sigma,T) \leqslant CT^{A(\sigma)(1-\sigma)+\epsilon}$. What is the smallest $\sigma_0 \geqslant \frac{1}{2}$ such that $A(\sigma) \leqslant 2$? This approach was taken up by Montgomery, who showed in 1971 that $\sigma_0 \leqslant 9/10$. Successive results by Huxley, Ramachandra, Forti & Viola, Jutila have brought the value of σ_0 closer to $\frac{1}{2}$. Today's best result is by Jutila (1977), who showed that for $\sigma_0 = 11/14$, then $A(\sigma_0) \leqslant 2$. There is still a long way to go.

G. Zero-Free Regions for $\zeta(s)$ and the Error Term in the Prime Number Theorem

The knowledge of larger zero-free regions for $\zeta(s)$ leads to better estimates of the various functions connected with the distribution of primes.

I have already indicated that de la Vallée Poussin determined a zero-free region, which he used in an essential way in his proof of the prime number theorem. There have been many extensions of his result, and the largest known zero-free region has been determined by Richert [and published in Walfisz's book (1963)]:

There exists $C > 0$ and $t_0 = t_0(C) > e^{e^{2C}}$ such that $\zeta(s) \neq 0$ for every $s = \sigma + it$ in the region

$$1 - \frac{C}{\lambda(t)} \leqslant \sigma \leqslant 1, \quad t_0 < |t|$$

where $\lambda(t) = (\log |t|)^{2/3} (\log \log |t|)^{1/3}$. The constant C may be taken to be $C = 1/8757$ and $t_0(C)$ is effectively computable.

It is to be remarked that up to now, no one has succeeded in knowing that $\zeta(s)$ has a zero-free region of the form $\{\sigma + it \mid \sigma \geqslant \sigma_0\}$ with $\frac{1}{2} < \sigma_0 < 1$.

The sharp evaluation of the error depends on the knowledge of large zero-free regions for $\zeta(s)$; the larger is the region, the better is the estimate of the differences $|\pi(x) - \mathrm{Li}(x)|$ and $|\psi(x) - x|$, appearing in the prime number theorem. The theory hinges on a formula due to Landau, which gives an estimate of the differences $|\psi(x) - x|$ in terms of the zeroes of $\zeta(s)$. All started, once more, with Riemann, who stated in 1859, what amounted to:

$$\psi(x) = x - \sum \frac{x^\rho}{\rho} - \log(2\pi) - \frac{1}{2}\gamma - \frac{1}{2}\log\left[1 - \frac{1}{x^2}\right]$$

(for $x > 1$, x not a prime power); γ denotes Euler's constant and the sum is over all non-trivial zeroes of ρ of $\zeta(s)$. This formula was proved in 1895 by von Mangoldt. Later, Landau gave the estimate that for $T > 0$ sufficiently large, then

$$\psi(x) = x - \sum_T \frac{x^\rho}{\rho} + O\left[x \frac{(\log xT)^2}{T}\right] + O(\log x).$$

Here, the sum is for all nontrivial zeroes $\rho = \sigma + it$ of $\zeta(s)$, with $|t| \leqslant T$.

Using whatever is known about zero-free regions for $\zeta(s)$ it is possible to deduce an estimate for the error in the prime number theorem. Thus, Littlewood shows that

$$\pi(x) = \text{Li}(x) + O\left[xe^{-C(\log x \, \log \log x)^{1/2}}\right];$$

later Tschudakoff obtained

$$\pi(x) = \text{Li}(x) + O\left[xe^{-C(\log x)^{\alpha}}\right];$$

with $\alpha < 4/7$, and $C > 0$.

Today's best known result on a zero-free region for $\zeta(s)$ leads to the better estimates

$$\pi(x) = \text{Li}(x) + O\left[xe^{-C\frac{(\log x)^{3/5}}{(\log \log x)^{1/5}}}\right];$$

here the constant C and the constant in O are explicitly computable.

In 1901, von Koch showed that Riemann's hypothesis is equivalent to the following form of the error

$$\pi(x) = \text{Li}(x) + O(x^{1/2} \log x).$$

The knowledge that many zeroes of $\zeta(s)$ are on the critical line leads also to better estimates. Thus, Rosser & Schoenfeld proved in 1975 that

$$0.998684x < \theta(x) < 1.001102x$$

(the lower inequality for $x \geqslant 1319007$, the upper inequality for all x).

H. The Growth of $\zeta(s)$

Now I shall focus a little bit on the growth of $\zeta(s)$.

The pivotal fact is that for every real σ there exists $\alpha > 0$ such that $|\zeta(\sigma + it)| = O(t^{\alpha})$ as $t > 0$ tends to infinity. So, along vertical lines, the function has finite order of growth.

Let $\omega(\sigma) = \inf\{\alpha > 0 \mid |\zeta(\sigma + it)| = O(t^{\alpha})\}$, so that $\omega(\sigma) \geqslant 0$. The function $\omega(\sigma)$ is continuous and nonincreasing.

It may be shown that $\zeta(1 + it) = O(\log t)$, hence $\omega(1) = 0$, and also $\omega(\sigma) = 0$ for $\sigma > 1$.

From the functional equation, it follows that $\omega(\sigma) = \frac{1}{2} - \sigma$, for $\sigma \leqslant 0$.

Concerning the values of $\omega(\sigma)$, for $0 < \sigma < 1$, it is important to estimate $\omega(\frac{1}{2})$.

The following hypothesis was made by Lindelöf: $\omega(\frac{1}{2}) = 0$. Equivalently,

$$\omega(\sigma) = \begin{cases} \frac{1}{2} - \sigma & \text{for } \sigma \leqslant \frac{1}{2} \\ 0 & \text{for } \sigma > \frac{1}{2}. \end{cases}$$

Already in 1921, the inseparable Hardy and Littlewood showed that $\omega(\frac{1}{2}) \leqslant \frac{1}{6} = 0.166...$. Successive work by Walfisz, Titchmarsh, Min, Chen, Ivić, and Kolesnik have brought down this estimate to $\omega(\frac{1}{2}) \leqslant 139/858 = 0.1620046...$ (see Ivić's book, page 196). Last Minute: In their forthcoming paper On the order of $\zeta(\frac{1}{2} + it)$, to appear in the *Annali della Scuola Normale Superiore di Pisa*, Bombieri & Iwaniec prove that $\omega(\frac{1}{2}) \leqslant \frac{9}{56}$.

Lindelöf's hypothesis is also equivalent to the following assertion:

For every $\sigma > \frac{1}{2}$: $N(\sigma, T + 1) - N(\sigma, T) = o(\log T)$.

And, it implies the density hypothesis already discussed before, as well as:

$N(\sigma, T) = O(T^{\epsilon})$ for every $\epsilon > 0$, when $\sigma \geqslant \frac{3}{4} + \delta$, $\delta > 0$.

It may be shown that if Riemann's hypothesis is true then so is Lindelöf's hypothesis.

Without any unwarranted assumptions, the following result holds:

There exist constants C, $a > 0$ such that for every σ, $\frac{1}{2} < \sigma < 1$, for every β, $\sigma \leqslant \beta \leqslant 1$, and for every $t \geqslant 3$ one has

$$|\zeta(\beta + it)| \leqslant C t^{a(1-\sigma)^{3/2}} (\log t)^{2/3}.$$

C, a may be taken to be 2100, 86, respectively.

II. The *nth* Prime and Gaps

The results in the preceding section concern the asymptotic behaviour of $\pi(x)$ – and its comparison with other known functions. But nothing was said about the behaviour of $\pi(x)$ in short intervals, nor about the size of the *n*th prime, the difference between consecutive' primes, etc. These are questions putting in evidence the fine points in the distribution of primes, and much more irregularity is expected.

A. Some Properties of $\pi(x)$

In this respect, historically the first statement is Bertrand's experimental observation (in 1845):

Between $n \geqslant 2$ and $2n$ there is always a prime number.

Equivalently, this may be stated as

$$\pi(2n) - \pi(n) \geqslant 1 \quad \text{(for } n \geqslant 2\text{),}$$

or also as

$$p_{n+1} < 2p_n \quad \text{(for } n \geqslant 1\text{).}$$

This statement has been known as "Bertrand's postulate" and it was proved by Tschebycheff in 1852 as a by-product of his estimates already indicated for $\pi(x)$. As a matter of fact, the following inequalities are easy to guarantee:

$$1 < \frac{1}{3}\frac{n}{\log n} < \pi(2n) - \pi(n) < \frac{7}{5}\frac{n}{\log n} \quad \text{(for } n \geqslant 5\text{),}$$

and clearly $\pi(4) - \pi(2) = 1$, $\pi(6) - \pi(3) = 1$, $\pi(8) - \pi(4) = 2$.

The proof of Bertrand's postulate and the above lower bound for the difference $\pi(2n) - \pi(n)$ are, of course, intimately connected. I noted in the literature, among the simplest proofs, one by Ramanujan (1919), which uses simple properties of the Γ-function, and one by Erdös (1930) stressing the divisibility properties of the middle binomial coefficient $\binom{2n}{n}$; it is reproduced in Hua's book (1982), quoted in the General References.

The simplest elementary proof of Bertrand's postulate, which I came across, is by Moser (1949).

More generally, Erdös proved in 1949 that for every $\lambda > 1$ there exists $C = C(\lambda) > 0$, and $x_0 = x_0(\lambda) > 1$ such that

$$\pi(\lambda x) - \pi(x) > C \frac{x}{\log x},$$

which is just a corollary of the prime number theorem.

The following result of Ishikawa (1934) is also a consequence of Tschebycheff's theorem:

If $x \geqslant y \geqslant 2$, $x \geqslant 6$, then $\pi(xy) > \pi(x) + \pi(y)$.

A proof may be found in Trost's book.

On the other hand, using deeper methods, Vaughan proved that

$$\pi(x + y) \leqslant \pi(x) + \frac{2y}{\log y} ;$$

by the result of Rosser & Schoenfeld (1975) already mentioned, it follows that $\pi(x + y) \leqslant \pi(x) + 2\pi(y)$.

The relation $\pi(2x) < 2\pi(x)$ for $x \geqslant 11$, was also proved by Rosser & Schoenfeld, as a consequence of their refined estimates of $\theta(x)$.

It was also conjectured by Hardy & Littlewood in 1923 that

$$\lim_{y \to \infty} \sup \ (\pi(x + y) - \pi(y)) \leqslant \pi(x).$$

Even though Segal checked in 1962 that the inequality holds for $x + y \leqslant 101,081$, this conjecture seems unlikely to be true, in view of results in opposite directions. For these questions, see Hensley & Richards (1973).

Two statements which are still waiting to be proved, or disproved, are the following:

In 1882, Opperman stated that $\pi(n^2 + n) > \pi(n^2) > \pi(n^2 - n)$ for $n > 1$.

In 1904, Brocard asserted that $\pi(p_{n+1}^2) - \pi(p_n^2) \geqslant 4$ for $n \geqslant 2$; that is, between the squares of two successive primes greater than 2 there are at least four primes.

B. The nth Prime

Now I shall consider specifically the nth prime.

The prime number theorem yields easily:

$$p_n \sim n \log n, \quad \text{that is,} \quad \lim_{n \to \infty} \frac{p_n}{n \log n} = 1.$$

In other words, for large indices n, the nth prime is about the size of $n \log n$. More precisely

$$p_n = n \log n + n(\log \log n - 1) + o\left(\frac{n \log \log n}{\log n} \right).$$

So, for large n, $p_n > n \log n$. But Rosser proved in 1938 that for every $n > 1$:

$$n \log n + n(\log \log n - 10) < p_n < n \log n + n(\log \log n + 8)$$

and also that for every $n \geqslant 1$: $p_n > n \log n$.

Rosser's estimates have been improved by Robin (1983) as follows:

If $n \geqslant 2$ then $p_n \geqslant n \log n + n (\log \log n - 1.0072629)$;

if $n \geqslant 2$ and $p_n \leqslant 10^{11}$ then $p_n \geqslant n \log n + n(\log \log n - 1)$;

if $n \geqslant 7022$ then $p_n \leqslant n \log n + n(\log \log n - 0.9385)$.

The following results by Ishikawa (1934) are also consequences of Tschebycheff's theorems (see Trost's book):

If $n \geqslant 2$, then $p_n + p_{n+1} > p_{n+2}$;

if $m,n \geqslant 1$, then $p_m p_n > p_{m+n}$.

In a very interesting paper, Pomerance considered in 1979 the prime number graph, consisting of all the points (n, p_n) of the plane (with $n \geqslant 1$). He proved Selfridge's conjecture: there exist infinitely many n with $p_n^2 > p_{n-i} p_{n+i}$ for all positive $i < n$. Also, there are infinitely many n such that $2p_n < p_{n-i} + p_{n+i}$ for all positive $i < n$.

C. Gaps between Primes

It is important to study the difference $d_n = p_{n+1} - p_n$ between consecutive primes, that is, the size of the gaps between consecutive primes. This gives an indication of the distribution of primes in short intervals.

It is quite easy to show that lim sup $d_n = \infty$, that is, for every $N > 1$ there exists a string of at least N consecutive composite integers; for example:

$$(N+1)! + 2, \quad (N+1)! + 3, \quad (N+1)! + 4, \quad ..., \quad (N+1)! + (N+1).$$

Actually, gaps of size N have been found experimentally between numbers much smaller than $(N+1)! + 1$.

[At this point, it is appropriate to report, as an aside, that this is readily generalized. Jarden showed in 1967, that in any arithmetic progression $\{kd + a \mid k = 1, 2, ...\}$, with $d \geqslant 1$, $a \geqslant 0$, for every $N > 1$ and $s \geqslant 1$, there exist N successive terms with at least s distinct prime factors, so arbitrarily long strings of highly composite integers.]

Record

In 1981 Weintraub found a string of 653 consecutive composite numbers following the prime 11,000,001,446,613,353. The previous record-holder was Brent (1980), with 651 consecutive composite numbers following the prime 2,614,941,710,599; he also established that there is no gap of this size between smaller primes.

To find lim inf d_n is quite another matter — a difficult problem which I shall discuss in more detail (see (c)). It is not even known if lim inf $d_n < \infty$; this would mean that there exist some integer $k > 0$ such that there exist infinitely many pairs of successive primes with difference $2k$ (see Polignac's conjecture, in the subsection on possible gaps between primes).

The next natural question is to estimate the rate of growth of the difference $d_n = p_{n+1} - p_n$, and compare it with various functions of p_n. For example, to compare d_n with p_n with powers p_n^θ, with $(\log p_n)^\theta$, etc. (where $\theta > 0$).

For example the old result of Tschebycheff on Bertrand's postulate tells that $d_n < p_n$ for every $n \geqslant 1$.

Using the methods of Tschebycheff, Nagura showed in 1952 that if $\epsilon = \frac{1}{5}$, for every $m \geqslant 25$, there exists a prime p satisfying $m < p < (1+\epsilon)m$. So, $d_n < \frac{1}{5} p_n$, for $n \geqslant 10$. More refined estimates of Tschebycheff's function $\theta(x)$ allows one to consider smaller values of ϵ and to calculate m_0.

For example, Rohrbach & Weis showed in 1964 that, if $\epsilon = 1/13$, then $m_0 = 118$; while Schoenfeld proved in 1976 that, if $\epsilon = 1/16597$, then $m_0 = 2010760$.

By the prime number theorem the above statement holds for every $\epsilon > 0$, with appropriate m_0, which means that

$$\lim_{n \to \infty} \frac{d_n}{p_n} = 0.$$

Clearly, since $d_n < p_n$ for every $n \geqslant 1$, then $d_n = O(p_n)$. This leads to the question of finding the best function $f(p_n)$ such that $d_n = O(f(p_n))$.

Before telling which are the results, I'll indicate what has been conjectured to be true.

It is believed that for every $\epsilon > 0$, $d_n = O(p_n^\epsilon)$ – a statement due to Piltz, in 1884, which has not been refuted up to now.

Using a probabilistic argument, Cramér conjectured in 1937 even that $d_n = O((\log p_n)^2)$, but this is very far from being established. Anyway, it is known that $d_n \neq O(\log p_n)$, as I will indicate soon.

Assuming the Riemann hypothesis, Cramér showed that $d_n = O(p_n^{1/2}\log p_n)$; a simpler proof is in Ivić's book (1985). If one assumes only the weaker density hypothesis on the nontrivial zeroes of $\zeta(s)$, it may be shown that $d_n = O(p^{\frac{1}{2}+\epsilon})$ for every $\epsilon > 0$. The estimates which can be deduced from unproved assumptions are goals to be attained.

Independently of any unproved assumption, the classical theorem was established by Hoheisel in 1930. It was a breakthrough not just a new record. As Turán said (1971) "in an age when pathbreakers are mostly forgotten and only world records and recorders are known, it is worth stressing the importance of the result."[*] For the first time the barrier 1 was broken. Hoheisel showed that there exists $\theta < 1$ such that

$$\pi(x + x^\theta) - \pi(x) \sim \frac{x^\theta}{\log x} \; ;$$

in fact, he took $\theta = 1 - \frac{1}{33000} + \epsilon$ (for any $\epsilon > 0$). It is easy to see that this asymptotic equality implies, with the same value of θ, that $d_n = O(p_n^\theta)$, or still that, for x sufficiently large, there is a prime p such that $x - x^\theta < p \leqslant x$.

In 1937, Ingham refined this result and showed that, for every $\epsilon > 0$, $d_n = O(p_n^{\frac{5}{8} + \epsilon})$.

There have been many improvements of this result, both by sharpening the estimates, and by introducing new methods. I shall only mention the papers which contain new methods, as well as the latest best values achieved.

In 1969 (see also 1971), Montgomery showed that one may take $\theta = \frac{3}{5} + \epsilon$. Huxley gave $\theta = \frac{7}{12} + \epsilon$ in 1972. With sieve methods, Iwaniec & Jutila reached in 1979, the value $\theta = \frac{13}{23} + \epsilon$, soon improved by Heath-Brown & Iwaniec (1979) to $\theta = \frac{11}{20} + \epsilon$. This was the previous record, while the next record was due to Iwaniec & Pintz (1984): $\theta = \frac{1}{2} + \frac{1}{21} = 0.5476...$ Thus, for large n :

$$p_{n+1} < p_n + p_n^{\frac{1}{2} + \frac{1}{21}} \; .$$

The race continued with papers by Pintz (1981, 1984) and Mozzochi (1986), who holds the record today: $\theta = \frac{11}{20} - \frac{1}{384} = 0.5473958...$.

Concerning the comparison of d_n and $\log p_n$, the following results have been established.

First, Erdös proved in 1940 that

$$\liminf \frac{d_n}{\log p_n} < C < 1.$$

---

[*]This is also my point of view; see the Conclusion.

In 1966, Bombieri & Davenport showed that C may be taken to be 0.467. This was somewhat improved by Huxley in 1977, but the best result is now by Maier (1985) where the constant is replaced by 0.248.

All the above results were obtained with rather sophisticated means. However, in 1984, using only the prime number theorem, Powell gave a proof of

$$\lim \inf \frac{d_n}{\log p_n} \leqslant 1$$

(problem proposed by Powell in 1980, and again in 1983, at the *American Mathematical Monthly*); see also Prachar's book (1957, p. 155).

On the other hand, Erdös (1935) and Rankin (1938, 1963) showed that

$$\lim \sup \frac{d_n}{\log p_n} \times \frac{(\log_3 p_n)^2}{(\log_2 p_n)(\log_4 p_n)} \geqslant e^{\gamma},$$

where $\gamma = 0.577...$ is Euler's constant, $e^{\gamma} = 1.78...$ and $\log_2 p_n = \log \log p_n$, $\log_3 p_n = \log \log \log p_n$, $\log_4 p_n = \log \log \log \log p_n$.

During the recent Conférence Internationale de Théorie des Nombres at Université Laval in Québec, Maier & Pomerance announced that they may slightly improve on Erdös and Rankin's results, replacing e^{γ} by ce^{γ}, where $c = 1.31256...$ satisfies the relation $4/c = 3 + e^{-4/c}$; the proof brings a new method into the study of the problem, so it is worth more than the admittedly modest improvement.

Erdös offers a prize of US$10,000 to anyone who will prove that e^{γ} may be replaced by ∞, in Erdös-Rankin inequality.

The set X of limit points of the sequence

$$\left[\frac{d_n}{\log p_n} \right]_{n \geqslant 1}$$

contains infinity, because $\lim \sup(d_n/\log p_n) = \infty$. (This consequence of Rankin's result had been proved by Westzynthius, in 1931.) Ricci proved in 1955 (and so claims Erdös) that X has positive Lebesgue measure and contains positive numbers less than 1; however no explicit number in X has yet been identified. Even more, in a forthcoming paper, Hildebrand & Maier show that a positive proportion of all positive real numbers are in the set X; thus there exists $C > 0$ such that if $x \geq 1$ the set $X \cap [0,x]$ contains a subset with measure at least Cx.

In a much lighter vein, a quick proof that

$$\limsup \frac{d_n}{\log p_n} \geq 1.7$$

was given by Shafer in 1984 (solution of the problem proposed by Powell, loc. cit.).

The sequence of differences $(d_n)_{n \geq 0}$ behaves rather erratically. In 1948, Erdös & Turán, proved that for every real number $t > 0$, there exist infinitely many $n \geq 2$ such that $p_{n+1}^t - p_n^t > p_n^t - p_{n-1}^t$ and also infinitely many $n \geq 2$ such that the opposite inequality $p_{n+1}^t - p_n^t < p_n^t - p_{n-1}^t$ holds; in particular, each of the inequalities $d_n > d_{n-1}$ and $d_n < d_{n-1}$ holds infinitely often. They have also proved that each of the inequalities $\log p_{n+1} - \log p_n > \log p_n - \log p_{n-1}$ and $\log p_{n+1} - \log p_n < \log p_n - \log p_{n-1}$ holds infinitely often.

But here is an assertion which ought to be false, yet Erdös & Turán, with their immense skill, could not disprove:

There exists an integer $n_0 \geq 1$ such that for every $k \geq 1$, the following inequalities hold:

$$d_{n_0+2k-1} < d_{n_0+2k} \quad \text{and} \quad d_{n_0+2k} > d_{n_0+2k+1}.$$

Erdös proved also, in 1949, that for every $A > 0$ there exist infinitely many n such that $d_n > A \log p_n$, $d_{n+1} > A \log p_n$, that is

$$\limsup \frac{\min \{d_n, d_{n+1}\}}{\log p_n} = \infty.$$

In 1981, with considerable deeper tools from sieve theory, Maier showed that for every $k \geq 1$:

$$\limsup \frac{\min\{d_n, d_{n+1}, \ldots, d_{n+k}\}}{\log p_n} \times \frac{(\log_3 p_n)^2}{(\log_2 p)(\log_4 p)} > 0.$$

D. The Possible Gaps between Primes

Now that the growth of d_n has been described, it is also natural to ask which values are possible for d_n; of course, for $n > 1$, $d_n = p_{n+1} - p_n$ is even.

The following result is an easy application of the prime number theorem, and was proposed by Powell as a problem in the *American Mathematical Monthly* (1983; solution by Davies in 1984):

For every natural number M, there exists an even number $2k$, such that there are more than M pairs of successive primes with difference $2k$.

Proof. Let n be sufficiently large, consider the sequence of primes

$$3 = p_2 < p_3 < \cdots < p_n$$

and the $n - 2$ differences $p_{i+1} - p_i$ ($i = 2, ..., n - 1$). If the number of distinct differences is less than

$$\left[\frac{n-2}{M}\right],$$

then one of these differences, say $2k$, would appear more than M times. In the alternative case,

$$p_n - p_2 \geqslant 2 + 4 + \cdots + 2\left[\frac{n-2}{M}\right].$$

But the right-hand side is asymptotically equal to n^2/M^2, while the left-hand side is asymptotically equal to $n \log n$, by the prime number theorem. This is impossible.   □

Of a totally different order of difficulty is Polignac's conjecture (1849): for every even natural number $2k$ there are infinitely many pairs of consecutive primes p_n, p_{n+1} such that $d_n = p_{n+1} - p_n = 2k$.

In particular, this conjecture includes as a special case the following open problem: are there infinitely many pairs of primes p, $p+2$?

A positive answer means that $\lim \inf d_n = 2$, where $d_n = p_{n+1} - p_n$. This question will be considered in the next section. I remark in this connection, that it is also not known that every even natural number is a difference of two primes (even without requiring them to be consecutive).

E. Interlude

This is a point when it is good to pause and to understand the above problems and conjectures as special cases of important and much more general questions.

Let $f(X)$ be a non-constant irreducible polynomial with integral coefficients. Assume:

(*)   there exists no integer $n > 1$ dividing all the values $f(k)$, for every integer k.

If condition (*) is satisfied, I shall say that $f(X)$ is strongly primitive, since this condition implies that the greatest common divisor of all the coefficients is equal to 1 (and such polynomials are called primitive).

This is a natural restriction to impose, in order that the problems considered below be nontrivial.

Questions:

(1)   Does there exist an integer $n \geq 1$ such that $f(n)$ is a prime?

(1_∞) Do there exist infinitely many integers $n \geq 1$ such that $f(n)$ is a prime [when $f(X)$ has positive leading coefficient]?

(2)   Does there exist a number p such that $f(p)$ is a prime?

(2_∞) Do there exist infinitely many primes p such that each $f(p)$ is a prime [when $f(X)$ has positive leading coefficient]?

I discuss some examples.

$f(X) = X + 2$: question (2_∞) will be considered in Section III; explicitly; do there exist infinitely many (consecutive) primes with difference 2?

$f(X) = 2k - X$ (with $2k \geq 4$): question (2) will be considered in Section VI; explicitly: is every even number $2k \geq 4$ equal to the sum of two primes?

$f(X) = 2X + 1$: question (2_∞) will be considered in Chapter 5, Section II; explicitly: do there exist infinitely many primes p such that $2p + 1$ is also a prime?

$f(X) = dX + a$ with $d \geqslant 1$, $1 \leqslant a \leqslant d - 1$ and $\gcd(a,d) = 1$: question (1_∞) will be considered in Section IV.

$f(X) = X^2 + 1$: question (1_∞) will be considered in Chapter 6.

In Chapter 6, I shall indicate many results that would follow if the answer to question (1) is assumed to be affirmative for every polynomial $f(X)$ satisfying condition (*).

It is of course implicit that, for an arbitrary polynomial, the question (1) remains open. Nevertheless, for any given polynomial $f(X)$, it is usually easy to determine numerically an integer n (and sometimes even a prime p) such that $f(n)$ [respectively, $f(p)$] is a prime. But ... look at Chapter 6, Section III to see that after all, it may not be so easy.

In view of the difficulty of these problems, it is reasonable to consider corresponding somewhat weaker questions, phrased by replacing prime numbers by "almost-primes," which I define now.

Let $k \geqslant 1$. An integer $n = \prod_{i=1}^{r} p_i^{e_i}$ (written as the product of its prime factors) is called a k-almost-prime when $\sum_{i=1}^{r} e_i \leqslant k$. The set of all k-almost-primes, shall be denoted by P_k.

Almost-primes may be handled by the methods of sieve theory. It is absolutely not my intention to enter into explanations concerning this theory; it has evolved from quantitative formulations of the sieve to Eratosthenes and it is now one of the most powerful methods to attack certain problems about prime numbers. The book to read is *Sieve Methods* by Halberstam & Richert (1974).

In order to develop a quantitative approach to fundamental questions, it is convenient to introduce the following notation. If $x \geqslant 1$ and $f(X)$ is a polynomial of the type indicated let

$$\pi_{f(X)}(x) = \#\{n \geqslant 1 \mid |f(n)| \leqslant x \text{ and } |f(n)| \text{ is a prime}\}.$$

More generally, let $f_1(X)$, ..., $f_s(X)$ be non-constant irreducible polynomials with integral coefficients. Assume that the following condition is satisfied:

(*) there does not exist any integer $n > 1$ dividing all the products $f_1(k)f_2(k)\cdots f_s(\kappa)$, for every integer k.

For every $x \geqslant 1$, let

$$\pi_{f_1(X),\dots,f_s(X)}(x) = \#\{n \geqslant 1 \mid |f_i(n)| \leqslant x \text{ and } |f_i(n)| \text{ is a prime,}$$

$$\text{for } i = 1, \dots, s\}.$$

If $f(X) = X$, clearly it is the function $\pi(x)$. Later, I shall be dealing explicitly with estimates of such prime counting functions, for example,

$$\pi_{dX+a}(x) = \#\{p \text{ prime} \mid p \leqslant x, \ p \equiv a \pmod{d}\},$$

$\pi_{X^2+1}(x)$ and also $\pi_{X,X+2}(x)$, $\pi_{X,2X+1}(x)$.

Similarly, if $k \geqslant 2$, $f(X)$ is a polynomial of the type indicated, it is natural to consider the k-almost-prime counting function denoted by

$$\pi^{(k)}_{f(X)}(x) = \#\{n \geqslant 1 \mid |f(n)| \leqslant x \text{ and } |f(n)| \in P_k\}.$$

III. Twin Primes

If p and $p + 2$ are primes, they are called twin primes.

The smallest pairs of twin primes are $(3,5)$, $(5,7)$, $(11,13)$ and $(17,19)$. Twin primes have been characterized by Clement in 1949 as follows.

Let $n \geqslant 2$. The integers n, $n + 2$ form a pair of twin primes if and only if

$$4[(n - 1)! + 1] + n \equiv 0 \quad [\bmod\ n(n + 2)].$$

Proof. If the congruence is satisfied, then $n \neq 2, 4$, and

$$(n - 1)! + 1 \equiv 0 \pmod{n},$$

so by Wilson's theorem n is a prime. Also

$$4(n - 1)! + 2 \equiv 0 \ (\text{mod } n + 2);$$

hence multiplying by $n(n + 1)$,

$$4[(n + 1)! + 1] + 2n^2 + 2n - 4 \equiv 0 \ (\text{mod } n + 2)$$

and then

$$4[(n + 1)! + 1] + (n + 2)(2n - 2) \equiv 0 \ (\text{mod } n + 2);$$

from Wilson's theorem $n + 2$ is also a prime.

Conversely, if n, $n + 2$ are primes, then $n \neq 2$ and

$$(n - 1)! + 1 \equiv 0 \ (\text{mod } n),$$

$$(n + 1)! + 1 \equiv 0 \ (\text{mod } n + 2).$$

But $n(n + 1) = (n + 2)(n - 1) + 2$, so $2(n - 1)! + 1 = k(n + 2)$, where k is an integer. From $(n - 1)! \equiv -1 \ (\text{mod } n)$, then $2k + 1 \equiv 0 \ (\text{mod } n)$ and substituting $4(n - 1)! + 2 \equiv -(n + 2) \ (\text{mod } n(n + 2))$, therefore $4[(n - 1)! + 1] + n \equiv 0 \ (\text{mod } n(n + 2))$. $\square$

However, this characterization has no practical value in the determination of twin primes.

It should be added here that Clement's characterization is an immediate consequence of the following variant of Wilson's theorem, given by Coblyn in 1913:

The integer $n \geqslant 2$ is a prime if and only if n divides each of the numbers $(r - 1)!(n - r)! + (-1)^{r-1}$ for $r = 1, 2, \ldots, n - 1$.

Thus, if p, $p + 2$ are primes then, taking $n = p$, $r = 1$, respectively $n = p + 2$, $r = 3$, gives easily that $p(p + 2)$ divides $4[(p - 1)! + 1] + p$.

Sergusov (1971) and Leavitt & Mullin (1981) proved the following elementary fact: $n = pp'$ where (p, p') is a pair of twin primes, if and only if $\phi(n)\sigma(n) = (n - 3)(n + 1)$. Here $\sigma(n)$ denotes the sum of all divisors of n.

The main problems is to ascertain whether there exist infinitely many twin primes.

For every $x > 1$, let $\pi_2(x)$ denote the number of primes p such that $p \leqslant x$ and $p + 2$ is also a prime. [With the notation previously introduced, $\pi_2(x) = \pi_{X,X+2}(x + 2)$.]

Brun announced in 1919 that there exists an effectively computable integer x_0 such that, if $x \geqslant x_0$, then

$$\pi_2(x) < \frac{100x}{(\log x)^2} .$$

The proof appeared in 1920.

In another paper of 1919, Brun had a weaker estimate for $\pi_2(x)$, but he also proved the famous result that

$$\Sigma \left[\frac{1}{p} + \frac{1}{p + 2} \right] \text{ (sum for all primes } p \text{ such that } p+2 \text{ is also a prime)}$$

converges, which expresses the scarcity of twin primes, even if there are infinitely many of them. The sum

$$B = \left[\frac{1}{3} + \frac{1}{5} \right] + \left[\frac{1}{5} + \frac{1}{7} \right] + \left[\frac{1}{11} + \frac{1}{13} \right] + \cdots + \left[\frac{1}{p} + \frac{1}{p + 2} \right] + \cdots$$

is now called Brun's constant. Based on heuristic considerations about the distribution of twin primes, B has been calculated, for example, by Shanks & Wrench (1974) and by Brent (1976):

$$B = 1.90216054... .$$

Brun also proved that for every $m \geqslant 1$ there exist m successive primes which are not twin primes.

The estimate of $\pi_2(x)$ has been refined by a determination of the constant and of the size of the error. This was done, among others, by Bombieri & Davenport, in 1966. It is an application of sieve methods and its proof may be found, for example, in the book of Halberstam & Richert.

Here is the result:

$$\pi_2(x) \leqslant 8 \prod_{p>2} \left[1 - \frac{1}{(p-1)^2}\right] \frac{x}{(\log x)^2} \left\{1 + O\left[\frac{\log \log x}{\log x}\right]\right\}.$$

The constant 8 has been improved to $\frac{68}{9} + \epsilon$ by Fouvry & Iwaniec (1983) and further by Bombieri, Friedlander & Iwaniec (1986) to $7 + \epsilon$. Actually, it was conjectured by Hardy & Littlewood that the factor should be 2 instead of 8.

The infinite product

$$C_2 = \prod_{p>2} \left[1 - \frac{1}{(p-1)^2}\right]$$

is called the twin-prime constant; its value is 0.66016... . It had been calculated by Wrench in 1961 and I shall explain the method in the similar calculation of Artin's constant, in Chapter 6.

To give a feeling for the growth of $\pi_2(x)$, I reproduce below part of the calculations of Brent (1975, 1976):

x	$\pi_2(x)$
10^3	35
10^4	205
10^5	1224
10^6	8169
10^7	58980
10^8	440312
10^9	3424506
10^{10}	27412679
10^{11}	224376048

Record

The largest exact value for the number of twin primes below a given limit has been published by Brent in 1976:

$$\pi_2(10^{11}) = 224,376,048.$$

At that time, with his well-known generosity, Shanks wanted to give one pair of twins to every American (see page 219 of his book, 2nd edition). To keep this goal a reality, he will need more help from Australia.

Any extension in the calculation of lists of primes automatically provides an improvement of this record.

Here is a list (compiled by Yates) of all known pairs of twin primes having at least 1000 digits.

For his accomplishment, Dubner used his low cost and very performing computer, especially adapted to number theory.

Twin Primes	Number of Digits	Discoverer	Year
$107570463 \times 10^{2250} \pm 1$	2259	Dubner (verified by Atkin)	1985
$43690485351513 \times 10^{1995} \pm 1$	2009	Same	1985
$520995090 \times 2^{6624} \pm 1$	2003	Atkin & Rickert	1984
$519912 \times 10^{1420} \pm 1$	1426	Dubner (verified by Atkin)	1984
$217695 \times 10^{1404} \pm 1$	1410	Same	1984
$219649815 \times 2^{4481} \pm 1$	1358	Atkin & Rickert	1983
$1639494 \times (2^{4423}-1) \pm 1$	1338	Keller	1983
$2445810 \times (2^{4253}-1) \pm 1$	1287	Same	1983
$218313 \times 10^{1068} \pm 1$	1074	Dubner (verified by Atkin)	1985
$499032 \times 10^{1040} \pm 1$	1046	Same	1984
$403089 \times 10^{1040} \pm 1$	1046	Same	1984
$256200945 \times 2^{3426} \pm 1$	1040	Atkin & Rickert	1980

Sieve theory has been used in attempts to prove that there exists infinitely many twin primes. Many authors have worked with this method.

To begin, in his famous paper of 1920, Brun showed that 2 may be written, in infinitely many ways, in the form $2 = m - n$, where m, n are 9-almost-primes (see the definition of almost-primes in Section II E). This was soon improved by Rademacher (1924), with 9 replaced by 7.

Later, Rényi showed in 1947 that there exists $k \geqslant 1$ such that 2 may be written in infinitely many ways, in the form $2 = m - p$, where p is a prime and m is a k-almost-prime.

The best result to date, with sieve methods, is due to Chen (announced in 1966, published in 1973, 1978); he proved that in Rényi's result k may be taken equal to 2; so $2 = m - p$, with $m \in P_2$ with p prime, in infinitely many ways.

The sieve methods used for the study of twin primes are also appropriate for the investigation of Goldbach's conjecture (see Section VI).

Addendum on Polignac's Conjecture

The general Polignac's conjecture can be, in part, treated like the twin-primes conjecture.

For every $k \geqslant 1$ and $x > 1$, let $\pi_{2k}(x)$ denote the number of integers $n > 1$ such that $p_n \leqslant x$ and $p_{n+1} - p_n = 2k$.

With Brun's method, it may be shown that there exists a constant $C_k > 0$ such that

$$\pi_{2k}(x) < C_k \frac{x}{(\log x)^2} .$$

For each x, the values of $2k$ for which $\pi_{2k}(x)$ is maximal are the most likely differences between consecutive primes, up to x.

Let $k(x)$ be the integer defined by:

$$k(x) = \min\{k \geqslant 1 \mid \pi_{2k}(x) \text{ is maximal}\}.$$

By the prime number theorem, there exists a constant $C' > 0$ such that

$$C' \frac{x}{(\log x)^2} < \pi_{2k(x)}(x).$$

In 1980, Erdös & Straus studied the behaviour of the function $k(x)$, as x tends to infinity. Their results depended on the following conjecture of Hardy & Littlewood (1923): there exists a constant $C > 0$ such that for every $k \geqslant 1$:

$$\pi_{2k}(x) \sim (C + o(1)) \frac{x}{(\log x)^2} \prod_{\substack{p>2 \\ p|k}} \frac{p-1}{p-2}.$$

This implies, in conjunction with the preceding estimate, that $\lim_{x\to\infty} k(x) = \infty$, which means that there is no most likely difference between consecutive primes.

IV. Primes in Arithmetic Progression

A. There Are Infinitely Many!

A classical and most important theorem was proved by Dirichlet in 1837. It states:

   If $d \geqslant 2$ and $a \neq 0$ are integers that are relatively prime, then the arithmetic progression

$$a, \quad a + d, \quad a + 2d, \quad a + 3d, \ ...$$

contains infinitely many primes.

   Many special cases of this theorem were already known — apart from Euclid's theorem (when $a = 1$, $d = 2$).

   Indeed, if $d = 4$ or $d = 6$ and $a = -1$, the proof is very similar to Euclid's proof.

   Using simple properties of quadratic residues, it is also easy to show that each of these arithmetic progressions contain infinitely many primes:

$d = 4$, $a = 1$;
$d = 6$, $a = 1$;
$d = 3$, $a = 1$;
$d = 8$, $a = 3$, or $a = 5$, or $a = 7$ (this includes the progressions with $d = 4$);

$d = 12$, $a = 5$, or $a = 7$, or $a = 11$ (this includes also the progressions with $d = 6$).

For $d = 8, 16, ...$, or more generally $d = 2^r$, and $a = 1$, the ingredients of a simple proof are: to consider $f(N)$ where

$$f(X) = X^{2^{r-1}} + 1, \quad N = 2p_1 p_2 \cdots p_n \,,$$

each p_i prime, $p_i \equiv 1 \pmod{2^r}$, and then to use Fermat's little theorem. These are hints for a reader wanting to find this proof by himself.

The proof when d is arbitrary and $a = 1$ is also elementary and provides a good pretext to mention the cyclotomic polynomials and some of their elementary properties.

For every $n \geq 2$, let

$$\zeta_n = \cos \frac{2\pi}{n} + i \sin \frac{2\pi}{n}.$$

Then $|\zeta_n| = 1$, $\zeta_n^n = 1$, so ζ_n is an nth root of 1. The argument of ζ_n is $2\pi/n$, hence $\zeta_n^k \neq 1$ for $k = 1, 2, ..., n-1$, and also $\zeta_n^k \neq \zeta_n^j$ for distinct exponents $k \neq j$, $1 \leq k < j \leq n-1$. Thus

$$X^n - 1 = \prod_{j=0}^{n-1} (X - \zeta_n^j).$$

A complex number ϵ is said to be a primitive nth root of 1 when $\epsilon^n = 1$ and $\epsilon, \epsilon^2, ..., \epsilon^{n-1}$ are different from 1. Thus, $\epsilon = \zeta_n^k$ where $\gcd(k,n) = 1$. The number of primitive nth roots of 1 is $\phi(n)$ (the Euler function at n).

The nth cyclotomic polynomial is, by definition,

$$\Phi_n(X) = \prod_{\substack{1 \leq k \leq n-1 \\ \gcd(k,n)=1}} (X - \zeta_n^k),$$

that is, $\Phi_n(X)$ is the polynomial of smallest degree having as roots all the primitive nth roots of 1. It is clear that

$$X^n - 1 = \prod_{m|n} \Phi_m(X).$$

Conversely, by using the Möbius function, the cyclotomic polynomials $\Phi_n(X)$ are expressible in terms of the binomials $X^m - 1$:

$$\Phi_n(X) = \prod_{m|n} (X^m - 1)^{\mu(n/m)}.$$

It is obvious that, if p is any prime, and $r \geqslant 1$, then

$$\Phi_p(X) = \frac{X^p - 1}{X - 1} = X^{p-1} + X^{p-2} + \cdots + X + 1,$$

$$\Phi_{p^r}(X) = \frac{X^{p^r} - 1}{X^{p^{r-1}} - 1} = X^{p^{r-1}(p-1)} + X^{p^{r-1}(p-2)} + \cdots + X^{p^{r-1}} + 1.$$

For every $n \geqslant 1$, $\Phi_n(X)$ is a monic polynomial with integer coefficients.

Also, it is not hard to show:

If $n > 1$, then $\Phi_n(0) = 1$,

$$\Phi_n(1) = \begin{cases} p & \text{when } n \text{ is a power of } p, \\ 1 & \text{otherwise,} \end{cases}$$

$|\Phi_n(a)| > 1$ for every $a > 1$.

The following identities turn out to be useful:

If a prime p divides m, then $\Phi_{pm}(X) = \Phi_m(X^p)$.
If a prime p does not divide m, and $r \geqslant 1$, then

$$\Phi_{mp^r}(X) = \frac{\Phi_m(X^{p^r})}{\Phi_m(X^{p^{r-1}})}.$$

Legendre showed in 1830:

If $a > 1$, $n \geqslant 2$, then the following three sets of primes coincide:

$E_1 = \{p \mid p$ divides $a^n - 1$, but p does not divide $a^m - 1$ for every m, $1 \leqslant m \leqslant n - 1\}$,
$E_2 = \{p \mid p \equiv 1 \pmod{n}, p$ divides $\Phi_n(a)\}$,
$E_3 = \{p \mid p$ does not divide n, p divides $\Phi_n(a)\}$.

That the set E_1 is contained in E_2 was already seen in Chapter 2,

Section II, so the only point to establish is that E_3 is contained in E_1. Please, do it. This is all that is required to prove Dirichlet's theorem in the special case $a = 1$:

For every $d \geqslant 2$ there exist infinitely many primes p in the arithmetic progression $1, 1 + d, 1 + 2d, ..., 1 + kd, ...$.

Proof. Let $p_1, ..., p_r$ be primes such that $p_i \equiv 1 \pmod{d}$ for $i = 1, 2, ..., r$. Let $N = dp_1 ... p_r$, so $|\Phi_d(N)| > 1$.

Let p be a prime dividing $\Phi_d(N)$. Since $\Phi_d(N) \equiv \Phi_d(0) \equiv 1 \pmod{N}$, p does not divide N; hence, p does not divide d. By Legendre's result above, $p \equiv 1 \pmod{d}$ and clearly $p \neq p_1, p_2, ..., p_r$. This shows that there must exist infinitely many primes in the arithmetic progression $\{1 + kd \mid k \geqslant 1\}$. $\square$

The above elegant proof goes back to Wendt (1895), but is of course subsumed, as a special case, in Dirichlet's theorem. Other proofs have been given by Nagell (1951), Rotkiewicz (1961), Estermann (1963), Sierpiński (1964), Niven & Powell (1976), and certainly many others.

Concerning the arithmetic progresssions $\{kd - 1 \mid k \geqslant 1\}$ there have also been special proofs. It is possible to give a proof based on properties of cyclotomic polynomials — however, this involves more than I want to present here and the proof is laborious, not transparent (see, for example, Nagell's or Landau's book).

On the other hand, I came across a charming proof by Powell (1977), which connects with the numbers of certain Lucas sequences. However, it is only good for the arithmetic progressions $\{kq - 1 \mid k \geqslant 1\}$, where q is any odd prime.

To say the truth, the proof for the arithmetic progression $\{kq^r - 1 \mid k \geqslant 1\}$ (q an odd prime, $r \geqslant 1$) is Theorem XXIX in Carmichael's paper of 1913 (quoted in Chapter 2). But it would be inconvenient to look it up there, as it comes at the end of a long paper. So, I prefer to present the proof rediscovered by Powell, which is self-contained.

As in Chapter 2, Section IV, let P, Q be nonzero relatively prime integers, such that $D = P^2 - 4Q \neq 0$. The roots of $X^2 - PX + Q$ are $\alpha, \beta = (P \pm \sqrt{D})/2$; so $\alpha + \beta = P$, $\alpha\beta = Q$, $\alpha - \beta = \sqrt{D}$.

Consider the associated Lucas sequences, defined recursively by $U_0 = 0$, $U_1 = 1$, $U_n = PU_{n-1} - QU_{n-2}$ (for $n \geqslant 2$) and by $V_0 = 2$, $V_1 = P$, $V_n = PV_{n-1} - QV_{n-2}$ (for $n \geqslant 2$). Or, equivalently,

$$U_n = \frac{\alpha^n - \beta^n}{\alpha - \beta}, \quad V_n = \alpha^n + \beta^n, \text{ for } n \geqslant 0.$$

Recall the following properties, which were given in Chapter 2, Section IV:

(1) If m divides n, then U_m divides U_n;

(2) $\gcd(U_n, Q) = 1$ for every $n \geqslant 1$;

(3) if $\gcd(m, 2Q) = 1$, there exists the smallest index $\rho(m) = r$ such that m divides U_r; then m divides U_n if and only if r divides n.

I shall also need the following congruences, where p is an odd prime:

(4) $U_p \equiv (P^2 - 4Q)^{(p-1)/2} \pmod{p}$;

(5) $V_p \equiv P \pmod{p}$;

(6) $2U_{p+1} \equiv P(U_p + 1) \pmod{p}$;

(7) $2bU_{p-1} \equiv P(U_p - 1) \pmod{p}$.

For convenience, I give their proofs:

$$(\alpha - \beta)^p = (\alpha^p - \beta^p) - \binom{p}{1}\alpha\beta(\alpha^{p-2} - \beta^{p-2})$$

$$+ \binom{p}{2}(\alpha\beta)^2(\alpha^{p-4} - \beta^{p-4}) - \cdots \pm \binom{p}{\frac{p-1}{2}}(\alpha\beta)^{(p-1)/2}(\alpha^2 - \beta^2),$$

so

$$(\alpha - \beta)^{p-1} = U_p - \binom{p}{1}QU_{p-2} + \binom{p}{2}Q^2U_{p-4} - \cdots \pm \binom{p}{\frac{p-1}{2}}Q^{(p-1)/2}U_2.$$

Hence $U_p \equiv (P^2 - 4Q)^{(p-1/2)} \pmod{p}$. Similarly,

$$(\alpha + \beta)^p = (\alpha^p + \beta^p) + \binom{p}{1}\alpha\beta(\alpha^{p-2} + \beta^{p-2}) + \binom{p}{2}(\alpha\beta)^2(\alpha^{p-4} + \beta^{p-4})$$

$$+ \cdots + \binom{p}{\frac{p-1}{2}}(\alpha\beta)^{(p-1)/2}(\alpha^2 + \beta^2),$$

so

$$P^P = V_p + \begin{bmatrix} p \\ 1 \end{bmatrix} QV_{p-2} + \begin{bmatrix} p \\ 2 \end{bmatrix} Q^2V_{p-4} + \cdots + \begin{bmatrix} p \\ \frac{p-1}{2} \end{bmatrix} Q^{(p-1)/2}V_2.$$

Hence

$$V_p \equiv P \pmod{p}.$$

Next,

$$U_{p+1} = \frac{\alpha^P(\alpha - \beta) + (\alpha^P - \beta^P)\beta}{\alpha - \beta} = \alpha^P + \beta U_p,$$

and also $U_{p+1} = \beta^P + \alpha U_p$; hence

$$2U_{p+1} = (\alpha^P + \beta^P) + (\alpha + \beta)U_p = V_p + PU_p \equiv P(1 + U_p) \pmod{p}.$$

Finally, $U_{p+1} = PU_p - QU_{p-1}$, hence

$$2QU_{p-1} = 2PU_p - 2U_{p+1} \equiv P(U_p - 1) \pmod{p}.$$

Now, I give Powell's proof of the proposition:

For every odd prime q, there exist infinitely many primes p in the arithmetic progression

$$q - 1, \quad 2q - 1, \quad 3q - 1, \, ..., \, kp - 1, \, ... \, .$$

Proof. Suppose that $p_1, p_2, \, ..., \, p_m$ ($m \geq 0$) are the only primes of the form $kp - 1$. Recall that if h is such that $1 \leq h \leq q - 1$, then the Legendre symbol (h/q) is given by

$$\left[\frac{h}{q} \right] \equiv h^{(q-1)/2} \pmod{q}.$$

There are exactly $(q-1)/2$ nonquadratic residues modulo q, so there exists h, $1 \leq h \leq q - 1$, such that $h^{(q-1)/2} \equiv -1 \pmod{q}$.

Let $t = h\Pi_{j=1}^{m} p_j^2$ and $P = 2$, $Q = 1 - 4t$.

So $P^2 - 4Q = 16t$, and $\alpha = 1 + 2\sqrt{t}$, $\beta = 1 - 2\sqrt{t}$. The sequence of numbers $(U_n)_{n \geq 0}$ defined by $U_0 = 0$, $U_1 = 1$, $U_n = 2U_{n-1} - (1-4t)U_{n-2}$, satisfies the following congruences:

$$U_q \equiv (P^2 - 4Q)^{(q-1)/2} \equiv (16t)^{(q-1)/2} = 4^{q-1}\left[\prod_{j=1}^{m} p_j\right]^{q-1} h^{(q-1)/2}$$

$$\equiv -1 \pmod{q},$$

and

$$U_q = \frac{(1 + 2\sqrt{t})^q - (1 - 2\sqrt{t})^q}{4\sqrt{t}} \equiv q \pmod{4t}.$$

If p is any prime dividing U_q, then $p \neq q$, p does not divide $4t$, so $p \neq 2, p_1, p_2, \ldots, p_m$; hence $p \not\equiv -1 \pmod{q}$; also, p does not divide h.

Now

$$U_p \equiv (P^2 - 4Q)^{(p-1)/2} \equiv (16t)^{(p-1)/2} \equiv 4^{p-1}\left[\prod_{j=1}^{m} p_j\right]^{p-1} h^{(p-1)/2}$$

$$\equiv h^{(p-1)/2} \equiv \epsilon \pmod{p},$$

where $\epsilon = 1$ or -1.

If $\epsilon = 1$, then $U_p \equiv 1 \pmod{p}$; so p divides $2QU_{p-1}$.

Since p divides U_q, then by (2) p does not divide Q, hence p divides U_{p-1}. By (3), q divides $p - 1$. This means that $p \equiv 1 \pmod{q}$.

If this happens for every prime p dividing U_q, then $U_q \equiv 1 \pmod{q}$, which is not true.

So there exists a prime p such that p divides U_q, the corresponding value of ϵ is $\epsilon = -1$. By (6), p divides $2U_{p+1}$, so p divides U_{p+1}. By (3), q divides $p + 1$, that is $p \equiv -1 \pmod{q}$. This is a contradiction and concludes the proof.   ☐

The preceding proofs of special cases of Dirichlet's theorem are elementary in the following sense: the proof is done by associating an appropriate polynomial and studying the prime divisors of the values of this polynomial. It is natural to ask for which values of a, d, $1 \leqslant a < d$, $\gcd(a,d) = 1$, is there an elementary proof of the existence of infinitely many primes in $\{a + kd \mid k \geqslant 1\}$.

First of all, this leads to the consideration of the set $P(f)$, of all primes p dividing $f(n)$, for some integer n, where $f(X)$ is any non-constant polynomial with integral coefficients.

It was easy for Schur to show (in 1912) — but anyone could have done it — that if $f(X) = a_0 X^n + a_1 X^{n-1} + \cdots + a_n$, with integral coefficients, $n \geqslant 1$, $a_0 \neq 0$, then $P(f)$ is an infinite set.

Indeed, since there exist infinitely many primes, the statement is true when $a_n = 0$. Now let $a_n \neq 0$ and assume that $p_1, ..., p_r$ are the only primes in the set $P(f)$. Choose an integer c, with $|c|$ sufficiently large such that $f(ca_np_1 \cdots p_r) > |a_n|$.

Since $(1/a_n)f(ca_np_1 \cdots p_r) \equiv 1 \pmod{p_1 \cdots p_r}$, then there exists a prime p dividing the integer $(1/a_n)f(ca_np_1 \cdots p_r)$, and thus $p \in P(f)$, $p \neq p_1, ..., p_r$.

Nagell also proved (in 1923 and again in 1969) that if $f_1, ..., f_m$ are nonconstant polynomials with integral coefficients, then $\cap_{i=1}^m P(f_i)$ is also an infinite set. This seems to have been a popular theorem, as I noted several other proofs of it, by Fjellstedt (1955), Grölz (1969) and the simplest one by Hornfeck (1970); see also the paper by Gerst & Brillhart (1971) with many other interesting related results.

Incidentally, for the cyclotomic polynomials, Legendre's result (quoted above) indicates that, for every $n > 1$,

$$P(\phi_n) \subseteq \{p \text{ prime} \mid p \text{ divides } n\} \cup \{p \text{ prime} \mid p \equiv 1 \pmod{n}\};$$

so, in particular, from Schur's result it follows again that there exist infinitely many primes p such that $p \equiv 1 \pmod{n}$.

The fact that $P(f)$ is infinite, is a companion to the result proved in Chapter 3, Section II, that $f(X)$ assumes infinitely many composite values. On the other hand, as I mentioned in the Interlude (Section II, (E)), the statement that *every nonconstant irreducible polynomial with integral coefficients [and no prime dividing all its values, condition (*)] has at least one prime value* will be amply discussed in Chapter 6. It has never been proved.

In the third of a series of papers, Wójcik gave an elementary proof (in 1969) that the arithmetic progression $\{a + kd \mid k \geqslant 0\}$ contains infinitely many primes, provided $a^2 \equiv 1 \pmod{d}$. Another proof was given by Ram Murty (unaware of Wójcik's result) in his Bachelor Thesis (Carleton University, Ottawa, 1975). Ram Murty also proved Bateman & Low's conjecture (1975): if there is an elementary proof that $\{a + kd \mid k \geqslant 0\}$ has infinitely many primes, then $a^2 \equiv 1 \pmod{d}$.

Thus, for example, there is an elementary proof for the progression $\{4 + 15k \mid k \geqslant 1\}$, which uses the polynomial $f(X) = X^4 - X^3 + 2X^2 + X + 1$.

From properties of Euler's function, it is easy to see that if $d \geqslant 2$ is such that there is an elementary proof for every progression $\{a + kd \mid k \geqslant 1\}$ with $1 \leqslant a < d$, $\gcd(a,d) = 1$, then necessarily $d = 2$, 4, 8, 6, 12, 24. In their paper quoted above, Bateman & Low give elementary proofs for all arithmetic progressions with $d = 24$.

After this excursus into special cases of Dirichlet's theorem, which may be proved with elementary methods, I should mention that in order to deal with arbitrary arithmetic progressions $\{a + kd \mid k \geqslant 1$ and $\gcd(a,d) = 1\}$, it is necessary to consider the characters associated to the finite abelian group of residue classes modulo d. They possess a so-called "separation property," which allows for the consideration of all the individual arithmetical progressions with the same given difference. Each such character may be lifted to a modular character χ with modulus d, which is a function defined over the integers, with values which are dth roots of 1, and such that:

$$\begin{cases} \chi(n) = 0 \text{ if and only if } \gcd(n,d) \neq 1; \\ \text{if } n \equiv m \pmod{d}, \text{ then } \chi(n) = \chi(m); \\ \chi(nm) = \chi(n)\chi(m); \text{ in particular, } \chi(1) = 1. \end{cases}$$

The principal character χ_0 with modulus d is defined by $\chi_0(n) = 1$ when $\gcd(n,d) = 1$.

With the modular characters χ, Dirichlet considered the L-series

$$L(s|\chi) = \sum_{n=1}^{\infty} \frac{\chi(n)}{n^s} \quad \text{(for} \quad \mathrm{Re}(s) > 1).$$

Thus if $d = 2$ (a degenerate case) this gives the Riemann zeta-series.

The theory of L-functions associated with characters is developed following the pattern of the theory of Riemann zeta function. In particular, these functions are extended to meromorphic functions of the whole complex plane, satisfying an appropriate functional equation. The study of the location of zeroes of the L-functions is, of course, very important, and leads to a generalized Riemann hypothesis. However, the discussion of these topics would lead too far away from my main purpose. The reader may wish to consult the book of Prachar (1957), listed in the General References at the Bibliography.

An essential point in the proof of Dirichlet's theorem is to establish that $L(1|\chi) \neq 0$ when $\chi \neq \chi_0$.

A detailed discussion of Dirichlet's theorem, with several variants of proofs, is in Hasse's book *Vorlesungen über Zahlentheorie* (now available in English translation, from Springer-Verlag).

In 1949, Selberg gave an elementary proof of Dirichlet's theorem, similar to his proof of the prime number theorem.

Concerning Dirichlet's theorem, de la Vallée Poussin established the following additional density result. For a, d as before, and $x \geqslant 1$ let

$$\pi_{d,a}(x) = \#\{p \text{ prime} \mid p \leqslant x, \ p \equiv a \pmod{d}\}.$$

[Observe that $\pi_{d,a}(x) = \pi_{dX+a}(x)$, with the notation previously introduced.]

Then

$$\pi_{d,a}(x) \sim \frac{1}{\phi(d)} \cdot \frac{x}{\log x}.$$

Note that the right-hand side is the same, for any a, such that $\gcd(a,d) = 1$.

It follows that

$$\lim_{x \to \infty} \frac{\pi_{d,a}(x)}{\pi(x)} = \frac{1}{\phi(d)},$$

and this may be stated by saying that the set of primes in the arithmetic progression $\{a + kd \mid k \geqslant 1\}$ has natural density $1/\phi(d)$ (with respect to the set of all primes).

Despite the fact that the asymptotic behavior of $\pi_{d,a}(x)$ is the same, for every a, $1 \leqslant a < d$, with $\gcd(a,d) = 1$, Tschebycheff had noted, already in 1853, that $\pi_{3,1}(x) < \pi_{3,2}(x)$ and $\pi_{4,1}(x) < \pi_{4,3}(x)$ for small values of x; in other words, there are more primes of the form $3k + 2$ than of the form $3k + 1$ (resp. more primes $4k + 3$ than primes $4k + 1$) up to x (for x not too large). Are these inequalities true for every x? The situation is somewhat similar to that of the inequality $\pi(x) < \text{Li}(x)$. Once again, as in Littlewood's theorem, it may be shown that these inequalities are reversed infinitely often. Thus, Leech computed in

1957 that $x_1 = 26861$ is the smallest prime for which $\pi_{4,1}(x) > \pi_{4,3}(x)$; see also Bays & Hudson (1978), who found that $x_1 = 608,981,813,029$ is the smallest prime for which $\pi_{3,1}(x) > \pi_{3,2}(x)$. About this question, see also Shanks' paper (1959) and Bays & Hudson (1979).

On the average $\pi_{3,2}(x) - \pi_{3,1}(x)$ is asymptotically $x^{1/2}/(\log x)$.

However, Hudson (with the help of Schinzel) showed in 1985 that

$$\lim_{x\to\infty} \frac{\pi_{3,2}(x) - \pi_{3,1}(x)}{\dfrac{x^{1/2}}{\log x}}$$

does not exist (in particular, it is not equal to 1).

For the computation of the exact number of primes $\pi_{d,a}(x)$, below x in the arithmetic progression $\{a + kd \mid k \geqslant 1\}$, Hudson derived in 1977, a formula similar to Meissel's formula for the exact value of $\pi(x)$. In the same year, Hudson & Brauer studied in more detail the particular arithmetic progressions $4k \pm 1$, $6k \pm 1$.

In 1964, Kapferer indicated the following corollary of Dirichlet's theorem:

If $\gcd(a,d) = 1$ but 2 divides ad, then there exists an infinite sequence of pairwise relatively prime integers $n_1, n_2, n_3, \ldots$, such that $a + n_i d = p_i$ is a prime (for every $i \geqslant 1$).

Proof. To begin, let $A_0 = a + d$, $D_0 = 2d(a - d)$. Then $\gcd(A_0, 2) = \gcd(A_0, d) = \gcd(A_0, a - d) = 1$, so $\gcd(A_0, D_0) = 1$. By Dirichlet's theorem, there exists $m_1 \geqslant 1$ such that $A_0 + m_1 D_0 = p_1$ is a prime. But $A_0 + m_1 D_0 = a + n_1 d$ with $n_1 = 1 + 2m_1(a - d)$, so $n_1 \equiv 1$ (mod $a{-}d$).

The proof proceeds by induction. Assume that $k \geqslant 1$ and $n_1, n_2, \ldots, n_k$ have already been found, so that they are pairwise relatively prime and $a + n_i d = p_i$ is a prime (for $i = 1, \ldots, k$).

Let $N_k = n_1 . n_2 \ldots n_k$ so $N_k \equiv 1$ (mod $a - d$). Let $A_k = a - d + 2dN_k$, $D_k = 2dN_k(a - d)$. Then $\gcd(A_k, 2) = 1$, $\gcd(A_k, d) = 1$, $\gcd(A_k, N_k) = \gcd(a - d, N_k) = 1$, and also $\gcd(A_k, a - d) = \gcd(A_k, 2dN_k) = 1$. So $\gcd(A_k, D_k) = 1$. By Dirichlet's theorem, there exists $m_{k+1} \geqslant 1$ such that $A_k + m_{k+1} D_k = p_{k+1}$ is a prime. But $A_k + m_{k+1} D_k = a + n_{k+1} d$

with $n_{k+1} = 2N_k - 1 + 2m_{k+1}N_k(a - d)$, so $n_{k+1} \equiv 1 \pmod{a - d}$ and $\gcd(n_{k+1},n_i) = 1$ for $i = 1, ..., k$. The proof may now be completed by induction. $\square$

A natural temptation is to try to prove:

If $\gcd(a,d) = 1$ and 2 divides ad, there exist infinitely many primes $q_1,q_2,q_3, ...,$ such that $p_i = a + q_i d$ is a prime (for $i \geqslant 1$).

This seems out of reach of present day methods. In fact, I don't even know how to prove that there is even one prime q such that $p = a + qd$ is also a prime.

Taking $a = 2$, $d = 1$, the statement is nothing short than the existence of infinitely many pairs of twin primes.

Taking $a = 1$, $d = 2$, the above statement reads:

There exist infinitely many primes q such that $2q + 1$ is also a prime.

I shall treat this question in Chapter 5, Section II.

Here are some other corollaries of Dirichlet's theorem, which are related to Fermat numbers, Mersenne numbers, and Polignac's conjecture. They dispel any hasty or optimistic thoughts about the primality of those numbers, or the validity of the conjecture. The proofs were given by Schinzel & Sierpiński, and are just a simple exercise (see Schinzel, 1959).

Given $m \geqslant 1$, if n is any integer such that $2^{n+1} \geqslant q$ for every prime factor of m, then each one of the Fermat numbers $F_n = 2^{2^n} + 1$ is congruent modulo m to infinitely many primes.

Yet, this does not suffice to conclude that F_n is a prime.

Similarly, given $m \geqslant 1$, if p is any prime such that $p \geqslant q$ for every prime factor of m, then each one of the Mersenne numbers $M_p = 2^p - 1$ is congruent modulo m to infinitely many primes.

Once more, this does not suffice to conclude that M_p is a prime.

Concerning Polignac's conjecture, given k, m, positive integers, there exist infinitely many pairs of primes (p,q) such that $2k \equiv p - q \pmod{m}$. But nothing may be inferred about the representation of $2k$ as the difference of two primes.

B. The Smallest Prime in an Arithmetic Progression

I return now to Dirichlet's theorem, to stress one aspect which was left untouched. Namely, with $d \geqslant 2$ and $a \geqslant 1$, relatively prime, let $p(d,a)$ be the smallest prime in the arithmetic progression $\{a + kd \mid k \geqslant 0\}$. Can one find an upper bound, depending only on a, d, for $p(d,a)$?

Let $p(d) = \max\{p(d,a) \mid 1 \leqslant a < d, \gcd(a,d) = 1\}$. Again, can one find an upper bound for $p(d)$ depending only on d? And how about lower bounds?

Estimates of the size of $p(d,a)$ are among the most difficult questions, and depend very closely on the zeroes of the L-functions associated with the characters modulo d. There are several conjectures, a number of results involving the generalized Riemann hypothesis for the zeroes of the L-functions — and there is, above all, a brilliant theorem of Linnik.

To begin, I shall indicate upper bounds. The results about $p(d,a)$ are, unfortunately, somewhat inconclusive.

First, just by looking at the prime factors of Mersenne numbers $M_q = 2^q - 1$ (with q prime), which must be of the form $p \equiv 1 \pmod{q}$, Chowla noted in 1934, that $p(q,1) < 2^q$.

Niven & Powell also noted (1976), from their proof of existence of infinitely many primes p, $p \equiv 1 \pmod{d}$, that $p(d,1) < d^{3d}$ (when $d > 2$).

Of course these results are very weak, but also very simple to prove.

In a more sophisticated tone, in 1937 Turán showed, using the generalized Riemann hypothesis for L-series:

For every $\epsilon > 0$, for every $d \geqslant 2$:
$$\#\{a \mid 1 \leqslant a < d, \gcd(a,d) = 1, p(d,a) < \phi(d)(\log d)^{2+\epsilon}\} = \phi(d) - o(\phi(d)).$$

Using Brun's sieve, Erdös showed in 1950 that for every $d \geqslant 2$ and for every $C > 0$ there exists $C' > 0$ (depending on C) such that
$$\frac{\#\{a \mid 1 \leqslant a < d, \gcd(a,d) = 1, p(d,a) < C\phi(d)\log d\}}{\phi(d)} \geqslant C';$$

that is, for a positive proportion of values of a, the inequality for $p(d,a)$ holds.

Elliott & Halberstam did not require the generalized Riemann hypothesis to show in 1971:

For every $\epsilon > 0$, for every $d \geqslant 2$, not belonging to a sequence of density 0:

$$\#\{a|\ 1 \leqslant a < d,\ \gcd(a,d) = 1,\ p(d,a) < \phi(d)(\log\ d)^{1+\epsilon}\} = \phi(d) - o(\phi(d)).$$

In the above results, the bound indicated for $p(d,a)$ is not valid for every a, $1 \leqslant a < d$, $\gcd(a,d) = 1$; so, it is not an upper bound for $p(d)$.

In 1970, Motohashi showed that for every $a \geqslant 1$ and for every $\epsilon > 0$ there exists $C = C(\epsilon) > 0$ such that for infinitely many primes q:

$$p(q,a) < C(\epsilon)q^{\theta+\epsilon},$$

where

$$\theta = \frac{2e^{1/4}}{2e^{1/4} - 1} \cong 1.63773 \ ... \ .$$

Last minute information (1987): Bombieri, Friedlander & Iwaniec have proved a very strong theorem, from which it may be deduced the following estimate for $p(d,a)$ (note that $a \geqslant 1$ is given): $p(d,a) < d^2/(\log\ d)^k$, for every $k > 0$, $d \geqslant 2$, $\gcd(d,a) = 1$ and d is outside a set of density 0.

Now, I shall indicate results and conjectures about upper bounds for $p(d)$.

In 1934, using the generalized Riemann hypothesis, Chowla showed that, for every $\epsilon > 0$ there exists $d(\epsilon) > 0$ such that $p(d) < d^{2+\epsilon}$ for every $d \geqslant d(\epsilon)$, and he also conjectured that $p(d) < d^{1+\epsilon}$, for every sufficiently large $d \geqslant 2$.

Again with the generalized Riemann hypothesis, Heath-Brown indicated that

$$p(d) \leqslant C(\phi(d))^2\ (\log\ d)^4;$$

this follows from the work of Titchmarsh (1930).

Linnik's theorem of 1944, which is one of the deepest results in analytic number theory asserts:

There exists $L > 1$ such that $p(d) < d^L$ for every sufficiently large $d \geqslant 2$.

Note that the absolute constant L, called Linnik's constant, is effectively computable.

It is clearly important to compute the value of L. Pan (Cheng-Dong) was the first to evaluate Linnik's constant, giving $L \leqslant 5448$ in 1957. Subsequently, a number of papers have appeared, where the estimate of the constant was improved:

777, by Chen in 1965,
550, by Jutila in 1970,
168, by Chen in 1977,
 80, by Jutila in 1977,
 36, by Graham in 1981,
 20, by Graham in 1982,
 17, by Chen in 1979.

Concerning the last two articles, see Graham's note added to his paper of 1981. Recently, Chen and Liu (Jianmin) have claimed (privately) that $L \leqslant 13.5$.

Schinzel & Sierpiński (1958) [see Chapter 6, Section II conjecture (S')] and Kanold (1963) have conjectured that $L = 2$, that is, $p(d) < d^2$ for every sufficiently large $d \geqslant 2$.

Explicitly, this means that if $1 \leqslant a < d$, $\gcd(a,d) = 1$, there is a prime number among the numbers $a, a + d, a + 2d, ..., a + (d - 1)d$.

Kanold elaborated on his conjecture in two notes, written in 1964 and 1965. His argument was based on the following function considered by Jacobsthal in 1960 (three articles). For every $m \geqslant 1$, consider the smallest integer $r \geqslant 1$ such that r consecutive integers contain at least one integer a, relatively prime to m. Define the Jacobsthal function by $g(m) = r$.

Thus $g(1) = 1$, $g(p) = 2$ for every prime, $g(2^e) = 2$ for every $e \geqslant 1$. Kanold showed also that if

$$m = 2^e \prod_{i=1}^{r} p_i^{e_i} \quad \text{with} \quad e \geqslant 1, \, e_1 \geqslant 1, \, ..., \, e_r \geqslant 1,$$

then $g(m) = 2g(p_1 p_2 \cdots p_r)$.

The growth of $g(m)$ is difficult to study. Erdös has estimated the Jacobsthal function in 1962, and among other results, he proved: except for a set of integers of density 0,

$$g(m) = \frac{m}{\phi(m)} \, \omega(m) + o(\log \log \log m)$$

[where $\omega(m)$ denotes the number of distinct prime factors of m].

The best estimate to date is by Iwaniec (1978):

$$g(m) < C \, \frac{m}{\phi(m)} \, \omega(m) \log^2 \omega(m) \quad \text{(for some constant } C > 0\text{)}.$$

Here are Kanold's results, relating Jacobsthal's function with the value of $p(d)$:

Let $d \geqslant 2$ and $m_d = \prod_{\substack{p < d \\ p \nmid d}} p$. If $g(m_d) < d$, then $p(d) < d^2$.

So the question becomes: for which values of d is $g(m_d) < d$? Kanold tried to show that this happens for d sufficiently large, using Erdös' estimate of $g(m_d)$. No wonder that he did not succeed, because Pomerance established in 1980 that $g(m_d) \geqslant d$ for all sufficiently large d.

In his 1965 note, Kanold showed how to deduce Linnik's theorem, and from this, how to prove Dirichlet's theorem, starting with the following conjecture about the Jacobsthal function:

There exists a constant C, $0 < C < 2$, such that for every n,

$$g(p_1 p_2 \cdots p_n) < p_n^{2-C}$$

(where $p_1 < p_2 < \cdots < p_n < \cdots$ is the increasing sequence of primes).

Needless to say that this conjecture, if true, should be difficult to prove.

This approach to Linnik's theorem will demand a more refined study of Jacobsthal's function.

Heath-Brown has advanced in 1978 the conjecture that $p(d) \leqslant Cd(\log d)^2$ and Wagstaff sustained in 1979 that $p(d) \sim \phi(d)(\log d)^2$ on heuristic grounds.

In his Bachelor's thesis (1977), Kumar Murty made use of the important sieve results of Bombieri and Vinogradov (see Halberstam & Richert's book, of 1974) to establish that $p(d) < d^{2+\epsilon}$ for every $\epsilon > 0$ and for every d not belonging to a sequence of density zero. This is the old result of Chowla, without appealing to the generalized Riemann hypothesis, but only valid for almost all $d \geqslant 2$.

Concerning the lower bounds for $p(d)$, I first mention the following result of Schatunowsky (1893), obtained independently by Wolfskehl[*] in 1901.

$d = 30$ is the large integer with the following property: if $1 \leqslant a < d$ and $\gcd(a,d) = 1$, then $a = 1$ or a is prime.

The proof is elementary, and may be found in Landau's book *Primzahlen* (1909), page 229. It follows at once that if $d > 30$ then $p(d) > d + 1$.

From the prime number theorem, it follows already that for every $\epsilon > 0$ and for all sufficiently large d:

$$p(d) > (1 - \epsilon)\phi(d)\log d.$$

This implies that

$$\lim \inf \frac{p(d)}{\phi(d)\log d} \geqslant 1.$$

In his 1950 paper in Szeged, Erdos showed that

---

[*] Paul Wolfskehl is usually remembered as the rich mathematician who endowed a substantial prize for the discovery of a proof of Fermat's last theorem. I tell the story in my book *13 Lectures on Fermat's Last Theorem*. Here, I wish to salute his memory, on behalf of all the young assistants, who in the last 80 years, have been thrilled finding mistakes in an endless and continuous flow of "proofs" of Fermat's theorem.

$$\lim \inf \frac{p(d)}{\phi(d)\log d} > 1.$$

Actually, Erdős proved more: for every $\epsilon > 0$ there exist infinitely many $d \geq 2$ such that there exists $C > 0$ (depending only on ϵ) for which

$$\frac{\#\{a \mid 1 \leq a < d,\ \gcd(a,d) = 1,\ \frac{p(d,a)}{\phi(d)\log d} > 1 + \epsilon\}}{\phi(d)} \geq C;$$

it is suggestive to say, that for those values of d there is a positive proportion of values of a for which the inequality holds.

Moreover, $p(d)/[\phi(d) \log d]$ is unbounded. But this follows also from the next lower estimates.

In 1961 and 1962, Prachar and Schinzel established: there exists an absolute constant $C > 0$ such that for every $a \geq 1$

$$p(d,a) > Cd(\log d)\frac{(\log_2 d)(\log_4 d)}{(\log_3 d)^2}$$

holds for infinitely many $d \geq 2$ with $\gcd(d,a) = 1$ (as already indicated, $\log_2 d = \log \log d$, $\log_3 d = \log \log \log d$, $\log_4 d = \log \log \log \log d$).

Wagstaff contributed to the subject with two papers. In 1978 he proved: for every $\epsilon > 0$ and every $a \geq 1$ there exists a set Q of primes, with $\Sigma_{p \in Q}(1/p) < \infty$, such that if q is any prime not in the set Q,

$$p(q,a) > q\frac{\log q}{(\log \log q)^{1+\epsilon}}.$$

In his second paper, Wagstaff proclaimed and gave a heuristic justification that, after disregarding a set of density zero, then $p(d) \sim \phi(d) \log d \log \phi(d)$ (as d increases to infinity).

The next step is from Pomerance.

Let Q be the set of all integers $d \geq 2$ with more than $e^{(\log_2 d)/(\log_3 d)}$ distinct prime factors. The set Q has density zero. In 1980, Pomerance proved that

$$\lim \inf \frac{p(d)}{\phi(d)\log d} \geq e^\gamma,$$

and that for every $d \geq 2$, outside the set Q:

$$p(d) > [e^\gamma + o(1)]\phi(d) \log d \, \frac{(\log_2 d)(\log_4 d)}{(\log_3 d)^2}.$$

In particular, for $d \notin Q$, then

$$\lim \frac{p(d)}{\phi(d)\log d} = \infty.$$

Another problem, which has been dealt with by Richert, concerns the existence of k-almost-primes in arithmetic progressions. If $d \geq 1$, $1 \leq a < d$, $\gcd(a,d) = 1$, and $k \geq 1$ let $P_k(d,a)$ be the smallest k-almost-prime in the arithmetic progression $\{a + nd \mid n \geq 1\}$; so $P_1(d,a)$ coincides with $p(d,a)$. Using Selberg weighted sieves, Richert proved in 1969 that, for $k \geq 2$ and d sufficiently large and a arbitrary:

$$P_k(d,a) \leq d^{1 + \frac{1}{k-(9/7)}}.$$

For special values $k = 2, 3, 4$, actually $P_2(d,a) \leq d^{11/5}$, $P_3(d,a) \leq d^{11/7}$, $P_4(d,a) \leq d^{11/8}$. For this matter, see Halberstam & Richert's book. In 1978, Heath-Brown improved the preceding result about 2-almost-primes. Namely, he proved that the number of 2-almost-primes n, such that $n \equiv a \pmod{d}$ and $n \leq d^{2-\delta}$ with $\delta \leq 0.035$, is at least

$$C \frac{d^{2-\delta}}{\log d} \quad \text{(with } C > 0\text{)}.$$

In particular, $P_2(d,a) \leq d^{2-\delta}$; this holds for all sufficiently large d.

C. Strings of Primes in Arithmetic Progression

Now I consider the question of existence of sequences of k primes $p_1 < p_2 < p_3 < \cdots < p_k$ with difference $p_2 - p_1 = p_3 - p_2 = \cdots = p_k - p_{k-1}$, so that these primes are in arithmetic progression.

In 1939, van der Corput proved that there exist infinitely many sequences of three primes in arithmetic progression (see Section VI); this was proved again by Chowla in 1944.

Even though it is easy to give examples of four (and sometimes more) primes in arithmetic progression, the following question remains open:

Is it true that for every $k \geqslant 4$, there exist infinitely many arithmetic progressions consisting of k primes?

In examining the first case, $k = 4$, one already discovers how very difficult this question is. In fact, to date the best result was obtained by Heath-Brown, who showed in 1981 that there exist infinitely many arithmetic progressions consisting of four numbers, of which three are primes and the other is of type P_2 (a prime, or the product of two, not necessarily distinct, prime factors).

It is conjectured that for every $k > 3$ there exists at least one arithmetic progression consisting of k prime numbers.

There have been extensive computer searches for long strings of primes in arithmetic progression.

Record

The longest known string of primes in arithmetic progression contains 19 terms; its difference is 4,180,566,390; its smallest term is 8,297,644,387, its largest term is 83,547,839,407.

It was discovered by Pritchard (1985) after truly enormous computations (14000 hours of computer time). The previous records were by Pritchard (1982) and Weintraub (1977), and contained 18 terms, resp. 17 terms.

Pritchard also listed the known strings of m primes in arithmetical progression with minimal last term, as well as the estimated size of the last term (for every $m = 2, 3, ..., 19$).

More work on this question has been done by Grosswald & Hagis (1979).

In this connection the following statement, due to M. Cantor (1861), and quoted in Dickson's *History of the Theory of Numbers*, Vol. I, p. 425, is easy to prove:

Let $d \geqslant 2$, let $a, a + d, ..., a + (n - 1)d$ be n prime numbers in arithmetic progression. Let q be the largest prime such that $q \leqslant n$. Then either $\prod_{p \leqslant q} p$ divides d, or $a = q$ and $\prod_{p < q}$ divides d.

Proof. First an easy remark. If p is a prime, not dividing d, and if

$a, a + d, ..., a + (p - 1)a$ are primes, then these numbers are pairwise incongruent modulo p and p divides exactly one of these numbers. Now assume that $\Pi_{p \leqslant q} p$ does not divide d, so that there exists a prime $p \leqslant n$ such that p does not divide d. Choose the smallest such prime p. By the remark, there exists j, $0 \leqslant j \leqslant p - 1$, such that p divides $a + jd$, so $p = a + jd$, since p is a prime number. But a is a prime; if $a \neq a + jd$, then a divides d (by the choice of p), so a divides p, that is $a = p = a + jd$. This proves that $p = a$. If $p < q$, then $p \leqslant n - 1$, so p divides $a + pd$, hence $p = a + pd = p(1 + d)$, which is an absurdity.

I have therefore established that if $\Pi_{p \leqslant q} p$ does not divide d then $q = a$ and $\Pi_{p < q} p$ divides d.   □

A particular case of this proposition had been established by Lagrange.

Taking a to be an odd prime p, one is tempted to try to find a sequence of p primes in arithmetic progression. I have already evoked this problem in Chapter 3, Section II. For example, I recall that this is easy when $p = 3, 5, 7$:

$$p = 3, 5, 7;$$
$$p = 5, 11, 17, 23, 29;$$
$$p = 7, 157, 307, 457, 607, 757, 907.$$

G. Löh, who has been told of this problem, determined the "smallest" instances for $p = 11$ and 13. A quick search showed that $p_n = 11 + nd$ is a prime for $n = 0,1, ..., 10$, where $d = 2 \times 3 \times 5 \times 7 \times 7315048 = 1536160080$, and this is the smallest possible value of d. For $p = 13$, a much more extensive computation was required to show that $p_n = 13 + nd$ is a prime for $n = 0,1, ..., 12$, where

$$d = 2 \times 3 \times 5 \times 7 \times 11 \times 4293861989 = 9918821194590.$$

This information is given here for the first time.

The same ought to be true for every p, but, in my opinion, this is so difficult that no one will prove, and no one will find a counter-example in the near future.

A related conjecture is the following: There exist arbitrary long arithmetic progressions of *consecutive* primes.

Record

The longest known string of consecutive primes in arithmetic progression contains six terms; its difference is 30 and the initial term is 121174811. It was discovered by Lander & Parkin (1967). A second example was discovered by Weintraub (1977); its difference is 30 and it begins with 999,900,067,719,989.

V. Primes in Special Sequences

A general problem worth considering is the following. Let $(a_k)_{k \geqslant 0}$ be a sequence of natural numbers. Does it contain infinitely many primes?

According to Dirichlet's theorem, this is true if $a_k = a + kd$, where $a \geqslant 1, d \geqslant 1 \gcd(a,d) = 1$. Note that $a_k = f(k)$, with $f(X) = a + dX$.

No similar theorem has ever been proved for irreducible polynomials $f(X) = c_0 + c_1 X + c_2 X^2 + \cdots + c_n X^n$ of degree $n \geqslant 2$, integer coefficients, and strongly primitive, in the sense indicated in Chapter 4, Interlude.

Thus, for example, it is not known whether there exist infinitely many primes of the form $k^2 + 1$.

I'll return to this question in Chapter 6.

Similarly, it is not known whether there exist infinitely many primes of the form $2^n + 1$ (since they must be Fermat primes), nor whether there exist infinitely many primes of the form $2^n - 1$ (since they must be Mersenne primes). Also still unsolved is the question of whether there exist infinitely many prime Fibonacci numbers (see Chapter 5, Section VIII).

On the other hand, rather definite assertions may be made about the primes of the form $[\alpha k]$, or $[k^\alpha]$, where $k = 1,2, \ldots$ and α is a real number, not an integer, $\alpha > 1$.

First I consider the sequence $[\alpha k]$ $(k = 1,2, \ldots)$. If $\alpha = a/b$, with a,b relatively prime integers, $b \geqslant 2$, then the set of all numbers $[(a/b)k]$ $(k = 1,2, \ldots)$ is the union of b arithmetic progressions with difference a, of which, at least one has initial term relatively prime to a. So, the sequence contains infinitely many primes, as guaranteed by

Dirichlet's theorem.

Now let α be irrational. The key to arrive at a conclusion concerns the uniform distribution of sequences of fractional parts $\alpha p - [\alpha p]$ (where $p = 2, 3, 5, 7, ...$).

It is not my intention to explain more about uniform distribution than it is absolutely necessary for my immediate purpose. It is a beautiful theory. A very elementary introduction may be found in Niven's book *Irrational Numbers* (1956); for a more detailed study, consult Hlawka's book (1979).

The name "uniform distribution" conveys perfectly well the idea. Let $A = \{\alpha_1, \alpha_2, \alpha_3, ...\}$ be a sequence of real numbers, which I assume first to be in the interval $[0,1]$, that is, $0 \leqslant \alpha_i \leqslant 1$ for all $i = 1, 2, ...$. The length of any closed interval $I = [a,b]$, with $0 \leqslant a < b \leqslant 1$, is $|I| = b - a$. For every $n \geqslant 1$ let $I(A,n) = \#\{i \mid 1 \leqslant i \leqslant n, \alpha_i \in I\}$, so $I(A,n) \leqslant n$.

The sequence A is said to be uniformly distributed if, for every interval I, as above, the following limit exists and is equal to $|I|$:

$$\lim_{n \to \infty} \frac{I(A,n)}{n} = |I| \ .$$

If the sequence A is uniformly distributed, then the set of numbers of the sequence is everywhere dense in the interval $[0,1]$.

If $A = \{\alpha_1, \alpha_2, \alpha_3, ...\}$ is any sequence of real numbers (not necessarily all belonging to the interval $[0,1]$), consider the sequence of fractional parts $A' = \{\alpha_1 - [\alpha_1], \alpha_2 - [\alpha_2], \alpha_3 - [\alpha_3], ...\}$.

If the sequence A' is uniformly distributed, then the sequence A is said to be uniformly distributed modulo 1.

The theory of sequences, uniformly distributed modulo 1, was developed in 1916 by Weyl and has its root in an old approximation theorem of Kronecker (see Hardy & Wright's book).

For example, Weyl proved that the real number α is irrational if and only if the sequence $\{\alpha, 2\alpha, 3\alpha, ...\}$ is uniformly distributed modulo 1.

For my present purpose, I need the following result of Vinogradov (1948) which I state in a special case:

If α is an irrational number, then the sequence $\{2\alpha, 3\alpha, 5\alpha, ..., p_n\alpha, ...\}$ is uniformly distributed modulo 1.

This was shown already by Turán in 1937, assuming the Riemann hypothesis. An accessible proof of this theorem may be found in Ellison & Mendès France's book listed in the general references (see p. 347).

This theorem is applied as follows.

If $x > 1$, let $\pi_\alpha(x) = \#\{p \text{ prime} \mid p \leqslant x \text{ and there exists } k \geqslant 1 \text{ such that } p = [k\alpha]\}$.

Then

$$\pi_\alpha(x) \sim \frac{1}{\alpha} \times \frac{x}{\log x} \qquad \text{(as } x \text{ tends to infinity).}$$

In particular, there exists infinitely many primes of the form $[k\alpha]$ (where $\alpha > 1$ is any given irrational number).

Proof. This is now quite easy to show. Indeed, let p be a prime, $p \leqslant x$. Then $p = [k\alpha]$ means that $k\alpha - 1 < p < k\alpha$, that is,

$$k - \frac{1}{\alpha} < \frac{1}{\alpha}p < k;$$

this is equivalent to

$$1 - \frac{1}{\alpha} < \frac{1}{\alpha}p - \left[\frac{1}{\alpha}p\right] < 1,$$

because $[(1/\alpha)p] = k - 1$. But $1/\alpha$ is irrational and the sequence $\{(1/\alpha)p_n\}_{n \geqslant 1}$ is uniformly distributed modulo 1. Thus, asymptotically

$$\pi_\alpha(x) \sim \frac{1}{\alpha}\pi(x) \sim \frac{1}{\alpha} \times \frac{x}{\log x},$$

by the prime number theorem. $\square$

Now I turn my attention to the sequence $[k^\alpha]$ ($k = 1,2,3, ...$). For every $x > 1$ let $\pi^\alpha(x) = \#\{p \text{ prime} \mid p \leqslant x \text{ and there exists } k \geqslant 1 \text{ such that } p = [k^\alpha]\}$.

It is conjectured that

$$\pi^\alpha(x) \sim C \frac{x^{1/\alpha}}{\log x}$$

(for some constant $C > 0$).

This has been shown to be true for $1 < \alpha < 10/9$ by Pyatetskii-Shapiro (1953) and Kolesnik (1967). Moreover, Deshouillers showed in 1972 (see 1976) that for every $\alpha > 1$,

$$\pi^\alpha(x) = O\left[\frac{x}{\log x}\right].$$

Also, for almost all α (in the sense of Lebesgue's measure), $\lim_{x\to\infty}\pi^\alpha(x) = \infty$.

VI. Goldbach's Famous Conjecture

In a letter of 1742 to Euler, Goldbach expressed the belief that:

(G) *Every integer $n > 5$ is the sum of three primes.*

Euler replied that this is easily seen to be equivalent to the following statement:

(G') *Every even integer $2n > 4$ is the sum of two primes.*

Indeed, if (G') is assumed to be true and if $2n \geqslant 6$, then $2n - 2 = p + p'$ so $2n = 2 + p + p'$, where p, p' are primes. Also $2n + 1 = 3 + p + p'$, which proves (G).

Conversely, if (G) is assumed to be true, and if $2n \geqslant 4$, then $2n + 2 = p + p' + p''$, with p, p', p'' primes; then necessarily $p'' = 2$ (say) and $2n = p + p'$.

Note that it is trivial that (G') is true for infinitely many even integers: $2p = p + p$ (for every prime).

It was also shown by Schinzel in 1959, using Dirichlet's theorem on primes in arithmetic progressions, that given the integers $k \geqslant 2$, $m \geqslant 2$, there exist infinitely many pairs of primes (p,q) such that $2k \equiv p + q \pmod{m}$. Of course, this falls short of proving Goldbach's conjecture, since it would still be necessary to show that if an even number is congruent, modulo every $m \geqslant 2$, to a sum of two primes, then it is itself a sum of two primes.

Very little progress was made in the study of this conjecture before the development of refined analytical methods and sieve theory.

And despite all the attempts, the problem is still unsolved.

There have been three main lines of attack, reflected, perhaps inadequately, by the keywords "asymptotic," "almost primes," "basis."

An asymptotic statement is one which is true for all sufficiently large integers.

The first important result is due to Hardy & Littlewood in 1923 — it is an asymptotic theorem. Using the circle method and a modified form of the Riemann hypothesis, they proved that there exists n_0 such that every odd number $n \geqslant n_0$ is the sum of three primes.

Later, in 1937, Vinogradov gave a proof of Hardy & Littlewood's theorem, without any appeal to the Riemann hypothesis — but using instead very sophisticated analytic methods. A simpler proof was given, for example, by Estermann in 1938. There have been calculations of n_0, which may be taken to be $n_0 = 3^{3^{15}}$.

The approach via almost-primes consists in showing that there exist h, $k \geqslant 1$ such that every sufficiently large even integer is in the set $P_h + P_k$ of sums of integers of P_h and of P_k. What is intended is, of course, to show that h, k can be taken to be 1.

In this direction, the first result is due to Brun (1919, *C.R. Acad. Sci. Paris*): every sufficiently large even number belongs to $P_9 + P_9$.

Much progress has been achieved, using more involved types of sieve, and I note papers by Rademacher, Estermann, Ricci, Buchstab, and Selberg, in 1950, who showed that every sufficiently large even integer is in $P_2 + P_3$.

While these results involved summands which were both composite, Rényi proved in 1947 that there exists an integer $k \geqslant 1$ such that every sufficiently large even integer is in $P_1 + P_k$. Subsequent work provided explicit values of k. Here I note papers by Pan, Pan (no mistake — two brothers, Cheng Dong and Cheng Biao), Barban, Wang, Buchstab, Vinogradov, Halberstam, Jurkat, Richert, Bombieri (see detailed references in Halberstam & Richert's book *Sieve Methods*).

The best result to date — and the closest one has come to establishing Goldbach's conjecture — is by Chen (announcement of results in 1966; proofs in detail in 1973, 1978). In his famous paper, Chen proved:

Every sufficiently large even integer may be written as $2n = p + m$, where p is a prime and $m \in P_2$.

As I mentioned before, Chen proved at the same time, the "conjugate" result that there are infinitely many primes p such that $p + 2 \in P_2$; this is very close to showing that there are infinitely many twin primes.

The same method is good to show that for every even integer $2k \geqslant 2$, there are infinitely many primes p such that $p + 2k \in P_2$; so $2k$ is the difference $m - p$ ($m \in P_2$, p prime) in infinitely many ways.

A proof of Chen's theorem is given in the book of Halberstam & Richert. See also the simpler proof given by Ross (1975).

The "basis" approach began with the famous theorem of Schnirelmann (1930), proved, for example, in Landau's book (1937) and in Gelfond & Linnik's book (translated in 1965):

There exists a positive integer S, such that every sufficiently large integer is the sum of at most S primes.

It follows that there exist a positive integer $S_0 \geqslant S$ such that every integer (greater than 1) is a sum of at most S_0 primes.

S_0 is called the Schnirelmann constant.

I take this opportunity to spell out Schnirelmann's method, which is also useful in a much wider context. Indeed, it is applicable to sequences of integers $A = \{0, a_1, a_2, \ldots\}$ with $0 < a_1 < a_2 < \cdots$.

Schnirelmann defined the density of the sequence A as follows. Let $A(n)$ be the number of $a_i \in A$, such that $0 < a_i \leqslant n$. Then the density is

$$d(A) = \inf_{n \geqslant 1} \frac{A(n)}{n}.$$

So if $1 \notin A$, then $d(A) = 0$. Also $d(A) \leqslant 1$ and $d(A) = 1$ exactly when $A = \{0, 1, 2, 3, \ldots\}$.

Some densities may be easily calculated:

If $A = \{0, 1^2, 2^2, 3^2, \ldots\}$, then $d(A) = 0$, if $A = \{0, 1^3, 2^3, 3^3, \ldots\}$, then $d(A) = 0$, and more generally, if $k \geqslant 2$ and $A = \{0, 1^k, 2^k, 3^k, \ldots\}$, then $d(A) = 0$.

Similarly, if $A = \{0, 1, 1 + m, 1 + 2m, 1 + 3m, \ldots\}$ (where $m > 0$ is an integer), then $d(A) = 1/m$.

If $A = \{0, 1, a, a^2, a^3, ...\}$ (where $a > 1$ is any integer), then $d(A) = 0$.

All the preceding examples were trivial. How about the sequence $P' = \{0,1,2,3,5,7, ..., p, ...\}$ of primes (to which 0, 1 have been added)? It follows from Tschebycheff's estimates for $\pi(n)$ that P' has density $d(P') = 0$.

Many problems in number theory are of the following kind: to express every natural number (or every sufficiently large natural number) as sum of a bounded number of integers from a given sequence. Because sums are involved, these problems constitute the so-called "additive number theory." Hardy & Littlewood studied such questions systematically, already in 1920, and called this branch "partitio numerorum."

Goldbach's problem is a good example: to express every integer $n > 5$ as sum of three primes.

If $A = \{a_0 = 0, a_1, a_2, ...\}$, $B = \{b_0 = 0, b_1, b_2, ...\}$, consider all the numbers of the form $a_i + b_j$ $(i,j \geq 0)$ and write them in increasing order, each one only once. The sequence so obtained is called $A + B$. This is at once generalized for several sequences

$$A_1 = \{0, a_{11}, a_{12}, ...\}, ..., \quad A_n = \{0, a_{n1}, a_{n2}, ...\}.$$

In particular, taking $A = B$, then $A + B$ is written $2A$; taking $A_1 = \cdots = A_n = A$, then $A_1 + \cdots + A_n$ is written nA.

A sequence $A = \{0,a_1,a_2, ...\}$ is called a basis of order $k \geq 1$ if $kA = \{0,1,2,3, ...\}$; that is, every natural number is the sum of k numbers of A (some may be 0). And A is called an asymptotic basis of order $k \geq 1$ if there exists $N \geq 1$ such that $\{N, N+1, N+2, ...\} \subseteq kA$.

For example, Lagrange showed that every natural number is the sum of 4 squares (some may be 0); in this terminology, the sequence $\{0,1^2,2^2,3^2, ...\}$ is a basis of order 4.

It is intuitively reasonable to say that the greater the density of a sequence is, the more likely it will be a basis. I proceed to indicate how this is in fact true.

Schnirelmann showed the elementary, but useful fact: If A, B are sequences as indicated, then

$$d(A + B) \geq d(A) + d(B) - d(A)d(B).$$

This is rewritten as

$$d(A + B) \geqslant 1 - (1 - d(A))(1 - d(B)),$$

and may be at once generalized to

$$d(A_1 + \cdots + A_n) \geqslant 1 - \prod_{i=1}^{n} (1 - d(A_i))$$

for any sequences $A_1, A_2, \ldots, A_n$.

Here is an immediate corollary.

If $d(A) = \alpha > 0$, then there exists $h \geqslant 1$ such that A is a basis of order h.

Proof. By the above inequality, if $k \geqslant 1$, then $d(kA) \geqslant 1 - (1 - \alpha)^k$. Since $\alpha > 0$, there exists k such that $d(kA) > \frac{1}{2}$, thus, for every $n \geqslant 1$, $s = (kA)(n) = \#\{a_i \in kA \mid 1 \leqslant a_i \leqslant n\} > n/2$. But $2s + 1 > n$, so the integers $0, a_1, \ldots, a_s, n - a_1, n - a_2, \ldots, n - a_s$, cannot all be distinct. So $n = a_{i_1} = a_{i_2}$. This shows that every natural number is a sum of two integers of kA, so A is a basis of order $h = 2k$. $\square$

Thus, from Schnirelmann's inequality and its corollary, the only sequences which remain to be considered are those with density 0, like $P' = \{0,1,2,3,5,7, \ldots\}$, or better, $P = \{0,2,3,5,7, \ldots\}$.

Even though $d(P') = 0$, Schnirelmann proved that $d(P' + P') > 0$ and therefore he concluded that P' is a basis, so P is an asymptotic basis (of some order S) for the set of natural numbers.

One extra digression, which will not be further evoked, is about the sharpening of Mann (1942) of Schnirelmann's density inequality:

$$d(A_1 + \cdots + A_n) \geqslant \min\{1, d(A_1) + \cdots + d(A_n)\}.$$

In his small and neat book (1947), Khinchin wrote an interesting and accessible chapter on Schnirelmann's ideas of bases and density of sequences of numbers, and he gave the proof by Artin & Scherk (1943) of Mann's inequality.

In respect to Goldbach's problem, I note that, according to the theorems of Hardy & Littlewood and Vinogradov (which avoids the Riemann hypothesis), Schnirelmann's theorem holds with $S = 3$. But the point is that the proof of Schnirelmann's theorem is elementary and based on totally different ideas.

With his method, Schnirelmann estimated that $S \leqslant 800,000$. Subsequent work, at an elementary level, has allowed Vaughan to show in 1976 that $S \leqslant 6$. As for S_0, the best values to date are by Deshouillers (1976): $S_0 \leqslant 26$, and later by Riesel & Vaughan (1983): $S_0 \leqslant 19$.

In 1949, Richert proved the following analogue of Schnirelmann's theorem: every integer $n > 6$ is the sum of distinct primes.

Here I note that Schinzel showed in 1959 that Goldbach's conjecture implies (and so, it is equivalent to) the statement:

Every integer $n > 17$ is the sum of exactly three distinct primes.

Thus, Richert's result will be a corollary of Goldbach's conjecture (if and when it will be shown true).

Now I shall deal with the number $r_2(2n)$ of representations of $2n \geqslant 4$ as sums of two primes, and the number $r_3(n)$ of representations of the odd number $n > 5$ as sums of three primes. A priori, $r_2(2n)$ might be zero (until Goldbach's conjecture is established), while $r_3(n) > 0$ for all n sufficiently large.

Hardy & Littlewood gave, in 1923, the asymptotic formula below, which at first relied on a modified Riemann's hypothesis; later work of Vinogradov removed this dependence:

$$r_3(n) \sim \frac{n^2}{2(\log n)^3} \left\{ \prod_{p} \left[1 + \frac{1}{(p-1)^2} \right] \prod_{\substack{p|n \\ p>2}} \left[1 - \frac{1}{p^2 - 3p + 3} \right] + o(1) \right\}.$$

With sieve methods, it may be shown that

$$r_2(2n) \leqslant C \, \frac{2n}{(\log 2n)^2} \, \log \log 2n.$$

On the other hand, Powell proposed in 1985 to give an elementary proof of the following fact (problem in *Mathematics Magazine*): for every $k > 0$ there exist infinitely many even integers $2n$, such that $r_2(2n) > k$. A solution by Finn & Frohliger was published in 1986.

For every $x \geqslant 4$ let

$$G'(x) = \#\{2n \mid 2n \leqslant x, \ 2n \text{ is not a sum of two primes}\}.$$

Van der Corput (1937), Estermann (1938), and Tschudakoff (1938) proved   independently   that $\lim \ G'(x)/x \ = \ 0$,   and   in   fact,

$G'(x) = O(x/(\log x)^{\alpha})$, for every $\alpha > 0$.

Using Tschebycheff's inequality, it follows that there exist infinitely many primes p such that $2p = p_1 + p_2$, where $p_1 < p_2$ are primes; so p_1, p, p_2 are in arithmetic progression (this result was already quoted in Section IV, while discussing strings of primes in arithmetical progressions).

The best result in this direction is the object of a deep paper by Montgomery & Vaughan (1975), and it asserts that there exists an effectively computable constant α, $0 < \alpha < 1$, such that for every sufficiently large x, $G'(x) < x^{1-\alpha}$. In 1980, Chen & Pan (Cheng Dong) showed that $1 - \alpha = \frac{1}{100}$ is a possible choice. In a second paper (1983), Chen succeeded in taking $\alpha = \frac{1}{25}$ (also done independently by Pan). The finding of a sharp upper bound for $G'(x)$ – a delicate problem, full of pitfalls – has eluded many gifted mathematicians (see Pintz, 1986).

Concerning numerical calculations about Goldbach's conjecture, Stein & Stein (1965) calculated $r_2(2n)$ for every even number up to 200,000 and conjectured that $r_2(2n)$ may take any integral value.

In 1964, Shen used a sieving process and verified Goldbach's conjecture up to 33,000,000.

Record

The record belongs to Stein & Stein (1965), and if it were not for Te Riele (thanks for the information), I would have missed it. Indeed, it is hidden in a note added at the end of the paper. Stein & Stein verified Goldbach's conjecture up to 10^8. Four other people (Light, Forrest, Hammond & Roe), unaware of the elusive note, have done the calculations again up to 10^8, but much later – in 1980.

To conclude this brief discussion of Goldbach's conjecture, I would like to mention the recent book of Wang (1984), which is a collection of selected important articles on this problem, (including the article by Chen), and contains a good bibliography.

VII. The Waring-Goldbach Problem

I will discuss the problem of expressing natural numbers as sums of kth powers of primes (where $k \geqslant 2$). This is a problem of the same kind as Goldbach's problem — which concerned primes, instead of their kth powers. It is also a problem of the same type as Waring's problem, which concerns kth powers of integers (whether prime or not) and how economically they may generate additively all integers. It is normal first to consider Waring's problem. Before I enter into any more explanation, I ask to be forgiven for going beyond the prime numbers; but to please the record lovers, I will include plenty of records from a very keen and challenging competition.

A. Waring's Problem

Waring's problem has a long and interesting history — see for example, Dickson's book, Vol. II.

Prehistory:

In 1770, Lagrange proved that every natural number is the sum of (at most) four squares. This result had been conjectured by Fermat, already one century before. Playing with squares was very much in favor at that time. Much fewer experiments involved cubes and higher powers. Nevertheless, in his book *Meditationes Algebricae* (third edition, 1782), Waring stated that every natural number is the sum of at most 9 (positive) cubes, at most 19 biquadrates, etc.

In other terms, for every $k \geqslant 2$ there exists a number $r \geqslant 1$ such that every natural number is the sum of at most r kth powers.

Waring's problem consists of two parts. First, to show that for every k the above number r exists. If this has been established, let $g(k)$ be the smallest possible value for r. The second part of Waring's problem consists in determining the value of $g(k)$ (which *a priori* might be infinity, if the first part would have negative solution). At worst, to determine lower and upper bounds for $g(k)$.

Since small numbers usually exhibit peculiarities, while phenomena tend to become more regular for sufficiently large numbers, it is appropriate to introduce the number $G(k)$, analogue to $g(k)$: $G(k)$ is the minimal value of r such that every sufficiently large integer is the sum of r kth powers.

Despite the notation, it is clear that $G(k) \leqslant g(k)$. Again, the problem arises to compute the exact value of $G(k)$, or at worst, lower and upper bounds for $G(k)$.

As in every difficult question, Waring's problem was treated first in special cases, that is, small exponents.

If $k = 2$, Lagrange had shown, as I indicated, that $g(2) \leqslant 4$. Noting that $a^2 \equiv 0$, 1 or 4 (mod 8), then every number of the form $8t + 7$ cannot be written as the sum of three squares; therefore $G(2)$ cannot be 3, hence $G(2) = g(2) = 4$. In this case, Waring's problem is completely solved.

The problem is much more difficult when $k > 2$. I shall indicate in tables below the main successive advances in the question, for special values of k. But first, I'll present the general results.

Euler − not Leonhard, but his son Johannes Albert (born in 1734, died in 1800, mathematician who became a member of Sankt Petersburg Academy) − made the following easy observation about the problem (around 1772). If $k \geqslant 2$, divide 3^k by 2^k to have

$$3^k = q \times 2^k + r \quad \text{with} \quad 1 \leqslant r < 2^k, \quad \text{so} \quad q = \left[\frac{3^k}{2^k} \right].$$

Let $s = q \times 2^k - 1 < 3^k$; if s is a sum of kth powers, it may only use 1 and 2^k; but $s = (q - 1) \times 2^k + (2^k - 1) \times 1$, so s is the sum of $q - 2 + 2^k$ (and not less) kth powers. Thus $g(k) \geqslant [(\frac{3}{2})^k] - 2 + 2^k$ and, therefore, if $k = 3, 4, 5,$ 6, 7, 8, 9, 10, ..., then respectively, $g(k) \geqslant 9,$ 19, 37, 73, 143, 279, 548, 1079, ...

A lower bound for $G(k)$ was obtained by Maillet in 1908: $G(k) \geqslant k + 1$. This means that there are arbitrarily large natural numbers which are not the sum of k kth powers (see, for example, Hardy & Wright's book, theorem 394).

Waring's problem for arbitrary exponent k [namely, the existence of $g(k) < \infty$] was first solved by Hilbert, in 1909. There have been also two other methods used to solve the problem, to which I'll refer afterwards.

Hilbert's complicated proof relied on a key result, which is the following interesting Hilbert-Waring's identity:

Given $k \geqslant 1$, $n \geqslant 1$, let

$$N = \begin{pmatrix} 2k + n - 1 \\ 2k \end{pmatrix}.$$

There exist N positive rational numbers $b_1, ..., b_N$, and integers $a_{1i}, ..., a_{ni}$ (for $i = 1,2, ..., N$), such that

$$(x_1^2 + \cdots + x_n^2)^k = \sum_{i=1}^{N} b_i(a_{1i}x_1 + \cdots + a_{ni}x_n)^{2k}$$

for all integers $x_1, ..., x_n$.

Stridsberg gave in 1912, a constructive proof of Hilbert-Waring's identity, from which it became possible to calculate an upper bound for $g(k)$. This was done by Rieger (1953):

$$g(k) \leqslant (2k + 1)^{260(k+3)^{3k+8}}.$$

A very poor estimate, which underlined the need for better techniques.

Hardy & Littlewood published in 1920, and the following years, several important papers on Waring's problem.   In particular, they developed a new analytical method that culminated in a new proof of Hilbert's theorem.   This led also to the estimate $G(k) \leqslant 2^{k-1}(k - 2) + 5$. In 1935, Vinogradov made many technical improvements on Hardy & Littlewood's proof, further simplified by Heilbronn (1936). Eventually, in 1947 Vinogradov obtained the estimate $G(k) \leqslant k(3 \ \log k + 11) -$ and even a better one in 1959, for $k \geqslant 170000$, which I will not quote, since it is too technical.   Chen arrived at the estimate (1958): $G(k) \leqslant k(3 \ \log k + 5.2)$; this was improved in 1984 by Balasubramanian and Mozzochi:

$$G(k) \leqslant [2A_2 + A_1 - 4],$$

where

$$A_1 = -\frac{\log(3k)}{\log \rho}, \quad A_2 = -\frac{\log(6k)}{\log \rho}, \quad \rho = \frac{k - 1}{k}.$$

Better lower bounds than Maillet's were found for special values of k, by Hardy & Littlewood. Thus, $G(2^t) \geqslant 2^{t+2}$ for $t \geqslant 2$, $G(p^t(p-1)) \geqslant p^{t+1}$ for odd primes p, and $t \geqslant 1$.

The third method to solve Waring's problem is based on the ideas of Schnirelmann, which I have explained in the section on Goldbach's problem. The proof was given by Linnik in 1943, and its framework is explained quite well in Ellison's paper. Linnik managed to prove that for every $k \geqslant 2$ there exists an integer s such that $sA^{(k)}$ has positive density; here $A^{(k)} = \{0, 1^k, 2^k, 3^k, ...\}$; so, by Schnirelmann's result, $sA^{(k)}$ – and therefore also $A^{(k)}$ – is a basis.

A proof of Waring's conjecture combining the Hardy & Littlewood method with Schnirelmann's was given by Newman in 1960.

Linnik did not try to determine $g(k)$. But in 1954, Rieger used Linnik's method to obtain $g(k) \leqslant 2^{2 \times 16 \, [(k+1)!]}$. Once more, this was a very poor bound.

The determination of $g(k)$ for $k \geqslant 7$ was possible, as the combined outcome of several papers of Dickson (1936), Rubugunday (1942), Niven (1944):

Let $3^k = q \times 2^k + r$, $1 \leqslant r < 2^k$. If $r + q \leqslant 2^k$, then $g(k) = [(\tfrac{3}{2})^k] - 2 + 2^k$.

If, however, $r + q > 2^k$, then $s = (q + 1)[(\tfrac{4}{3})^k] + q \geqslant 2^k$, and

$$
g(k) = \begin{cases} [(\tfrac{3}{2})^k] - 2 + 2^k + [(\tfrac{4}{3})^k] & \text{if} \ \ 2^k = s \\ [(\tfrac{3}{2})^k] - 3 + 2^k + [(\tfrac{4}{3})^k] & \text{if} \ \ 2^k < s. \end{cases}
$$

The point is that $r + q \leqslant 2^k$ for every $k \leqslant 200{,}000$ (as verified by Stemmler, in 1964). On the other hand, Mahler showed, in 1957, that there exist at most finitely many k such that $r + q > 2^k$; in fact, no such value of k has ever been found!

To delight the record lovers, but also not to forget the efforts of many obstinate mathematicians, the following tables give results for small exponents k, obtained with particular methods.

$k = 3$

Year	Author	$g(3)$	$G(3)$	Remark
1862	J. A. Euler	$\geqslant 9$		
1895	E. Maillet	$\leqslant 21$	$\geqslant 4$	$g(3)$ exists
1906	A. Fleck	$\leqslant 13$		
1909	A. Wieferich	$\leqslant 9$		$g(3)$ best possible
1909	E. Landau		$\leqslant 8$	
1943	Yu. V. Linnik		$\leqslant 7$	

The proof of Wieferich was incomplete; the missing case was dealt by A. J. Kempner in 1912.

In 1939, Dickson made effective Landau's result, by showing that all integers, except 23 and 239 are sums of eight cubes. A more accessible proof of Linnik's result was given by Watson (1951).

Present status: $4 \leqslant G(3) \leqslant 7$; the exact value of $G(3)$ is unknown, however recent computations of Bohman & Fröberg (1981) as well as of Romani (1982) point to the likelihood that $G(3) = 4$.

$k = 4$

Year	Author	$g(4)$	$G(4)$	Remark
1859	J. Liouville	$\leqslant 53$		$g(4)$ exists
1862	J. A. Euler	$\geqslant 19$		
1878	S. Réalis	$\leqslant 47$		
1878	E. Lucas	$\leqslant 41$		
1895	E. Maillet		$\geqslant 5$	
1906	A. Fleck	$\leqslant 39$		
1907	E. Landau	$\leqslant 38$		
1909	A. Wieferich	$\leqslant 37$		
1912	A. J. Kempner		$\geqslant 16$	
1921	G. H. Hardy & J. E. Littlewood		$\leqslant 21$	
1939	H. Davenport		$\leqslant 16$	$G(4)$ best possible
1971	F. Dress	$\leqslant 30$		
1974	J. R. Chen	$\leqslant 27$		
1974	H. E. Thomas	$\leqslant 22$		
1979	R. Balasubramanian	$\leqslant 21$		
1986	R. Balasubramanian, F. Dress & J. M. Deshouillers	$\leqslant 19$		$g(4)$ best possible

Present status: $G(4) = 16$, $g(4) = 19$, the problem is solved.

A rare occasion: Flags up!

$k = 5$

Year	Author	$g(5)$	$G(5)$	Remark
1862	J. A. Euler	$\geqslant 37$		
1892	E. Maillet	$\leqslant 192$	$\geqslant 6$	$g(5)$ exists
1907	A. Fleck	$\leqslant 156$		
1907	A. Wieferich	$\leqslant 59$		
1913	W. S. Baer	$\leqslant 58$		
1922	G. H. Hardy & J. E. Littlewood		$\leqslant 53$	
1933	L. E. Dickson	$\leqslant 54$		
1942	H. Davenport		$\leqslant 23$	
1964	J. R. Chen	$\leqslant 37$		$g(5)$ best possible
1985	K. Thanigasalam		$\leqslant 22$	
1986	R. C. Vaughan		$\leqslant 21$	

Present status: $g(5) = 37$, $6 \leqslant G(5) \leqslant 21$.

$k = 6$

Year	Author	$g(6)$	$G(6)$	Remark
1862	J. A. Euler	$\geqslant 73$		
1895	E. Maillet		$\geqslant 7$	
1907	A. Fleck	$\leqslant 184g(3) + 59$ $\leqslant 2451$		$g(6)$ exists
1912	A. J. Kempner	$\leqslant 970$		
1913	W. S. Baer	$\leqslant 478$		
1922	G. H. Hardy & J. E. Littlewood			
1922	G. H. Hardy & J. E. Littlewood		$\leqslant 133$ $\geqslant 9$	
1940	S. S. Pillai	$\leqslant 73$		$g(6)$ best possible
1942	H. Davenport		$\leqslant 36$	
1985	K. Thanigasalam		$\leqslant 34$	
1986	R. C. Vaughan		$\leqslant 31$	

Present status: $g(6) = 73$, $9 \leqslant G(6) \leqslant 31$.

$k = 7$

Year	Author	$g(7)$	$G(7)$	Remark
1862	J. A. Euler	⩾ 143		
1895	E. Maillet		⩾ 8	
1909	A. Wieferich	⩽ 3806		$g(7)$ exists
1922	G. H. Hardy & J. E. Littlewood		⩽ 325	
1942	H. Davenport		⩽ 53	
1985	K. Thanigasalam		⩽ 50	
1986	R. C. Vaughan		⩽ 45	

The paper where K. Sambasiva Rao (1941) claims to prove that $G(7) ⩽ 52$, is incorrect.

Present status: $143 ⩽ g(7) ⩽ 3806,\ 8 ⩽ G(7) ⩽ 45$.

$k = 8$

Year	Author	$g(8)$	$G(8)$	Remark
1862	J. A. Euler	⩾ 279		
1895	E. Maillet		⩾ 9	
1908	E. Maillet	< ∞		$g(8)$ exists
1908	A. Hurwitz	⩽ 840 × $g(4)$+273 ⩽ 36119		$g(8)$ exists
1922	G. H. Hardy & J. E. Littlewood		⩽ 773 ⩾ 32	
1941	V. Narasimhamurti		⩽ 73	
1985	K. Thanigasalam		⩽ 68	
1986	R. C. Vaughan		⩽ 62	

Present status: $279 ⩽ g(8) ⩽ 36119,\ 32 ⩽ G(8) ⩽ 62$.

$k = 9$

Year	Author	$g(9)$	$G(9)$	Remark
1862	J. A. Euler	$\geqslant 548$		
1895	E. Maillet		$\geqslant 10$	
1922	G. H. Hardy & J. E. Littlewood		$\geqslant 13$	
1941	V. Narasimhamurti		$\leqslant 99$	
1973	R. J. Cook		$\leqslant 96$	
1977	R. C. Vaughan		$\leqslant 91$	
1985	K. Thanigasalam		$\leqslant 87$	
1986	R. C. Vaughan		$\leqslant 82$	

The existence of $g(9)$ followed from Hilbert's theorem.
Present status: $g(9) \geqslant 548$, $13 \leqslant G(9) \leqslant 82$.

$k = 10$

Year	Author	$g(10)$	$G(10)$	Remark
1862	J. A. Euler	$\geqslant 1079$		
1895	E. Maillet		$\geqslant 11$	
1909	J. Schur	$< \infty$		$g(10)$ exists
1922	G. H. Hardy & J. E. Littlewood		$\geqslant 12$	
1941	V. Narasimhamurti		$\leqslant 122$	
1973	R. J. Cook		$\leqslant 121$	
1977	R. C. Vaughan		$\leqslant 107$	
1984	R. Balasubramanian & C. J. Mozzochi		$\leqslant 106$	
1985	K. Thanigasalam		$\leqslant 102$	

Present status: $g(10) \geqslant 1079$, $12 \leqslant G(10) \leqslant 102$.

After this discussion of Waring's problem, with the rich ideas involved, I wish to consider a natural generalization. As a matter of fact, the following situation had been already considered by Fermat.

Triangular numbers are 1, 3, 6, 10, ...; so they are the values of

$$f_3(X) = \frac{X(X-1)}{2} + X$$

for $X = 1, 2, 3, 4, \dots$.

Square numbers are, of course, 1, 4, 9, 16, ...; so they are the values of

$$f_4(X) = X^2 = 2 \times \frac{X(X-1)}{2} + X.$$

Pentagonal numbers are 1, 5, 12, 22, ...: so they are the values of

$$f_5(X) = 3 \times \frac{X(X-1)}{2} + X.$$

More generally, r-gonal numbers are the values of

$$f_r(X) = (r-2)\frac{X(X-1)}{2} + X \quad \text{for} \quad X = 1, 2, 3, 4, \dots .$$

Commenting on an empirical observation by Bachet, on Diophantus (Vol. IV, 31), Fermat stated that every natural number is the sum of at most three triangular numbres, of at most four square numbers, of at most five pentagonal numbers, ... and more generally, of at most r r-gonal numbers.

Fermat also wrote:

> I cannot provide here the proof, which depends on numerous and abstruse mysteries of the Science of numbers; I intend to devote to this subject a whole book, thus causing this branch of Arithmetic to progress in an astonishing way, far beyond its formerly known bounds.

As I have already mentioned, Lagrange proved in 1770 the assertion about squares.

In Gauss' *Tagebuch* entry 18, 1796, it is written:

EYPHKA! num = Δ + Δ + Δ .

This expressed the delight of the young Gauss, aged 19, having found the proof that every natural number is the sum of three triangular numbers. The proof involves, of course, the theory of ternary quadratic forms. An elegant modern presentation is in Serre's *Cours d'Arithmétique*, 1970 (it fits in a hand, and makes the mind travel ...).

Fermat's statement was finally completely proved by Cauchy, in 1815, who showed that every natural number is a sum of r r-gonal numbers, of which all but four are equal to 0 or to 1.

More generally, let $f(X)$ be a polynomial with integral coefficients, such that $\lim_{m\to\infty} f(m) = \infty$. Let S be the set consisting of 0, 1 and all positive integers which are values $f(m)$, for $m > 0$. Kamke showed in 1921 that there exists a natural number g such that every natural number is a sum of g numbers of S.

Taking $f(X) = X^k$ ($k \geqslant 1$), this result becomes Hilbert's theorem on Waring's problem.

B. The Waring-Goldbach Problem

And now, back to prime numbers!

The Waring-Goldbach problem asks for the possibility of expressing natural numbers as sums of kth powers of primes, with a bounded number of summands.

In 1937 and 1938, Vinogradov showed that for every $k \geqslant 1$ there exists an integer $V(k)$, such that every sufficiently large natural number is the sum of at most $V(k)$ kth powers of prime numbers. For $k = 1$, I have already reported this result in the section about Goldbach's problem. This was the first theorem concerning the question, which is fit to call the Waring-Goldbach problem.

In Hua's book (1957) it is shown that $V(5) \leqslant 25$; moreover, Hua also stated in 1959 that $V(6) \leqslant 37$, $V(7) \leqslant 55$, $V(8) \leqslant 75$.

Some progress has since been made. Thus, in 1985, Thanigasalam showed that $V(5) \leqslant 23$, $V(6) \leqslant 35$, $V(7) \leqslant 51$, $V(8) \leqslant 69$; he claimed even better, and the proof is in his new paper (1987): $V(6) \leqslant 33$, $V(7) \leqslant 47$, $V(8) \leqslant 63$, $V(9) \leqslant 83$, $V(10) \leqslant 107$.

I hope you didn't forget the meaning of $V(5) \leqslant 23$: every sufficiently large integer is the sum of at most 23 fifth powers of prime numbers. [This is not too far from the best known estimate $G(5) \leqslant 21$, for the representation as sums of fifth powers of any natural numbers.]

More generally, Hua proved the following theorem (see his book from 1957):

Let $f(X)$ be a polynomial with integral coefficients such that $\lim_{m \to \infty} f(m) = \infty$. Let S be the set consisting of 0, 1 and the values $f(p) > 0$, for all primes p. Then there exists a natural number h, such that every natural number n may be written in the form $n = s_1 + \cdots + s_h$, with $s_1, ..., s_h$ in the set S.

Hua's proof was analytical. However, following Linnik's approach to Waring's problem, Burr (in an unpublished manuscript) indicated an elementary proof of Hua's result.

I recommend the following expository articles on Waring's problem and variants: (1) a short survey which is easy and interesting to read, by Small (1977); (2) a more substantial overview, containing well presented proofs, by Ellison (1971); the appropriate chapters in the second volume of Ostmann's book (1956), especially in connection with Schnirelmann's method and the Kamke, Goldbach variants of Waring's problem.

VIII. The Distribution of Pseudoprimes and of Carmichael Numbers

Now I shall indicate results on the distribution of pseudoprimes and of Carmichael numbers.

A. Distribution of Pseudoprimes

Let $P\pi(x)$ denote the number of pseudoprimes (to the base 2) less than or equal to x.

Let $(psp)_1 < (psp)_2 < \cdots < (psp)_n < \cdots$ be the increasing sequence of pseudoprimes.

In 1949 and 1950, Erdös gave the following estimates

$$C \log x < P\pi(x) < \frac{x}{e^{\frac{1}{3}(\log x)^{1/4}}}$$

(for x sufficiently large and $C > 0$). They were later to be much improved, as I shall soon indicate.

From these estimates, it is easy to deduce that $\sum_{n=1}^{\infty} 1/(psp)_n$ is convergent (first proved by Szymiczek in 1967), while $\sum_{n=1}^{\infty} 1/(\log(psp)_n$ is divergent (first shown by Mąkowski in 1974).

Indeed, taking $x = (psp)_n$ and n sufficiently large, then

$$n < \frac{(psp)_n}{e^{(\log(psp)_n)^{1/4}/3}} < \frac{(psp)_n}{(\log n)^{4/3}} \;.$$

Hence

$$\sum_{n=1}^{\infty} \frac{1}{(psp)_n} < \sum_{n=1}^{\infty} \frac{1}{n(\log n)^{4/3}} \;,$$

and since this series is convergent, so is $\sum_{n=1}^{\infty} 1/(psp)_n$.

Rotkiewicz & Wasén gave in 1979 the following easy proof of Mąkowski's result:

$C/n < 1/\log(psp)_n$, by the Erdös result above; since $\sum_{n=1}^{\infty} 1/n$ is divergent, it follows that $\sum_{n=1}^{\infty} 1/\log(psp)_n$ is also divergent.

Erdös conjectures that $P\pi(x) \sim x^{1-\delta(x)}$ where $\lim_{x \to \infty} \delta(x) = 0$. From this it follows that $\sum 1/(psp)_n^{1-\epsilon}$ is divergent for every $\epsilon > 0$.

The early estimates of $P\pi(x)$ given by Erdös have been sharpened and the methods apply to arbitrary bases.

It is convenient to introduce the following notation for the counting functions of pseudoprimes, Euler and strong pseudoprimes in arbitrary bases $a \geqslant 2$:

$$P\pi_a(x) = \#\{n \mid 1 \leqslant n \leqslant x, n \text{ is } psp(a)\}, \qquad P\pi(x) = P\pi_2(x),$$
$$EP\pi_a(x) = \#\{n \mid 1 \leqslant n \leqslant x, n \text{ is } epsp(a)\}, \qquad EP\pi(x) = EP\pi_2(x),$$
$$SP\pi_a(x) = \#\{n \mid 1 \leqslant n \leqslant x, n \text{ is } spsp(a)\}, \qquad SP\pi(x) = SP\pi_2(x).$$

Clearly, $SP\pi_a(x) \leqslant EP\pi_a(x) \leqslant P\pi_a(x)$.

And now, the estimates of upper and lower bounds for these functions.

For the upper bound of $P\pi(x)$, improving on a previous result of Erdös (1956), Pomerance showed in 1981, that for all large x:

$$P\pi(x) \leqslant \frac{x}{\mathcal{l}(x)^{1/2}},$$

with

$$\mathcal{l}(x) = e^{\log x \, \log \log \, \log x \, x/\log \log x}.$$

The same bound is also good for $P\pi_a(x)$, with arbitrary basis a.

Concerning lower bounds, Pomerance, Selfridge & Wagstaff (1980) gave the following estimate for $SP\pi_a(x)$:

$$C_a \log x < SP\pi_a(x),$$

where $C_a = 1/(4a \log a)$, valid for $x \geqslant a^{15a} + 1$. This is of the same order as the lower bound for $P\pi(x)$ given by Erdös, which I have already mentioned. But the best result to date is by Pomerance, in 1982 (see Remark 3 of his paper):

$$e^{(\log x)^{\alpha}} \leqslant SP\pi_a(x),$$

where $\alpha = 5/14$. With the sieve methods of Fouvry (1985), so successful in dealing with the first case of Fermat's last theorem, α may be taken to be $\alpha > 0.68/1.68 > 2/5$. With less sophisticated methods (just an analysis of the average of $B_{psp}(n) = \#\{a \mid 1 < a < n, \gcd(a,n) = 1,$ such that n is $psp(a)\}$). Erdös & Pomerance (1976), quoted in Chapter 2, can take $\alpha = 15/38$, which is slightly smaller than 2/5.

Pomerance conjectures that each of the above counting functions is $x/\mathcal{l}(x)^{1+o(1)}$, where $o(1)$ depends on which function is considered.

Summarizing:

	$P\pi_a(x) \geqslant$	$EP\pi_a(x) \geqslant$	$SP\pi_a(x)$	Given by
Lower bound			$e^{(\log x)^{\alpha}}$	Pomerance, 1982
Upper bound	$\dfrac{x}{\mathcal{l}(x)^{1/2}}$			Pomerance, 1981

The tables of pseudoprimes suggest that for every $x > 170$ there exists a pseudoprime between x and $2x$. However, this has not yet been proved. In this direction, I note the following results of Rotkiewicz (1965):

If n is an integer, $n > 19$, there exists a pseudoprime between n and n^2.

Also, for every $\epsilon > 0$ there exists $x_0 = x_0(\epsilon) > 0$ such that, if $x > x_0$, then there exists a pseudoprime between x and $x^{1+\epsilon}$.

Halberstam & Rotkiewicz pointed out in 1968 that more can be said, by using information about gaps between consecutive primes. Recall (see Chapter 2, Section VIII) that $(2^{2p} - 1)/3$ is a pseudoprime, for every prime $p > 5$. As already seen, there exists θ, $0 < \theta < 1$, such that for every sufficiently large x, there exists a prime p between x and $x + x^\theta$. Hence, there exists a pseudoprime between x and $xe^{C(\log x)^\theta}$, for every sufficiently large x. In particular, θ may be taken to be equal to $5/8$ (Ingham), or even equal to $1/2 + 1/21$ (Iwaniec & Pintz).

Concerning pseudoprimes in arithmetic progression, Rotkiewicz proved in 1963 and 1967:

If $a \geqslant 1$, $d \geqslant 1$ and $\gcd(a,d) = 1$, there exist infinitely many pseudoprimes in the arithmetic progression $\{a + kd \mid k \geqslant 1\}$. Let $psp(d,a)$ denote the smallest pseudoprime in this arithmetic progression. Rotkiewicz showed in 1972:

For every $\epsilon > 0$ and for every sufficiently large d, $\log psp(d,a) < d^{4L^2+L+\epsilon}$, where L is Linnik's constant (seen in Section IV).

The use of a strong form of Riemann's hypothesis allows to obtain the estimate: $\log psp(d,a) < d^{18+\epsilon}$.

The above results have been extended by van der Poorten & Rotkiewicz in 1980: if a, $d \geqslant 1$, $\gcd(a,d) = 1$, then the arithmetic progression $\{a + kd \mid k \geqslant 1\}$ contains infinitely many odd strong pseudoprimes for each base $b \geqslant 2$.

B. Distribution of Carmichael Numbers

Now I turn my attention to the distribution of Carmichael numbers.

Let $CN(x)$ denote the number of Carmichael numbers n such that $n \leqslant x$.

I recall that it is believed that there exist infinitely many Carmichael numbers, but this assertion has still to be proved.

In 1956, Erdös showed that there exists a constant $c > \frac{1}{2}$ such that, for every sufficiently large x,

$$CN(x) \leqslant \frac{x}{L(x)^c},$$

where $L(x)$ was defined above.

Pomerance, Selfridge & Wagstaff improved this estimate in 1980: for every $\epsilon > 0$ there exists $x_0(\epsilon) > 0$ such that, if $x \geqslant x_0(\epsilon)$, then

$$CN(x) \leqslant \frac{x}{L(x)^{1-\epsilon}}.$$

It is much more difficult to provide lower bounds for $CN(x)$. In the above paper the following conjecture is proposed:

For every $\epsilon > 0$ there exists $x_0(\epsilon) > 0$ such that, if $x \geqslant x_0$, then

$$CN(x) \geqslant \frac{x}{L(x)^{2+\epsilon}}.$$

However, Pomerance conjectures for $CN(x)$, like for the pseudoprime counting functions, that $CN(x) = x/L(x)^{1+o(1)}$.

I refer now to tables of pseudoprimes, Carmichael numbers.

Poulet determined, already in 1938, all the (odd) pseudoprimes (in base 2) up to 10^8. Carmichael numbers were starred in Poulet's table.

In 1975, Swift compiled a table of Carmichael numbers up to 10^9.

The largest table of pseudoprimes and Carmichael numbers, including Euler and strong pseudoprimes (in base 2), is by Pomerance, Selfridge & Wagstaff (1980) and extends up to 25×10^9.

x	$P\pi(x)$	$EP\pi(x)$	$SP\pi(x)$	$CN(x)$
10^3	3	1	0	1
10^4	22	12	5	7
10^5	78	36	16	16
10^6	245	114	46	43
10^7	750	375	162	105
10^8	2057	1071	488	255
10^9	5597	2939	1282	646
10^{10}	14887	7706	3291	1547
25×10^9	21853	11347	4842	2163

C. Distribution of Lucas Pseudoprimes

In Chapter 2, Section X the Lucas pseudoprimes were studied. Recall that if P, Q are nonzero integers, $D = P^2 - 4Q$, and the Lucas sequence is defined by

$$U_0 = 0, \quad U_1 = 1, \quad U_n = PU_{n-1} - QU_{n-2} \quad (\text{for } n \geqslant 2),$$

then the composite integer n, relatively prime to D, is a Lucas pseudoprime (with parameters (P,Q)) when n divides $U_{n-(D/n)}$.

Since the concept of Lucas pseudoprime is quite recent, much less is known about the distribution of such numbers. My source is the paper of Baillie & Wagstaff (1980) quoted in Chapter 2. Here are the main results:

If x is sufficiently large, the number $L\pi(x)$ of Lucas pseudoprimes (with parameters (P,Q)), $n \leqslant x$, is bounded as follows:

$$L\pi(x) < \frac{x}{e^{Cs(x)}}$$

where $C > 0$ is a constant and $s(x) = (\log x \log \log x)^{1/2}$.

It follows (like in Szymiczek's result for pseudoprimes) for any given parameters (P,Q), $\Sigma(1/U_n)$ is convergent (sum for all Lucas pseudoprimes with these parameters).

Similarly, there is the following lower bound for the number $SL\pi(x)$ of strong Lucas pseudoprimes (with parameters (P,Q)) $n \leqslant x$ (see the definition in Chapter 2, Section X):

$SL\pi(x) > C'\log x$ (valid for all x sufficiently large) where $C' > 0$ is a constant.

Chapter 5

WHICH SPECIAL KINDS OF PRIMES HAVE BEEN CONSIDERED?

We have already encountered several special kinds of primes, for example, those which are Fermat numbers, or Mersenne numbers (see Chapter 2). Now I shall discuss other families of primes, among them the regular primes, the Sophie Germain primes, the Wieferich primes, the Wilson primes, the prime repunits, the primes in second-order linear recurring sequences.

Regular primes, Sophie Germain primes, and Wieferich primes have directly sprung from attempts to prove Fermat's last theorem. The interested reader may wish to consult my book *13 Lectures on Fermat's Last Theorem*, where these matters are considered in more detail. In particular, there is an ample bibliography including many classical papers, which will not be listed in the bibliography of this book.

I. Regular Primes

Regular primes appeared in the work of Kummer in relation with Fermat's last theorem. In a letter to Liouville, in 1847, Kummer stated that he had succeeded in proving Fermat's last theorem for all primes p satisfying two conditions. Indeed, he showed that if p satisfies these conditions, then there do not exist integers $x, y, z \neq 0$

such that $x^p + y^p = z^p$. He went on saying that "it remains only to find out whether these properties are shared by all prime numbers."

To describe these properties I need to explain some concepts that were first introduced by Kummer.

Let p be an odd prime, let

$$\zeta = \zeta_p = \cos \frac{2\pi}{p} + i \sin \frac{2\pi}{p}$$

be a primitive pth root of 1. Note that $\zeta^{p-1} + \zeta^{p-2} + \cdots + \zeta + 1 = 0$, because $X^p - 1 = (X - 1)(X^{p-1} + X^{p-2} + \cdots + X + 1)$ and $\zeta^p = 1$, $\zeta \neq 1$. Thus, ζ^{p-1} is expressible in terms of the lower powers of ζ. Let K be the set of all numbers $a_0 + a_1\zeta + \cdots + a_{p-2}\zeta^{p-2}$, with a_0 , ..., a_{p-2} rational numbers. Let A be the subset of K consisting of those numbers for which $a_0, a_1,$ ..., a_{p-2} are integers. Then K is a field, called the field of (p-) cyclotomic numbers. A is a ring, called the ring of (p-) cyclotomic integers. The units of A are the numbers $\alpha \in A$ which divide 1, that is $\alpha\beta = 1$ for some $\beta \in A$. The element $\alpha \in A$ is called a cyclotomic prime if α cannot be written in the form $\alpha = \beta\gamma$ with $\beta, \gamma \in A$, unless β or γ is a unit.

I shall say that the arithmetic of p-cyclotomic integers is ordinary if every cyclotomic integer is equal, in unique way, up to units, to the product of cyclotomic primes.

Kummer discovered as early as 1847, that if $p \leqslant 19$ then the arithmetic of the p-cyclotomic integers is ordinary; however it is not so for $p = 23$.

To find a way to deal with nonuniqueness of factorization, Kummer introduced ideal numbers. Later, Dedekind considered certain sets of cyclotomic integers, which he called ideals. I'm refraining from defining the concept of ideal, assuming that it is well known to the reader. Dedekind ideals provided a concrete description of Kummer ideal numbers, so it is convenient to state Kummer's results in terms of Dedekind ideals. A prime ideal P is an ideal which is not equal to 0 nor to the ring A, and which cannot be equal to the product of two ideals, $P = IJ$, unless I or J is equal to P. Kummer showed that for every prime $p > 2$, every ideal of the ring of p-cyclotomic integers is equal, in a unique way, to the product of prime ideals.

In this context, it was natural to consider two nonzero ideals I, J to be equivalent if there exist nonzero cyclotomic integers α, $\beta \in A$ such that $A\alpha I = A\beta J$. The set of equivalence classes of ideals forms a commutative semigroup, in which the cancellation property holds; there is therefore a smallest group containing it. This is called the group of ideal classes. Kummer showed that this group is finite. The number of its elements is called the class number, and denoted by $h = h(p)$. It is a very important arithmetic invariant. The class number $h(p)$ is equal to 1 exactly when every ideal of A is principal, that is, of the form $A\alpha$, for some $\alpha \in A$. Thus $h(p) = 1$ exactly when the arithmetic of the p-cyclotomic integers is ordinary. So, the size of $h(p)$ is one measure of the deviation from the ordinary arithmetic.

Let it be said here that Kummer developed a very deep theory, obtained an explicit formula for $h(p)$ and was able to calculate $h(p)$ for small values of p.

One of the properties of p needed by Kummer in connection with Fermat's last theorem was the following: p does not divide the class number $h(p)$. Today, a prime with this property is called a regular prime.

The second property mentioned by Kummer concerned units, and he showed later that it is satisfied by all regular primes. This was another beautiful result of Kummer, today called Kummer's lemma on units.

In his regularity criterion, Kummer established that the prime p is regular if and only if p does not divide the numerators of the Bernoulli numbers $B_2, B_4, B_6, \ldots, B_{p-3}$ (already defined in Chapter 4, Section I, A).

Kummer was soon able to determine all irregular primes less than 163, namely, 37, 59, 67, 101, 103, 131, 149, 157. He maintained the hope that there exist infinitely many regular primes. This is a truly difficult problem to settle, even though the answer should be positive, as it is supported by excellent numerical evidence.

On the other hand, it was somewhat of a surprise when Jensen proved, in 1915, that there exist infinitely many irregular primes. The proof was actually rather easy, involving some arithmetical properties of the Bernoulli numbers.

In Chapter 6, I shall return to a conjecture about the distribution of regular primes. Let $\pi_{reg}(x)$ be the number of regular primes $p \leqslant x$ (including $p = 2$),

$$\pi_{ii}(x) = \pi(x) - \pi_{reg}(x).$$

For each irregular prime p, the pair $(p, 2k)$ is called an irregular pair if $2 \leqslant 2k \leqslant p - 3$ and p divides the numerator of B_{2k}. The irregularity index of p, denoted $ii(p)$, is the number of irregular pairs $(p, 2k)$.

For $s \geqslant 1$ let $\pi_{iis}(x)$ be the number of primes $p \leqslant x$ such that $ii(p) = s$.

Record

The most extensive calculations about regular primes (and the validity of Fermat's last theorem) are due to Wagstaff (1978) and Tanner & Wagstaff (1987), who extended previous computations done by Johnson (1975). Here are the results with $N = 150,000$[*]:

$$\pi(N) = 13848$$
$$\pi_{reg}(N) = 8399$$
$$\pi_{ii}(N) = 5449$$
$$\pi_{ii1}(N) = 4215 \quad \text{(the smallest is 37)}$$
$$\pi_{ii2}(N) = 1020 \quad \text{(the smallest is 157)}$$
$$\pi_{ii3}(N) = 192$$
$$\pi_{ii4}(N) = 20$$
$$\pi_{ii5}(N) = 2 \qquad \text{(these are the primes 78233, 94963)}$$
$$\pi_{iis}(N) = 0 \qquad \text{for all } s \geqslant 6.$$

---

[*]This is the place to tell a little anecdote and a last minute development. In one of my reports on Fermat's last theorem, it was written in my preprint that Fermat's theorem holds for every prime p less than 125,080 — instead of 125,000, as indicated by Wagstaff. This was only due to a typing mistake. Nevertheless, it worried Wagstaff sufficiently that he undertook to extend his computations and test six more irregular primes up to that limit. This was done in June 1985, on a CYBER 205 computer. I believe that the record for the exponent in Fermat's last theorem is, for a while, safely in the hands of Tanner & Wagstaff.

The longest known string of consecutive regular primes has 27 primes and begins with 17881.

The longest known string of consecutive irregular primes has 11 primes and begins with 8597.

The only known irregular pairs which are "successive" $(p,2k)$, $(p,2k + 2)$, are $p = 491$, $2k = 336$ or $p = 587$, $2k = 90$. There are no known triples $(p,2k)$, $(p,2k + 2)$, $(p,2k + 4)$ of irregular pairs.

It is conjectured, but it has never been proved, that there are primes with arbitrarily high irregularity index.

Up to now, I have not touched the following question: when is the arithmetic of the p-cyclotomic field ordinary, that is, for which values of p is $h(p) = 1$?

Owing to the observed growth of $h(p)$, it was long suspected that $h(p) = 1$ only for $p \leqslant 19$. This is true and was proved by Uchida (and is also attributed to Montgomery) in 1971. The problem was similar to the determination of the imaginary quadratic fields with class number 1, on which I aleady reported in Chapter 3, Section II.

If $m > 1$, $m \not\equiv 2 \pmod 4$ (to avoid duplication) and ζ_m is a primitive mth root of 1, it is natural to ask when the class number of the cyclotomic field $Q(\zeta_m)$ is equal to 1. This was settled by Masley in his thesis (1972) and jointly published with Montgomery in 1976. The answer is $m = 3$, 4, 5, 7, 8, 9, 11, 12, 13, 15, 16, 17, 19, 20, 21, 24, 25, 27, 28, 32, 33, 35, 36, 40, 44, 45, 48, 60, 84 − and no others.

Regular primes may be better understood, when the subject of the class number of cyclotomic fields is considered from a broader point of view. This is provided by Iwasawa's theory.

Let p be a fixed odd prime, $\zeta_{p^{n+1}}$ a primitive root of unity of order p^{n+1} (for every $n \geqslant 0$). Let $K_n = Q(\zeta_{p^{n+1}})$, thus K_0 is the p-cyclotomic field considered before. Let $K_n^+ = Q(\zeta_{p^{n+1}} + \zeta_{p^{n+1}}^{-1})$ be the maximal real subfield of K_n. Just like the p-cyclotomic field, or for that matter, any algebraic number field (as shown by Minkowski), the group of classes of ideals of K_n, resp. K_n^+, is finite. Since p is now fixed, I shall use the notation:

h_n = class number of K_n,
h_n^+ = class number of K_n^+ .

It may be shown that h_n^+ divides h_n; let $h_n^- = h_n/h_n^+$. Denote by $p^{e(n)}$ the exact power of p dividing h_n.

Iwasawa proved (see, for example, his book of 1972, or Washington's book, 1980): there exist integers μ, $\lambda \geqslant 0$ and ν, such that for every n sufficiently large, $e(n) = \mu p^n + \lambda n + \nu$. Moreover, p is a regular prime if and only if $\mu = \lambda = \nu = 0$.

The question whether $e(n)$ could grow like p^n was important. This was ruled out by a famous theorem of Ferrero & Washington (1979), which stated that in this, and also in some other more general situations not treated in this book, that $\mu = 0$. Thus

$$e(n) = \lambda n + \nu.$$

With similar notation, considering the tower of real cyclotomic fields K_n^+, then also $e^+(n) = \lambda^+ n + \nu^+$, for all n sufficiently large.

Metsänkylä gave in 1975 and 1978 a criterion for the irregularity index of p to be λ. Namely, $ii(p) = \lambda$ if and only if the following two conditions are satisfied:

(i) $\lambda^+ = 0$

(ii) for every irregular pair $(p,2k)$,

$$\frac{B_{2k}}{2k} \not\equiv \frac{B_{2k+p-1}}{2k + p - 1} \pmod{p^2}.$$

Note that Kummer had proved the fundamental congruence, valid for any even index $2k$:

$$\frac{B_{2k}}{2k} \equiv \frac{B_{2k+p-1}}{2k + p - 1} \pmod{p}.$$

There is still much mystery surrounding the determination of the irregularity index of a prime, and it is probable that methods based on the theory of p-adic zeta functions will be helpful. For example, Báyer used this method in 1979 to give a new proof of Metsänkylä's criterion quoted above. The recent work of Skula (1980, 1986), with methods of the algebraic theory of cyclotomic fields is also noteworthy.

II. Sophie Germain Primes

I have already encountered the Sophie Germain primes, in Chapter 2, in connection with a criterion of Euler about divisors of Mersenne numbers.

I recall that p is a Sophie Germain prime when $2p + 1$ is also a prime. They were first considered by Sophie Germain and she proved the beautiful theorem:

If p is a Sophie Germain prime, then there are no integers x, y, z, different from 0 and not multiples of p, such that $x^p + y^p = z^p$.

In other words, for Sophie Germain's primes the "first case of Fermat's last theorem" is true. For a more detailed discussion, see my own book (1979).

It is not known whether there are infinitely many Sophie Germain primes, and I shall return to this matter in Chapter 6, where I'll discuss conjectures about their distribution.

Now I wish to explain in more detail the relations between the first case of Fermat's last theorem and primes like Sophie Germain primes.

Sophie Germain's theorem was extended by Legendre as follows: if $p > 2$ and $4p + 1$, or $8p + 1$, or $10p + 1$, or $14p + 1$, or $16p + 1$, are primes, then the first case of Fermat's last theorem is true for the exponent p. This was further extended by Dénes, as follows (1951):

If p is a prime, m is not a multiple of 3, $m \leqslant 53$, and $2mp + 1$ is also a prime, then the first case of Fermat's last theorem is true for the exponent p.

A better theorem had been proved by Krasner already in 1940: the first case is true for p, provided 3 does not divide m, $2mp + 1$ is a prime, $3^{m/2} < 2mp + 1$ and $2^{2m} \not\equiv 1 \pmod{2mp + 1}$; I showed in 1985 that the last condition is superfluous, provided also $3^{\phi(2m)} < 2mp + 1$.

The gist of the condition is that in the arithmetic progression of first term 1 and difference p, there should exist another prime $2mp + 1$, with m sufficiently small with respect to p (and not a multiple of 3). As already discussed in Chapter 4, Section IV, this cannot be guaranteed by any theorem, and one is not entitled to expect a prime so soon.

Undeterred by this difficulty, Adleman and Heath-Brown considered simultaneously the primes p in appropriate intervals, so as to average the phenomenon, and using a deep result in sieve theory by Fouvry (1985), they were able to show that the first case of Fermat's last theorem is true for infinitely many prime exponents.

Now I indicate estimates for the number of Sophie Germain primes less than any number $x \geqslant 1$.

More generally, let $a, d \geqslant 1$, with ad even and $\gcd(a,d) = 1$.

For every $x \geqslant 1$, let

$$S_{d,a}(x) = \#\{p \text{ prime} \mid p \leqslant x, a + pd \text{ is a prime}\}.$$

[In the notation proposed previously, $S_{d,a}(x) = \pi_{x,dX+a}(dx + a)$.]

If $a = 1, d = 2$, then $S_{2,1}(x)$ counts the Sophie Germain primes $p \leqslant x$.

The same sieve methods of Brun, used to estimate the number $\pi_2(x)$ of twin primes less than x, yield here the similar bound

$$S_{d,a}(x) < \frac{Cx}{(\log x)^2}.$$

By the prime number theorem,

$$\lim_{x \to \infty} \frac{S_{d,a}(x)}{\pi(x)} = 0.$$

It is then reasonable to say that the set $S_{d,a}$ has density 0. In particular, the set of Sophie Germain primes, and by the same token, the set of twin primes, have density 0.

In 1980, Powell gave a proof that $S_{d,a}$ has density 0, avoiding the use of sieve methods.

Concerning the numbers $8p + 1$, where p is a prime, here is a quite interesting and unexpected result of Vaughan (1973): either there are infinitely many primes p such that $8p + 1$ is a prime or the product of two (not necessarily distinct) primes, or there are infinitely many primes p such that $8p + 1$ is the product of three distinct primes. In the second case $\tau(8p + 1) = \tau(8p) = 8$ for infinitely many primes; here $\tau(n)$ denotes the number of divisors of n. This result is related with the following conjecture of Erdös & Mirsky (1952): there exist infinitely many n such that $\tau(n + 1) = \tau(n)$.

In 1973, Vaughan's result was very intriguing, but now it is super-seded. Indeed, Chen's method (developed in connection with Goldbach's conjecture) allowed to prove that, for infinitely many primes p, $8p + 1$ (and also $2p + 1$) is either a prime, or the product of two (not necessarily distinct) primes.

Concerning the conjecture of Erdös & Mirsky, significant progress was made by Spiro in her thesis (1981):

$$\tau(n) = \tau(n + 5040) \text{ for infinitely many integers } n.$$

Finally, in 1984, Heath-Brown fully proved the conjecture of Erdös & Mirsky. Moveover, he showed that for large x, there are at least $x/(\log x)^7$ integers $n \leqslant x$ such that $\tau(n) = \tau(n + 1)$. See also Spiro's paper (1986), where it is given a rather short proof that, for x large

$$\#\{n < x \mid \tau(n) = \tau(n + 2520)\} \geqslant \frac{x}{(\log x)^7} \, .$$

Last minute: Erdös has kindly informed me that the best results to date on this question are

$$\frac{x}{(\log \log x)^3} \leqslant \#\{n \mid n \leqslant x, \tau(n) = \tau(n + 1)\} \leqslant \frac{Cx}{(\log \log x)^{1/2}}$$

(for x large); the lower bound is due to Hildebrand, the upper bound to Erdös, Pomerance & Sárközy (still unpublished).

III. Wieferich Primes

A prime p satisfying the congruence

$$2^{p-1} \equiv 1 \pmod{p^2}$$

ought to be called a Wieferich prime.

Indeed, it was Wieferich who proved in 1909 the difficult theorem:

If the first case of Fermat's last theorem (which I abbreviate First FLT) is false for the exponent p, then p satisfies the congruence indicated.

It should be noted that, contrary to the congruence $2^{p-1} \equiv 1$ (mod p) which is satisfied by every odd prime, the Wieferich congruence is very rarely satisfied.

Before the computer age(!), Meissner discovered in 1913, and Beeger in 1922, that the primes $p = 1093$ and 3511 satisfy Wieferich's congruence. If you have not been a passive reader, you must have already calculated that $2^{1092} \equiv 1$ (mod 1093^2), in Chapter 2, Section III. It is just as easy to show that 3511 has the same property.

Record

Lehmer has shown in 1981 that, with the exceptions of 1093 and 3511, there are no primes $p < 6 \times 10^9$, satisfying Wieferich's congruence.

According to results already quoted in Chapter 2, Sections III and IV, the computations of Lehmer say that the only possible factors p^2 (where p is a prime less than 6×10^9) of any pseudoprime, must be 1093 or 3511. This is confirmed by the numerical calculations of Pomerance, Selfridge & Wagstaff, quoted at the end of Chapter 4, Section VIII.

In 1910, Mirimanoff proved the following theorem, analogous to Wieferich's theorem:

If the first case of Fermat's last theorem is false for the prime exponent p, then $3^{p-1} \equiv 1$ (mod p^2).

It may be verified that 1093 and 3511 do not satisfy Mirimanoff's congruence. These two results, together with the computations of Lehmer, ensure that the first case of Fermat's last theorem is true for every exponent having a prime factor less than 6×10^9. But this is not the record for the first case of Fermat's last theorem. Wait and see how much bigger it is!

More generally, for any base $a \geqslant 2$ (where a may be prime or composite) one may consider the primes p such that p does not divide a and $a^{p-1} \equiv 1$ (mod p^2). In fact, it was Abel who first asked for such examples (in 1828); these were provided by Jacobi, who indicated the following congruences, with $p \leqslant 37$:

$$3^{10} \equiv 1 \pmod{11^2}$$
$$9^{10} \equiv 1 \pmod{11^2}$$
$$14^{28} \equiv 1 \pmod{29^2}$$
$$18^{36} \equiv 1 \pmod{37^2}.$$

I will give soon a more extensive table.

The quotient

$$q_p(a) = \frac{a^{p-1} - 1}{p}$$

has been called the Fermat quotient of p, with base a. The residue modulo p of the Fermat quotient behaves somehow like a logarithm (this was observed already by Eisenstein in 1850): if p does not divide ab, then

$$q_p(ab) \equiv q_p(a) + q_p(b) \pmod{p}.$$

Also

$$q_p(p - 1) \equiv 1 \pmod{p}, \qquad q_p(p + 1) \equiv -1 \pmod{p}.$$

In my article "1093" (1983), I indicated many truly interesting properties of the Fermat quotient.

As an illustration, I quote the following congruence, which is due to Eisenstein (1850):

$$q_p(2) \equiv \frac{1}{p} \left[1 - \frac{1}{2} + \frac{1}{3} - \frac{1}{4} + \cdots - \frac{1}{p-1} \right] \pmod{p}.$$

The following problems are open:

(1)   Given $a \geqslant 2$, do there exist infinitely many primes p such that $a^{p-1} \equiv 1 \pmod{p^2}$?

(2)   Given $a \geqslant 2$, do there exist infinitely many primes p such that $a^{p-1} \not\equiv 1 \pmod{p^2}$?

The answer to (1) should be positive, why not? But I'm stating it with no base whatsoever, since the question is no doubt very difficult. However, as I shall indicate in Chapter 6, Section III, there is a heuristic reason to believe that the answer should be "yes."

I shall now describe unexpected connections between concepts and facts which have no apparent immediate reason to be related: Fermat quotients, primitive factors of binomials, Fermat numbers, Mersenne numbers, powerful numbers (I will define them below) — all this having a bearing with the first case of Fermat's last theorem.

It is convenient to introduce some notation.

If $k \geq 1$ and ℓ is any prime, let

$$\mathfrak{W}_\ell^{(k)} = \{p \text{ prime} \mid \ell^{p-1} \equiv 1 \pmod{p^k}\}.$$

Clearly, $\mathfrak{W}_\ell^{(1)}$ consists of all the primes different from ℓ .

As I have already mentioned, the results of Wieferich and Mirimanoff say that if the (First FLT) is false for the exponent p, then p belongs to the sets $\mathfrak{W}_2^{(2)}$, $\mathfrak{W}_3^{(2)}$. Later, Pollaczek (in 1917, continuing work of Vandiver and Frobenius) extended the above conclusion: p must belong to $\mathfrak{W}_\ell^{(2)}$, for all primes $\ell \leq 31$. Pollaczek's paper is indeed very interesting despite some flaws. He had to stop at $\ell = 31$ because his method led to the computation of very large determinants and the factorization of quite large integers, so it could not be continued indefinitely. In 1986, Granville understood and reworked Pollaczek's proof. With clever computational tricks and the help of MAPLE at the University of Waterloo (G. Gonnet, M. B. Monagan), he extended Pollaczek's result as follows: if (First FLT) is false for p, then p belongs to $\mathfrak{W}_\ell^{(2)}$, for every prime $\ell \leq 89$. This result is part of Granville's Ph.D. thesis (1987, Queen's University in Kingston) and is written in a preprint co-authored by Monagan. As I have already mentioned, there are rather few primes in $\mathfrak{W}_2^{(2)}, \mathfrak{W}_3^{(2)}$; thus, to require that $p \in \mathfrak{W}_\ell^{(2)}$ for all primes $\ell \leq 89$ is an extremely restrictive condition. In 1948, Gunderson quantified this restriction, by showing the following estimate:

If $p \in \mathfrak{W}_\ell^{(2)}$ for every prime $\ell \leq p_N$ (the Nth prime, $N \geq 3$), then

$$p - 1 \geq \binom{2N-2}{N-1} \times \frac{4}{N!} \times \frac{[\log(p/\sqrt{2})]^N}{(\log 2)(\log 3)\cdots(\log p_N)}.$$

This inequality holds for all primes p which are sufficiently large, say $p \geq G(N)$. In 1981, Shanks & Williams decided to calculate $G(N)$, and they made the surprising discovery that $G(N)$ grows to a maximum, at $G(29) = 4{,}408{,}660{,}978{,}137{,}503$, and then it decreases.

According to Granville's result, taking $p_N = 89$, so $N = 24$, it follows that the first case of Fermat's last theorem holds for every $p < G(24) = 714{,}591{,}416{,}091{,}389$.

Right now, due to difficulties stemming from the amount of computations, this is the record attained. If the calculations will be continued up to $p_{29} = 109$, then it will be ensured that the first case of FLT holds for all primes less than $G(29)$. The last drop of blood from Pollaczek and Gunderson's method!

This calls for an improvement of Gunderson's inequality.

I return to the fundamental theorem of Wieferich. It follows at once, as noted by Landau in 1913: If p is a prime such that $pc = 2^e \pm 1$, for some $e > 1$ and p not dividing c, then (First FLT) is true for p.

Thus, if p is a Mersenne prime, then (First FLT) is true for the exponent p, so the (First FLT) is true for the largest prime known today(!), namely,

$$M_{216091} = 2^{216091} - 1.$$

Combining the various criteria involving Fermat quotients (from Wieferich, to Pollaczek, to Granville), it is equally easy to prove:

If p is a prime, if there exists a natural number c, not a multiple of p, such that $pc = u \pm v$, where all prime factors of uv are at most 89, then (First FLT) is true for p.

This leads to the consideration of the following sets. If $L \geqslant 1$, let

> $\aleph_L = \{p$ prime $\mid$ there exists c, not a multiple of p, such that $pc = u \pm v$, where all prime factors of uv are at most $L\}$.

Is $\aleph_{89}$ an infinite set? If this is true, then (First FLT) is true for infinitely many primes p. This conclusion is exactly the difficult theorem proved by Adleman & Heath-Brown and Fouvry, which I discussed in the preceding section. It is therefore likely to be very difficult − but also interesting − to establish that $\aleph_{89}$ is infinite.

To further analyze the situation, I introduce other sets of primes.

If $k \geqslant 1$ and ℓ is any prime, let

$$\mathbb{N}_{\ell}^{(k)} = \{p \text{ prime} \mid \text{there exists } s \geqslant 1 \text{ such that } p \text{ divides } \ell^s + 1,$$
$$\text{but } p^{k+1} \text{ does not divide } \ell^s + 1\}.$$

For example, if $\ell \leqslant L$, then $\mathbb{N}_{\ell}^{(1)} \subseteq \mathbb{N}_L$. So, in order to show that $\mathbb{N}_L$ is infinite (say with $L = 89$), it suffices to find a prime $\ell \leqslant L$ such that $\mathbb{N}_{\ell}^{(1)}$ is infinite. In other words, consider the sequence of integers $\{\ell + 1, \ell^2 + 1, \ell^3 + 1, ...\}$. By Bang's theorem (see Chapter 2, Section II), there are infinitely many primes p dividing some number of the sequence because (with the only exception $\ell = 2$, $s = 3$) each number $\ell^s + 1$ has a primitive prime factor. Are there still infinitely many such primes belonging to $\mathbb{N}_{\ell}^{(1)}$?

This is true if there would exist infinitely many primitive prime factors whose squares are not factors. A hard question to settle — but once again, full of important consequences.

A result of 1968 by Puccioni, has been improved as follows (see my own paper, 1987):

If $k \geqslant 1$ and ℓ is any prime, then $\mathbb{N}_{\ell}^{(k)} \cup \mathbb{W}_{\ell}^{(k+2)}$ is an infinite set.

It suffices therefore to show that $\mathbb{W}_{\ell}^{(3)}$ is a finite set (for some prime ℓ) to conclude that $\mathbb{N}_{\ell}^{(1)}$ is infinite.

For example, if $\ell = 2$, no integer in $\mathbb{W}_{2}^{(3)}$ is known. There are heuristic reasons to believe that $\mathbb{W}_{\ell}^{(3)}$ is finite, but no proof has been found as yet (see Chapter 6, Section III).

The following elementary result is due to Powell (see the problem proposed to the *American Mathematical Monthly* in 1982, and its solution by Mattics in 1984):

The set $S = \bigcup\limits_{k \text{ odd}} (\mathbb{W}_{\ell}^{(k)} \backslash \mathbb{W}_{\ell}^{(k+1)})$ is infinite.

Proof. It is convenient to use the notation $q^e \parallel a$ (when q is a prime, $e \geqslant 0$) to express that q^e divides a, but q^{e+1} does not divide a. Then q^e is called the exact power of q dividing a.

Note that $q \in S$ when, and only when, the exact power q^e dividing

$\ell^{q-1} - 1$ has odd exponent e.

Assume that $q_1, \ldots, q_n$ $(n \geqslant 0)$ are the only odd primes in the above set S. Let $s = \prod_{i=1}^{m}(q_i - 1)^2$ when $n \geqslant 1$, or $s = 1$, otherwise. Since each q_i divides $\ell^{4s} - 1$, then q_i does not divide $\ell^{4s} + 1$.

Consider now the number $\ell^{4s} + 1$. If p is any of its odd prime divisors, then $p \neq q_1, \ldots, q_n$. So the exact power p^d dividing $\ell^{dq-1} - 1$ has even exponent d.

Let $p^e \| \ell^{4s} + 1$ with $e \geqslant 1$. Note that p does not divide $\ell^{4s} - 1$, so $p^e \| \ell^{8s} - 1$.

If r is the order of ℓ modulo p, then r divides $p - 1$ and $8s$. Write $p - 1 = rk$, so $p \nmid k$. Note that $\ell^r - 1$ divides $\ell^{p-1} - 1$ and

$$\frac{\ell^{p-1} - 1}{\ell^r - 1} = \ell^{r(k-1)} + \ell^{r(k-2)} + \cdots + \ell^r + 1 \equiv k \pmod{p};$$

so p does not divide the above quotient, hence $p^d \| \ell^r - 1$.

Write $8s = rhp$, with $f \geqslant 0$ and p not dividing h. Since $p \nmid r$, $p \neq 2$ and s is a square, then f is even.

Note that

$$\frac{\ell^{rh} - 1}{\ell^r - 1} = \ell^{r(h-1)} + \cdots + \ell^r + 1 \equiv h \pmod{p}$$

so p does not divide this quotient, and therefore $p^d \| \ell^{rh} - 1$.

I make use of the following easy fact: if $a, b \geqslant 1$ and $m \geqslant 1$, then $p^a \| m - 1$ if and only if $p^{a+b} \| m^{p^b} - 1$.

Then $p^{d+f} \| \ell^{8s} - 1$, so $e = d + f$ is even.

But p is an arbitrary odd prime factor of $\ell^{4s} + 1$. So $\ell^{4s} + 1 = c^2$ or $2c^2$ for some integer c.

Fermat had shown, with his famous method of infinite descent that the equations $X^4 + Y^4 = Z^2$, $X^4 + Y^4 = 2Z^2$ have no solution in nonzero integers. This is a contradiction, and the set S is infinite. $\square$

For the case when $\ell = 2$, it is possible to give a proof using Fermat numbers.

Since $\mathfrak{w}_\ell^{(3)}$ is likely to be finite, the above result means that it is probable that there exist infinitely many primes p, such that $\ell^{p-1} \not\equiv 1 \pmod{p^2}$.

In 1985, Granville published an elementary proof of the following interesting result:

Let ℓ be a prime, assume that there exists n_0 such that if p is a prime, $p \geqslant n_0$, $p \equiv 1 \pmod 4$, then $\ell^{p-1} \equiv 1 \pmod{p^2}$. Then the set $\mathfrak{W}_\ell^{(3)}$ is infinite.

The conclusion being unlikely, there should exist infinitely many primes p such that $p \equiv 1 \pmod 4$ and $\ell^{p-1} \not\equiv 1 \pmod{p^2}$.

Now I turn my attention to powerful numbers. If $n = \prod_{i=1}^{k} p_i^{e_i}$ (each p_i prime) with each $e_i \geqslant 2$, then n is called a powerful number. Thus, powerful numbers are exactly those of the form $a^2 b^3$, with a,b integers.

The theory of powerful numbers is quite interesting. I have summarized the main results in my paper *Impuissants devant les puissances* (1987). Here, I want only to discuss the conjecture of Erdös (1975 and 1976):

(E)   There do not exist three consecutive powerful numbers.

Earlier in 1970, Golomb had raised this question. This conjecture was also stated independently by Mollin & Walsh (1986), who proved:

The conjecture (E) is equivalent to the following statement:

(MW) If m is a square-free integer, $m \equiv 7 \pmod 8$, if $t_1 + u_1\sqrt{m}$ is the fundamental unit of the quadratic field $Q(\sqrt{m})$, writing $t_k + u_k\sqrt{m} = (t_1 + u_1\sqrt{m})^k$ for every integer $k \geqslant 1$, if there exists k odd such that t_k is even and powerful, then either u_k is odd or m does not divide u_k.

In this form, the conjecture may be tested with a computer to see whether it is false. For the moment, it stands like a rock. Mollin & Walsh used $m = 7$, $t_1 + u_1\sqrt{m} = 8 + 3\sqrt{7}$, and $k \leqslant 114,254,287$.

Here are other related conjectures (when $a \geqslant 2$):

(E'_a) There exist at most finitely many integers $k \geqslant 1$ such that $a^k - 1$ is powerful.

(E''_a) There exist at most finitely many even integers $2k \geqslant 2$ such that $a^{2k} - 1$ is powerful.

The next statement (which is also conjectural) concerns the sequence of binomials $a^m - 1$ (with $m \geq 1$), where $a \geq 2$:

(B_a) There exist infinitely many numbers $a^m - 1$ having a primitive prime factor p_m such that p_m^2 does not divide $a^m - 1$.

Finally, consider the statement:

(W_a) There exist infinitely many primes p such that $a^{p-1} \not\equiv 1 \pmod{p^2}$.

If $a = 2$, this means that there exist infinitely primes which are not Wieferich primes.

Note the following implications between these statements (for every even $a \geq 2$):

$$(E_a') \quad (B_a)$$
$$\downarrow \qquad \downarrow$$
$$(E) \to (E_a'') \to (W_a)$$

Clearly, $(E) \to (E_a'')$ and $(E_a') \to (E_a'')$ are trivial.

Proof that $(B_a) \to (W_a)$. The proof follows at once from De Leon's lemma (1978, solution to a problem proposed by Powell, 1977):

If p is a prime, $p \mid a^m - 1$ and $p^2 \mid a^{p-1} - 1$, then $p^2 \mid a^m - 1$.

Indeed, let r be the order of a modulo p, so $r \mid p - 1$, $r \mid m$. Then $p - 1 = rk$, $a^r = 1 + cp$, $a^{p-1} = (1 + cp)^k \equiv 1 + kcp \pmod{p^2}$; since $p \nmid k$, then by hypothesis, $p \mid c$; therefore letting $m = rh$ then $a^m = (1+cp)^h \equiv 1 + hcp \equiv 1 \pmod{p^2}$.

Now, if $a^m - 1$ has a primitive prime factor p_m such that $p_m^2 \nmid a^m - 1$, by the lemma, $a^{p_m-1} \not\equiv 1 \pmod{p_m^2}$; this shows property (W_a). $\square$

Proof that $(E_a'') \to (W_a)$. Assume that (W_a) is not true, so there exists p_0 such that if p is a prime, $p > p_0$, then $a^{p-1} \equiv 1 \pmod{p^2}$.

Let $t = \prod_{p \leq p_0} p$, hence $\phi(t) = \prod_{p \leq p_0}(p - 1)$. For every $h \geq 1$, let $a_h = a^{ht\phi(t)}$. Then $a_h - 1$ is a powerful number, as I proceed to show.

Note that $2 \nmid a_h - 1$, when a is even, while if a is odd, since $\phi(t)$ is even, then $4 \mid a_h - 1$. If p is a prime such that $2 < p \leq p_0$, then $p(p - 1)$ divides $t\phi(t)$; from $a^{p-1} \equiv 1 \pmod{p}$, it follows that

$a^{p(p-1)} \equiv 1 \pmod{p^2}$, therefore, $p^2 \mid a^{ht\phi(t)} - 1 = a_h - 1$. Finally, if $p > p_0$ and $p \mid a_h - 1$, then by hypothesis $a^{p-1} \equiv 1 \pmod{p^2}$; hence by De Leon's lemma, $p^2 \mid a_h - 1$. Since h is arbitrary, this is contrary to the hypothesis (E''_a). □

The above proof is just a slight modification of the proof, given by Granville (1986), that Erdös' conjecture implies (W_2). According to Wieferich's theorem (W_2) implies that there exist infinitely many prime exponents p for which (First FLT) is true. Thus, if Erdös' conjecture is established, the theorem of Adleman & Heath-Brown and Fouvry will follow easily. I find this connection between powerful numbers and Fermat's theorem rather startling.

The next implication holds for special values of a:

If a is even and $a - 1$ is powerful, then $(E'_a) \rightarrow (B_a)$. In particular, $(E'_2) \rightarrow (B_2)$.

Proof. If (B_a) is false, there exists m_0 such that for every $m > m_0$ and for every primitive prime divisor p_m of $a^m - 1$, $p_m^2 \mid a^m - 1$. Choose a prime $q > m_0$; if $s \geqslant 1$ and ℓ is a prime dividing $a^{q^s} - 1$, then there exists h, $0 \leqslant h \leqslant s$, such that ℓ is a primitive prime divisor of $a^{q^h} - 1$. If $h = 0$, then $\ell^2 \mid a - 1$, by hypothesis. If $h \geqslant 1$, again $\ell^2 \mid a^{q^h} - 1$, because $q > m_0$. Hence $\ell^2 \mid a^{q^s} - 1$, This shows that $a^{q^s} - 1$ is a powerful number for every $s \geqslant 1$, contradicting the hypothesis. □

Now, the Fermat and Mersenne numbers enter the scene!
Consider the statements:

(F') There exist infinitely many Fermat numbers which are not powerful.

(M') There exist infinitely many Mersenne numbers which are not powerful.

Note that (F') and (M') are weaker than the current conjectures that there exist infinitely many square-free Fermat numbers (resp.

Mersenne numbers).

Consider also the statement:

(W'_2) There exists only finitely many primes p such that $2^{p-1} \equiv 1$ (mod p^2).

I will show the following implications:

$$(F')$$

$$(W'_2) \qquad\qquad (B_2)$$

$$(M')$$

(F') $\rightarrow$ (B_2): Since the Fermat numbers are pairwise relatively prime, it suffices to show that, if p is a prime such that $p \mid F_n$, $p^2 \nmid F_n$ (for some F_n), then p is a primitive prime factor of $2^{2^{n+1}} - 1$ and $p^2 \nmid 2^{2^{n+1}} - 1$.

Indeed, $p \nmid 2^{2^n} - 1$ (because p divides $F_n = 2^{2^n} + 1$), so $p \mid 2^{2^{n+1}} - 1$ and $p^2 \nmid 2^{2^{n+1}} - 1$. Let $e \geqslant 1$ be the smallest integer such that p divides $2^{2^e} - 1$. If $e < n + 1$, then $p \nmid F_{e-1} = 2^{2^{e-1}} + 1$, because Fermat numbers are pairwise relatively prime; so $p \mid 2^{2^{e-1}} - 1$, which is a contradiction. So p is a primitive prime factor of $2^{2^{n+1}} - 1$. $\square$

(M') $\rightarrow$ (B): This is trivial, because if p divides the Mersenne number $M_q = 2^q - 1$ (q prime), then it must be a primitive prime factor. $\square$

If anyone is daring enough to assume that there exist infinitely many prime Fermat numbers, respectively, prime Mersenne numbers, then again (W_2) follows. Indeed, let p be a prime Fermat or Mersenne number, so $p = 2^r \pm 1$. Hence

$$r q_p(2) \equiv q_p(2^r) \equiv q_p(p \mp 1) \equiv \pm 1 \pmod{p};$$

therefore,

$$q_p(2) \not\equiv 0 \pmod{p}, \text{ that is, } 2^{p-1} \not\equiv 1 \pmod{p^2}.$$

Of course what I have just shown is weaker than the implications $(F') \to (B_2)$, $(M') \to (B_2)$.

The proofs of the implications $(W_2') \to (F')$ and $(W_2') \to (M')$ require a preliminary result about the sets of primes dividing the sequences of Fermat numbers and Mersenne numbers. Let

$\mathcal{P}(F) = \{p \text{ prime} \mid \text{there exists } n \text{ such that } p \text{ divides } F_n\}$,

$\mathcal{P}(M) = \{p \text{ prime} \mid \text{there exists a prime } q \text{ such that } p \text{ divides } M_q\}$,

$\mathcal{P}^{(2)}(F) = \{p \text{ prime} \mid \text{there exists } n \text{ such that } p^2 \text{ divides } F_n\}$,

$\mathcal{P}^{(2)}(M) = \{p \text{ prime} \mid \text{there exists a prime } q \text{ such that } p^2 \text{ divides } M_q\}$.

Since Fermat numbers, respectively Mersenne numbers, are pairwise relatively prime, then the sets $\mathcal{P}(F)$ and $\mathcal{P}(M)$ are infinite.

Rotkiewicz (1965) and Warren & Bray (1967) showed:

1) $\mathcal{P}(F) \cap \mathfrak{W}_2^{(2)} = \mathcal{P}^{(2)}(F)$,

2) $\mathcal{P}(M) \cap \mathfrak{W}_2^{(2)} = \mathcal{P}^{(2)}(M)$.

Proof. (1) If $p \mid F_n$, then $2^{2^n} \equiv -1 \pmod{p}$, so the order of 2 modulo p is 2^{n+1}, hence $2^{n+1} \mid p - 1$. So $2^{2^{n+1}} - 1$ divides $2^{p-1} - 1$. It follows from $p^2 \mid F_n$ that $p^2 \mid 2^{2^{n+1}} - 1$, hence $p^2 \mid 2^{p-1} - 1$.

Conversely, if $p \mid F_n$, then $p \nmid 2^{2^n} - 1$ and $p \nmid 2^{2^{n+1}} - 1$. If $p^2 \mid 2^{p-1} - 1$, by De Leon's lemma, $p^2 \mid 2^{2^{n+1}} - 1$, hence $p^2 \mid F_n$.

(2) If $p^2 \mid M_q$, then q is the order of 2 modulo p, so $q \mid p - 1$, hence $p^2 \mid 2^{p-1} - 1$.

Conversely, if $p \mid M_q$ and $p^2 \mid 2^{p-1} - 1$, by De Leon's lemma, $p^2 \mid M_q$. $\square$

$(W_2') \to (F')$. Assume that there exists n_0 such that, if $n > n_0$, then F_n is powerful. Since the Fermat numbers are pairwise relatively prime, then $\mathcal{P}^{(2)}(F)$ is an infinite set, hence so is the set $\mathfrak{W}_2^{(2)}$, by the above result. $\square$

$(W_2') \to (M')$. The proof is similar. $\square$

The next question concerns a fixed prime and a variable base:

(3) If p is an odd prime, does there exist one or more bases a, $2 \leqslant a < p$, such that $a^{p-1} \equiv 1 \pmod{p^2}$?

Few results are known.

Kruyswijk showed in 1966 that there exists a constant C such that for every odd prime p:

$$\#\{a \mid 2 \leqslant a < p, \ a^{p-1} \equiv 1 \pmod{p^2}\} < p^{1/2 \, + \, C/(\log\log p)}.$$

So, not too many bases are good for each prime p.

On the other hand, Gandhi claimed (but I have not been able to locate this in any published article):

$$\#\{q \text{ prime} \mid 2 \leqslant q < p, \ q^{p-1} \equiv 1 \pmod{p^2}\} < p^{1/2},$$

and

$$\#\{q \text{ prime} \mid 2 \leqslant q < p, \ q^{p-1} \not\equiv 1 \pmod{p^2}\} \geqslant \pi(p) - p^{1/2}.$$

Powell showed that if $p \not\equiv 7 \pmod 8$, then there is at least one prime $q < \sqrt{p}$ such that $q^{p-1} \not\equiv 1 \pmod{p^2}$ (problem posed in 1982 to the *American Mathematical Monthly*, solution published in 1986, by Tzanakis). In this respect, the best result — and this time I have checked the proof, even though it is still unpublished (no wonder, he is my student and I have priorities) — is by Granville (1987). He showed that if u is any positive integer and p any prime such that $p \geqslant u^{2u}$ then

$$\#\{q \text{ prime} \mid 2 \leqslant q \leqslant u^{1/u}, \ q^{p-1} \equiv 1 \pmod{p^2}\} < up^{1/2u}.$$

Record

Besides the computations of Lehmer for the Fermat quotient with base 2, Brillhart, Tonascia & Weinberger considered (in 1971) bases up to 99. Below is an extension of the table by Kloss (1965) for odd prime bases up to 97:

Base	Range up to	Prime solutions p of $a^{p-1} \equiv 1 \pmod{p^2}$
13	2^{30}	11 1006003
5	2^{29}	20771 40487 53471161
7	2^{28}	5 491531
11	2^{28}	71
13	2^{28}	863 1747591
17	2×10^8	3 46021 48947
19	2×10^8	3 7 13 43 137 63061489
23	2×10^8	13 2481757 13703077
29	2^{28}	None
31	2×10^8	7 79 6451 2806861*
37	2×10^8	3 77867
41	2×10^8	29 1025273 138200401*
43	2×10^8	5 103
47	2×10^8	None
53	2×10^8	3 47 59 97
59	2×10^8	2777
61	2×10^8	None
67	2×10^8	7 47 268573
71	2×10^8	3 47 331
73	2×10^8	3
79	2×10^8	7 263 3037 1012573* 60312841*
83	2×10^8	4871 13691
89	2×10^8	3 13
97	2×10^8	7 2914393*

The solutions marked with an asterisk have been obtained by Keller, and kindly communicated to me.

IV. Wilson Primes

This is a very short section — almost nothing is known.

Wilson's theorem states that if p is a prime, then $(p - 1)! \equiv -1$ (mod p), thus the so-called Wilson quotient

$$W(p) = \frac{(p - 1)! + 1}{p}$$

is an integer.

p is called a Wilson prime when $W(p) \equiv 0$ (mod p), or equivalently $(p - 1)! \equiv -1$ (mod p^2).

For example, $p = 5, 13$ are Wilson primes.

It is not known whether there are infinitely many Wilson primes. In this respect, Vandiver wrote:

> This question seems to be of such a character that if I should come to life any time after my death and some mathematician were to tell me it had been definitely settled, I think I would immediately drop dead again.

Record

There is only one other known Wilson prime, besides 5, 13. It is 563, which was discovered by Goldberg in 1953 (one of the first successful computer searches, involving very large numbers).

The search for Wilson primes was continued by Pearson (in 1963, up to 200,183), by Kloss (in 1965, up to 1,017,000) and by Keller (communicated by letter, up to 3×10^6).

V. Repunits and Similar Numbers

There has been a great curiosity about numbers all of whose digits (in the base 10) are equal to 1: 1, 11, 111, 1111, ... They are called repunits. When are such numbers prime?

The notation Rn is commonly used for the number

$$11...1 = \frac{10^n - 1}{9},$$

with n digits equal to 1. If Rn is a prime, then n must be a prime, because if $n,m > 1$ then

$$\frac{10^{nm} - 1}{9} = \frac{10^{nm} - 1}{10^m - 1} \times \frac{10^m - 1}{9}$$

and both factors are greater than 1.

It is very easy to see that a repunit (different from 1) cannot be a square. It is more delicate to show that it cannot be a cube, and it is not known whether it may be a kth power, where k is not a multiple of 2 or 3 (see Obláth, 1956).

The following prime repunits were known for a long time: $(10^p-1)/9$ with $p = 2, 19, 23$.

Brillhart has completely factored the numbers $(10^p-1)/9$ with $p = 31, 37, 41, 43$ (this was first published by Angell & Godwin, in 1974).

The method to show that a repunit Rp is composite is just an application of Fermat's little theorem; it is a routine calculation when p is not too large.

In particular, it was easy to establish that R71 is composite. However, a major accomplishment was to determine its complete factorization especially as it turned out that it had no small prime factors. Indeed, R71 is the product of a P30 (prime number with 30 digits) by a P41. The factorization was done by Davis & Holdridge in 1984, using the quadratic sieve developed theoretically by Pomerance. Incidentally, for a while R71 was the largest number factored by a "general" factoring method. This has now been superseded by Silverman, who factored with a general method the 81-digit number $(2^{269}+1)/3$ and found it to be the product of a P24 by a P57, as well as the 87-digit number

$$\frac{5^{128} + 1}{2 \times 257} = (P26) \times (P62),$$

done in 1986.

Record

The only known prime repunits are R2, R19, R23, and more recently R317 (discovered by Williams in 1978) and R1031 (discovered by Williams & Dubner in 1986). Moreover, Dubner has determined that for every other prime p less than 10000, the repunit Rp is composite. Actual factorizations are now known for every $p \leqslant 89$.

Open problem: are there infinitely many primes repunits?

For more information about repunits see, for example, the book by Yates (1982).

Williams & Seah considered also in 1979 numbers of the form $(a^n-1)/(a-1)$, where $a \neq 2, 10$. It is usually difficult to establish the primality of large numbers of this type. Yet, as stated in their paper, in the last ten years much progress has been made both in technology and in the theory of factorization and primality testing.

$$\text{Table of primes of the form } \frac{a^n - 1}{a - 1}$$

$$\text{for } n \leqslant 1000, \quad a = 3, 5, 6, 7, 11, 12;$$

$$\text{or } n \leqslant 10000, \quad a = 10 \text{ (repunits)}$$

a	n									
3	3	7	13	71	103	541				
5	3	7	11	13	47	127	149	181	619	929
6	2	3	7	29	71	127	271	509		
7	5	13	131	149						
10	2	19	23	317	1031					
11	17	19	73	139	907					
12	2	3	5	19	97	109	317	353		

VI. Primes with Given Initial and Final Digits

Here is a curious, but rather easy fact, established by Sierpiński in 1959:

Let $a_1, a_2, ..., a_m, b_1, b_2, ..., b_n$ be any digits $(0 \leqslant a_i, b_j \leqslant 9)$, satisfying $b_n = 1, 3, 7$ or 9. Then there exist infinitely many prime numbers p which are written in base 10, with the a_i as initial digits and the b_j as final digits:

$$p = a_1 a_2 \; ... \; a_m \; ... \; b_1 b_2 \; ... \; b_n.$$

This result is an application of Dirichlet's theorem on primes in arithmetic progressions and easy estimates.

As indicated by Borucki & Diaz in 1974, this result extends immediately to arithmetic progressions. Provided the arithmetic progression $\{\ell + dk \mid k = 1,2, ...\}$ with $\gcd(d,\ell) = 1$, contains one number terminating in $b_1 b_2 \; ... \; b_n$, then it contains infinitely many primes with the initial digits a_i and the final digits b_j.

VII. Numbers $k \times 2^n \pm 1$

As I had said in Chapter 2, the factors of Fermat numbers are of the form $k \times 2^n + 1$. This property brought these numbers into focus, so it became natural to investigate their primality.

Besides the Mersenne numbers, other numbers of the form $k \times 2^n - 1$, as well as numbers of the form $k^2 \times 2^n + 1$, $k^4 \times 2^n + 1$, $k \times 10^n + 1$, have also been tested for primality.

From Dirichlet's theorem on primes in arithmetic progressions, given $n \geqslant 1$, there exist infinitely many integers $k \geqslant 1$, such that $k \times 2^n + 1$ is a prime, and also infinitely many integers $k' \geqslant 1$, such that $k' \times 2^n - 1$ is a prime.

On the other hand, it is interesting to ask the following question: Given $k \geqslant 1$, does there exist some integer $n \geqslant 1$, such that $k \times 2^n + 1$ is prime?

This question was asked by Bateman and answered by Erdös & Odlyzko (in 1979). I quote special cases of their results.

For any real number $x \geqslant 1$, let $N(x)$ denote the number of odd integers k, $1 \leqslant k \leqslant x$, such that there exists $n \geqslant 1$ for which $k \times 2^n + 1$ is a prime.   Then there exists $C_1 > 0$, C_1 being effectively computable, such that $N(x) \geqslant C_1 x$ (for every $x \geqslant 1$).

The method developed serves also to study the sequences $k \times 2^n - 1$ ($n \geqslant 1$) and other similar sequences.

Before discussing the records, note that it is not interesting to ask for the largest prime of the form $k \times 2 + 1$; just take $k = 2^{q-1} - 1$, with $q = 216091$, to obtain the largest Mersenne prime.   So, n should be greater than 1.

Record

The largest known prime of the form $k \times 2^n + 1$ with $n \geqslant 2$, is $7 \times 2^{54486} + 1$; it has 16402 digits.   It was discovered in July 1987 by J. Young, and I owe this information to my special agent SSWJr. The second largest known prime of the form $k \times 2^n + 1$ with $n \geqslant 2$, is $39079 \times 2^{26506} + 1$; it has 7984 digits and was discovered by Keller in 1985 (kindly communicated to me by letter).

Previously, the largest known prime of the form $k \times 2^n + 1$ was $5 \times 2^{23473} + 1$, with 7067 digits, also discovered by Keller in 1984 (announced in 1985); this number is a factor of the Fermat number F_{23471}, as I have previously indicated.

The largest known prime of the form $k^2 \times 2^n + 1$ was discovered by Keller in 1984: $17^2 \times 2^{18502} + 1 = (17 \times 2^{9251})^2 + 1$.   This is also the largest known prime of the form $n^2 + 1$.

The largest known prime of the form $k^4 \times 2^n + 1$ is $6952^4 \times 2^{9952} + 1$ (Atkin & Rickert, private communication).   It is equal to $(869 \times 2^{2491})^4 + 1$ and it is also the largest known prime of the form $n^4 + 1$.

The largest known prime of the form $k \times 10^n + 1$ is $150093 \times 10^{8000} + 1$, with 8006 digits.   It was discovered by Dubner on March 12, 1986; it took approximately 60 days of computing (24 hours per day) over a period of 7 months!

Young's number is also the largest known prime which is not a Mersenne number.   In this category, the next largest are Dubner's $150093 \times 10^{8000} + 1$ (8006 digits), followed by Keller's $39079 \times 2^{26506} + 1$ (7984 digits), and Dubner's $217833 \times 10^{7150} + 1$

(7156 digits), $6006 \times 10^{7090} + 1$ (7094 digits).

Even though a positive proportion of integers $k \geqslant 1$ has the property that $k \times 2^n + 1$ is a prime (for some n), Sierpiński proved in 1960 the following interesting theorem:

There exist infinitely many odd integers k such that $k \times 2^n + 1$ is composite (for every $n \geqslant 1$).

The numbers k with the above property are called Sierpiński numbers.

The proof of Sierpiński used the method of covering congruences, namely: there is a finite set of primes $q_1, q_2, ..., q_s$, such that if k is odd and belongs to a particular arithmetic progression modulo $2q_1 q_2 ... q_s$ then for every $n \geqslant 1$, $k \times 2^n + 1$ is divisible by one of the primes $q_1, q_2, ..., q_s$. It follows that there exists a positive number $C_2 < \frac{1}{2}$ such that $N(x) \leqslant C_2 x$, for x sufficiently large. Erdös & Odlyzko conjectured that $N(x) \sim Cx$ (for some $C > 0$).

Record

The smallest known Sierpiński number is $k = 78557$ (Selfridge, 1963).

In 1983, superseding work done by Jaeschke, Keller showed that if there exists a smaller Sierpiński number, it must be one of a list of 69 odd numbers. In a letter of September 1986, Keller writes that he has now reduced further this list to only 57 odd numbers.

It follows from Dirichlet's theorem on primes in arithmetic progressions and Sierpiński's result that there exists infinitely many Sierpiński numbers which are primes. For such primes p, the equation $\phi(x) = 2^n p$ has no solution in integers x (for every positive integer n). This was shown by Mendelsohn in 1976.

Aigner extended the results of Sierpiński in 1961, and considered sequences of the type $kr^n + h$. After excluding some trivial cases, he arrived at a similar theorem.

Sierpiński's problem may be posed similarly for sequences $k \times 2^n - 1$, $n \geqslant 1$. These are also infinitely many odd integers k such that $k \times 2^n - 1$ is composite for every n. The smallest known value of k, found by Keller (and communicated by letter) is $k = 509203$.

Addendum on Cullen Numbers

The numbers of the form $Cn = n \times 2^n + 1$ are known as Cullen numbers. In 1958, Robinson showed the $C141$ is a prime, the only known Cullen prime for 25 years, apart from $C1 = 3$. In 1984, Keller showed the Cn is also prime for $n = 4713, 5795, 6611, 18496$, and for every other $n \leqslant 20000$, Cn is composite (private communication).

VIII.   Primes and Second-Order Linear Recurrence Sequences

In this short section I shall consider sequences $T = (T_n)_{n \geqslant 0}$, defined by second-order linear recurrences.

Let P, Q be given nonzero integers such that $D = P^2 - 4Q \neq 0$. These integers P, Q are the parameters of the sequence T, to be defined now.

Let T_0, T_1 be arbitrary integers and for every $n \geqslant 2$, let $T_n = PT_{n-1} - QT_{n-2}$.

The characteristic polynomial of the sequence T is $X^2 - PX + Q$; its roots are

$$\alpha = \frac{P + \sqrt{D}}{2}, \quad \beta = \frac{P - \sqrt{D}}{2}.$$

So $\alpha + \beta = P$, $\alpha\beta = Q$, $\alpha - \beta = \sqrt{D}$.

The sequences $(U_n)_{n \geqslant 0}$, $(V_n)_{n \geqslant 0}$, with parameters (P,Q) and $U_0 = 0$, $U_1 = 1$, resp. $V_0 = 2$, $V_1 = P$, are precisely the Lucas sequences, already considered in Chapter 2, Section IV.

Let $\gamma = T_1 - T_0\beta$, $\delta = T_1 - T_0\alpha$, then

$$T_n = \frac{\gamma\alpha^n - \delta\beta^n}{\alpha - \beta} = T_1 \frac{\alpha^n - \beta^n}{\alpha - \beta} - QT_0 \frac{\alpha^{n-1} - \beta^{n-1}}{\alpha - \beta},$$

for every $n \geqslant 0$.

If $U = (U_n)_{n \geqslant 0}$ is the Lucas sequence with the same parameters, then $T_n = T_1 U_n - QT_0 U_{n-1}$ (for $n \geqslant 2$).

It is also possible to define the companion sequence $W = (W_n)_{n \geqslant 0}$. Let

$$W_0 = 2T_1 - PT_0, \quad W_1 = T_1 P - 2QT_0$$

and

$$W_n = PW_{n-1} - QW_{n-2}, \quad \text{for } n \geq 2.$$

Again, $W_n = \gamma\alpha^n + \delta\beta^n = T_1 V_n - QT_0 V_{n-1}$, where $V = (V_n)_{n \geq 0}$ is the companion Lucas sequence with parameters (P,Q).

Now, I could proceed establishing many algebraic relations and divisibility properties of the sequences, just like for the Lucas sequences, in Chapter 2, Section IV. However, my aim is just to discuss properties related with prime numbers.

First, consider the set

$$\textbf{P}(T) = \{p \text{ prime} \mid \text{there exists } n \text{ such that } p \mid T_n\}.$$

The sequence T is called degenerate if $\alpha/\beta = \eta$ is a root of unity. Then $\beta/\alpha = \eta^{-1}$ is also a root of unity; thus $|\eta + \eta^{-1}| \leq 2$. But

$$\eta + \eta^{-1} = \frac{\alpha^2 + \beta^2}{\alpha\beta} = \frac{P^2 - 2Q}{Q},$$

so, if T is degenerate then $P^2 - 2Q = 0, \pm Q, \pm 2Q$.

All the possibilities are listed below where n is any non-zero integer:

$P^2 - 2Q$	P	Q	D	α	β	η	Order of root η
0	$2n$	$2n^2$	$-4n^2$	$(1+i)n$	$(1-i)n$	i	4
Q	$3n$	$3n^2$	$-3n^2$	$-i\omega\sqrt{3}n$	$i\omega^2\sqrt{3}n$	$-\omega^2$	6
$-Q$	n	n^2	$-3n^2$	$-\omega n$	$-\omega^2 n$	ω^2	3
$2Q$	$2n$	n^2	0	n	n	1	1
$-2Q$	0	n	$4n$	$2\sqrt{n}$	$-2\sqrt{n}$	-1	2

The last two cases have been excluded.

From this analysis, it follows at once that if T is degenerate, if k is the order of the root of unity $\eta = \alpha/\beta$, if $p \in \textbf{P}(T)$, then p divides $6\eta T_0 T_1 \cdots T_{k-1}$.

Indeed, if $m = qk + r$, $0 \leq r < k$, then

$$T_m = T_1 \frac{\alpha^{qk+r} - \beta^{qk+r}}{\alpha - \beta} - QT_0 \frac{\alpha^{qk+r-1} - \beta^{qk+r-1}}{\alpha - \beta} = \alpha^{qk} T_r,$$

and since $\alpha^3 = -n^3$, $\alpha^4 = -4n^4$, $\alpha^6 = -3^3 n^6$, the assertion follows.

Therefore, if T is degenerate then $P(T)$ is finite.

In 1954, Ward proved that the converse is true:

If T is any nondegenerate sequence, then $P(T)$ is infinite.

Earlier, in 1937, Hall had given a criterion for a prime p to belong to $P(T)$. His result involved the auxiliary Lucas sequences $U = (U_n)_{n \geq 0}$, $S = (S_n)_{n \geq 0}$ with $U_0 = 0$, $U_1 = 1$ and parameters P, Q, and $S_0 = 0$, $S_1 = 1$, with parameters $P' = 2T_1 - PT_0$, $Q' = QT_0^2 + T_1^2 - PT_0 T_1$. Since $\gamma + \delta = P'$, $\gamma\delta = Q'$ then $S_n = (\gamma^n - \delta^n)(\gamma - \delta)$ for every $n \geq 0$.

Hall showed that $p \in P(T)$ if and only if the ranks of appearance $\rho_p(U)$, $\rho_p(S)$ satisfy: $\rho_p(S)$ divides $\rho_p(U)$.

This is not difficult to prove. Henceforth, Ward attempted to prove his theorem as a corollary of Hall's result. But, this implication depends on a difficult problem, which I have alluded to in Chapter 2, Section II: if $a \neq \pm 1$, a not a square, can it be a primitive root modulo infinitely many primes? See Chapter 6, Section I, where I will discuss this problem.

The natural ensuing problem is to ask whether $P(T)$ has necessarily a positive density (in the set of all primes) and, if possible, to compute it.

The pioneering work was done by Hasse (1966). He wanted to study the set of primes p such that the order of 2 modulo p is even. This means that there exists $n \geq 1$ such that p divides $2^{2n} - 1$, but p does not divide $2^m - 1$, for $1 \leq m < 2n$. Hence $2^n \equiv -1 \pmod{p}$, so p divides $2^n + 1$, and conversely.

The sequence $H = (H_n)_{n \geq 0}$, with $H_n = 2^n + 1$, is the companion Lucas sequence with parameters $P = 3$, $Q = 2$.

Let $\pi_H(x) = \#\{p \in P(H) \mid p \leq x\}$ for every $x \geq 1$.

Hasse showed that

$$\lim_{x \to \infty} \frac{\pi_H(x)}{\pi(x)} = \frac{17}{24}.$$

The number 17/24 represents the density of primes dividing the sequence H.

However, if $Q \geqslant 3$ and Q is square-free and $H' = (H_n)_{n \geqslant 0}$ is the companion Lucas sequence with parameters $(Q+1, Q)$, then $H'_n = Q^n + 1$. For this sequence, the density of $P(H')$ is $\frac{2}{3}$.

In 1985, Lagarias reworked Hasse's method and showed, among other results, that for the sequence $V = (V_n)_{n \geqslant 0}$ of Lucas numbers, $P(V)$ has density $\frac{2}{3}$. Previously, in 1961, Ward had shown that $P(V)$ has density at least $\frac{1}{2}$ and at most $\frac{2}{3}$.

The prevailing conjecture is that for every nondegenerate sequence T, the set $P(T)$ has a positive density.

Many special cases have confirmed the conjecture and in fact the statement may be proved under the assumption of a generalized form of Riemann's hypothesis (see Stephens, 1976).

Now, I turn to another very interesting and difficult problem.

Let $T = (T_n)_{n \geqslant 0}$ be a second-order linear recurring sequence. For example, the Fibonacci or Lucas sequences. These sequences do contain prime numbers, but it is not known, and certainly difficult to decide, whether they contain infinitely many prime numbers.

It follows from Chapter 2, Section IV, relations (IV.15), resp. (IV.16):

If U_m is a prime (and $m \neq 4$) then m is a prime.
If V_m is a prime (and m is not a power of 2) then m is a prime.

Of course, the converse need not be true.

For $3 \leqslant n < 1000$, the only Fibonacci numbers which are primes are the U_n, with:

$$n = 3, 4, 5, 7, 11, 13, 17, 23, 29, 43, 47, 83, 131, 137, 359,$$
$$431, 433, 449, 509, 569, 571.$$

The above list was indicated by Brillhart (1963), completing work by previous authors, including Jarden (1958). To my knowledge, the largest Fibonacci number recognized to be a prime, is U_{2971}, discovered by Williams.

Concerning the Lucas numbers, if $0 \leqslant n \leqslant 500$, L_n is a prime exactly when

$$n = 0, 2, 4, 5, 7, 8, 11, 13, 16, 17, 19, 31, 37, 41, 47, 53, 61,$$
$$71, 79, 113, 313, 353$$

(as listed by Brillhart). More prime Lucas numbers, discovered by Williams, are V_{503}, V_{613}, V_{617}, V_{803}, the latter being, according to the information at my disposal, the largest known prime Lucas numbers.

It should be noted that, if T is neither a Lucas sequence nor a companion Lucas sequence, then T need not to contain any prime number. The following example was found by Graham in 1964: let $P = 1$, $Q = -1$ and

$$T_0 = 1786772701928802632268715130455793,$$

$$T_1 = 1059683225053915111058165141686995.$$

Then T contains no primes.

Recall that if $U = (U_n)_{n \geqslant 0}$ is a Lucas sequence, by the law of appearance, any prime p, not dividing $2Qd$, must divide $U_{p-(D/p)}$. Hence the number $\frac{1}{p} U_{p-(D/p)}$ is an integer. A natural question is to study the distribution of the residue classes of these integers modulo p.

For example, considering the sequence $(U_n)_{n \geqslant 0}$ of Fibonacci numbers, Wall asked (in 1960) whether there exists any prime p, $p \neq 2$, 5 such that $\frac{1}{p} U_{p-(5/p)} \equiv 0 \pmod{p}$, that is, p^2 divides $U_{p-(5/p)}$. Williams has checked that no prime $p < 10^9$ has this property. In his study (1982), Williams obtained the congruences:

$$5U_p \equiv \left(\frac{5}{p} \right)(5S - 2^P - 2) \pmod{p^2}$$

where

$$S = \sum_{k=0}^{[p/5]} \left(\begin{array}{c} p \\ 5k \end{array} \right),$$

and

$$\frac{1}{p} U_{p-(5/p)} \equiv \frac{2}{5} \sum_{k=1}^{p-1-[p/5]} \frac{(-1)^k}{k} \pmod{p}.$$

For the sequence of Pell numbers, namely, $P = 2$, $Q = -1$, Williams showed that if $p < 10^8$ and p^2 divides $U_{p-(D/p)} = U_{p-(2/p)}$, then $p = 13$

or 31 or 1546463.

These developments are parallel to some considerations and computations about Fermat's quotients $(2^{p-1}-1)/p$, in Section III.

IX. The NSW-Primes

NSW does not mean North-South-West, nor New South Wales, but stands for Newman, Shanks & Williams, and I had the privilege of getting acquainted with their paper, while still in preprint form. This was just after Dan Shanks' visit to Queen's University, memorable for more than one reason — in 197?.

The paper deals with numbers related to the order of simple groups. As it will be pointed out, these numbers have a behaviour which reminds one of the Mersenne numbers.

The NSW numbers are

$$S_{2m+1} = \frac{(1 + \sqrt{2})^{2m+1} + (1 - \sqrt{2})^{2m+1}}{2}$$

for $m \geqslant 0$.

This sequence begins with

$$S_1 = 1, \quad S_3 = 7, \quad S_5 = 41, \quad S_7 = 239, \quad S_9 = 1393, \ldots .$$

For the Mersenne numbers $M_{2m+1} = 2^{2m+1} - 1$, the following classical elementary results hold (see Chapter 2, Section IV):

(M1) If M_{2m+1} is a prime, then $2m + 1$ is a prime.

(M2) If p is an odd prime, if $d > 0$ divides M_p, then $d \equiv \pm 1 \pmod 8$ and $d = 2kp + 1$ (for some $k \geqslant 1$).

(M3) If p, q are distinct odd primes, then M_p, M_q are relatively prime.

(M4) If p is a prime such that $p \equiv 3 \pmod 4$ and $q = 2p + 1$ is a prime, then q divides M_p.

For the NSW numbers, the similar results hold. They are established considering properties of the rank of appearance of primes with respect to the sequence $W = (W_n)_{n \geqslant 0}$, defined by $W_0 = 0$, $W_1 = 1$,

$W_n = 2W_{n-1} + W_{n-2}$ (for $n \geqslant 2$), and noting that $W^2_{2m+1} = \frac{1}{2}(S^2_{2m+1} + 1)$.
Let us quote these results explicitly:

(NSW1)   If S_{2m+1} is a prime, then $2m + 1$ is a prime.

(NSW2)   If p is an odd prime, if $d > 0$ divides S_p, then $d \equiv \pm 1$
(mod 8) and $d = 2kp + 1$ (for some $k \geqslant 1$).

(NSW3)   If p, q are distinct odd primes, then S_p, S_q are
relatively prime.

(NSW4)   If p is a prime such that $p \equiv 3$ (mod 4) and $q = 2p + 1$ is
a prime, then q divides S_p.

The following problems are open:

Does there exist infinitely many primes p such that S_p is prime?
Does there exist infinitely many primes p such that S_p is
composite?

Even though the answer to both questions should be affirmative,
no one seems to know how to approach this problem.

The connection between the NSW-numbers and simple groups is
the following.

Let P be a prime, F_P the field with P elements, V a vector space
of dimension $2n$ over F_P, let $f: V \times V \to F_P$ be a bilinear form which
is nondegenerate and skew-symmetric. Let G be the group of all
linear transformations $\sigma: V \to V$ such that σ has determinant 1 and σ
leaves the form f invariant: $f(\sigma(x), \sigma(y)) = f(x,y)$ for all $x,y \in V$.
Then G is a group, called the symplectic group of dimension $2n$ over
F_P. It is denoted by $\mathrm{Sp}(2n, F_P)$.

The order of the symplectic group is equal to

$$\frac{1}{3} P^{n^2} \prod_{i=1}^{n} (P^{2i} - 1).$$

The first result states:

If $S_{2m+1} = P$ is a prime then the order of $\mathrm{Sp}(4, F_P)$ is a square,
namely $[P^2(P^2 - 1)W_{2m+1}]^2$ where W_{2m+1} was defined above.

The second result, which is much more difficult to establish, is
just the converse:

If the order of a symplectic group $Sp(2n,F_P)$ is a square, then the prime P is a NSW-number, $P = S_{2m+1}$ and $2n = 4$.

Thus, for example, taking $S_3 = 7$, gives the symplectic group $Sp(4,7)$ of order 11760^2.

The following values of $p < 1000$ yield prime NSW-numbers:

$$p = 3, 5, 7, 19, 29, 47, 59, 163, 257, 421, 937, 947.$$

Since there is no known fast algorithm to test the primality of these numbers (like the Lucas-Lehmer test for Mersenne numbers), the determination of the prime NSW-numbers was rather laborious.

The following question remained untouched: is there any finite simple group, other than a symplectic group, having order which is a square? This should be possible to answer, now that the simple groups have been completely classified.

Chapter 6

HEURISTIC AND PROBABILISTIC RESULTS ABOUT PRIME NUMBERS

The word "heuristic" means: based on, or involving, trial and error. Heuristic results are formulated following the observation of numerical data from tables or from extended calculations. Sometimes these results express the conclusions of some statistical analysis.

There are also probabilistic methods. The idea is quite well explained in Cramér's paper (1937) already quoted in Chapter 4:

> In investigations concerning the asymptotic properties of arithmetic functions, it is often possible to make an interesting use of probability arguments. If, e.g., we are interested in the distribution of a given sequence S of integers, we then consider S as a member of an infinite class C of sequences, which may be concretely interpreted as the possible realizations of some game of chance. It is then in many cases possible to prove that, *with a probability equal to* 1, a certain relation R holds in C, i.e., that in a definite mathematical sense, "almost all" sequences of C satisfy R. Of course we cannot, in general, conclude that R holds for the particular sequence S, but results suggested in this way may sometimes afterwards be rigorously proved by other methods.

Heuristic and probabilistic methods, if not properly handled with caution and intelligence, may give rise to "dream mathematics," far removed from the reality. Hasty conjectures, misinterpretation of numerical evidence have to be avoided.

I'll be careful and restrict myself to only a few of the reliable contributions. I have in mind, Hardy & Littlewood, with their famous conjectures in *Partitio Numerorum*. Dickson, Bouniakowsky, Schinzel & Sierpiński, with their intriguing hypotheses and Siegel's work concerning the distribution of regular primes. But I'll also discuss the distribution of Mersenne primes and some other topics.

I. Prime Values of Linear Polynomials

Once, more, the starting point is Dirichlet's theorem on primes in arithmetic progressions. It states that if $f(X) = bX + a$, with integers a, b such that $a \neq 0$, $b \geqslant 1$, $\gcd(a,b) = 1$, then there exist infinitely many integers $m \geqslant 0$ such that $f(m)$ is a prime.

In 1904, Dickson stated the following conjecture, concerning the simultaneous values of several linear polynomials:

(D)   *Let* $s \geqslant 1$, $f_i(X) = b_i X + a_i$ *with* a_i, b_i *integers*, $b_i \geqslant 1$ *(for* i = 1, ..., s). *Assume that the following condition is satisfied*:

(*)   *There does not exist any integer* $n > 1$ *dividing all the products* $f_1(k)f_2(k) \cdots f_s(k)$, *for every integer k*.

Then there exist infinitely many natural numbers m *such that all numbers* $f_1(m)$, $f_2(m)$, ..., $f_s(m)$ *are primes*.

The following statement looks weaker than (D):

(D$_0$)   *Under the same assumptions for* $f_1(X)$, ..., $f_s(X)$, *there exists a natural number* m *such that the numbers* $f_1(m)$, ..., $f_s(m)$ *are primes*.

Whereas, at first sight, one might doubt the validity of (D), the statement (D$_0$) is apparently demanding so much less, that it may be more readily accepted. But, the fact is that (D) and (D$_0$) are equivalent.

Indeed, if (D_0) is true, there exists $m_1 \geqslant 0$ such that $f_1(m_1)$, ..., $f_s(m_1)$ are primes. Let $g_i(X) = f_i(X + 1 + m_1)$ for $i = 1, ..., s$. Then (*) is satisfied by $g_1(X)$, ..., $g_s(X)$, hence by (D_0) there exists $k_1 \geqslant 0$ such that $g_1(k_1)$, ..., $g_s(k_1)$ are primes; let $m_2 = k_1 + 1 + m_1 > m_1$, so $f_1(m_2)$, ..., $f_s(m_2)$ are primes. The argument may be repeated, and shows that (D_0) implies (D).

Dickson did not explore the consequences of his conjecture. This was the object of a paper by Schinzel & Sierpiński (1958), which I find so interesting that I'll devote a rather large portion of Chapter 6 to it.

As a matter of fact, Schinzel had proposed a more embracing conjecture [the hypothesis (H)] dealing with polynomials not necessarily linear. But, before I discuss the hypothesis (H) and its consequences, I shall indicate the many interesting results which Schinzel & Sierpiński proved, under the assumption that the conjecture (D) is valid. Alas, despite the sophisticated weighted sieve methods employed, there is no hope of proving (D); this state of affairs will be more convincing when I will tell the marvelous consequences of hypothesis (D).

On the other hand, there are definite results if the expectations are lowered, and the question is about k-almost-primes which are values of $\prod_{i=1}^{s} f_i(X)$.

In Chapter 10 of Halberstam & Richert's book, it is shown, among more technical results, that if the polynomials $f_i(X)$ ($i = 1, 2, ..., s$) verify the assumption made in hypothesis (D), there exist infinitely many natural numbers m such that $\prod_{i=1}^{s} f_i(m)$ is a k-almost-prime, where

$$k > s - 1 + s \sum_{j=1}^{s} \frac{1}{j} + s \log 2.$$

Thus, if $s = 2, 3, 4, 5, 6$, then k may be taken respectively equal to $k = 6, 10, 15, 19, 24$.

Now I will show various consequences of Dickson's hypothesis (D).

(D_1) *Let $s \geqslant 1$, let $a_1 < a_2 < \cdots a_s$ be nonzero integers and assume that $f_1(X) = X + a_1$, ..., $f_s(X) = X + a_s$ satisfy condition* (*) *in* (D).

Then there exist infinitely many integers $m \geqslant 1$ such that $m+a_1$, $m+a_2$, ..., $m+a_s$ are consecutive primes.

Proof that (D) implies (D$_1$). By (D) there exists infinitely many integers $h \geqslant 0$ such that $h+a_1$, $h+a_2$, ..., $h+a_s$ are primes. Choose one such h which is sufficiently large; all I need is that $h \geqslant a_s - 2a_1 + 2$. Let

$$b = \frac{(h+a_s)!}{(h+a_1)(h+a_2)\cdots(h+a_s)}$$

and let $g_i(X) = bX + h + a_i$ $(i = 1, ..., s)$.

Note that $2(h+a_i) > h + a_s$, so each prime $h + a_i$ does not divide b, that is, $\gcd(b, h+a_i) = 1$.

I will now show that $g_1(X)$, ..., $g_s(X)$ satisfy condition (*). If there exists a prime p dividing the products $g_1(n)g_2(n)\cdots g_s(n)$, say for $n = 0, 1, ..., p - 1$, then p divides $(h+a_1)(h+a_2)\cdots(h+a_s)$, so there exists k, $1 \leqslant k \leqslant s$, such that $p = h + a_k$. But $h + a_s < 2(h+a_k) = 2p$ so p does not divide b. Hence for every $i = 1, ..., s$ there exists a unique integer m_i, $0 \leqslant m_i \leqslant p - 1$, such that $bm_i \equiv -(h+a_i) \pmod{p}$, that is, p divides $bm_i + h + a_i = g_i(m_i)$. If $s \leqslant p - 1$, there would be an integer n, $0 \leqslant n \leqslant p - 1$, such that p would not divide $\prod_{i=1}^{s} g_i(n)$, which is a contradiction. So $h + a_k = p \leqslant s$. But $h + a_k \geqslant a_s - a_1 + 2 \geqslant s + 1$, which is absurd. So, condition (*) is indeed satisfied.

By (D), there exist infinitely many integers $n \geqslant 1$ such that $g_i(n)$ are primes $(i = 1, ..., s)$; so there exist infinitely many integers $m = bn + h$ such that $m + a_i$ $(i = 1, ..., s)$ are primes.

It remains to show that $m + a_1 < m + a_2 < \cdots < m + a_s$ are consecutive primes. Suppose there exists a prime q such that $m + a_i < q < m + a_{i+1}$ (with $1 \leqslant i \leqslant s - 1$). Then $q = m + a_i + d = bn + h + a_i + d$ is a prime. Here $0 < d < a_{i+1} - a_i$, hence $h + a_i < h + a_i + d < h + a_{i+1} \leqslant h + a_s$, so $h + a_i + d$ divides b, therefore $h + a_i + d$ divides q. But $bn > 0$ so $q > h + a_i + d > 1$, and this is absurd. $\square$

The conjecture of Polignac (1849), discussed in Chapter 4, Sections II and III is a consequence of (D$_1$):

(D$_{1,1}$) *For every even integer $2k \geqslant 2$ there exist infinitely many pairs of consecutive primes with difference equal to $2k$. In particular, there* *exist infinitely many pairs of twin primes.*

Proof that (D₁) implies (D₁,₁). Let $f_1(X) = X + 1$, $f_2(X) = X + 2k + 1$. Then the products $f_1(0)f_2(0) = 1 + 2k$, $f_1(1)f_2(1) = 2(2 + 2k)$ are relatively prime. So, condition (*) of (D) is satisfied, and, by (D₁), there exist infinitely many integers $m \geqslant 1$ such that $m + 1$ and $m + 2k + 1$ are consecutive primes.   □

Here is an interesting consequence concerning the abundance of twin primes:

(D₁,₂) *For every integer $m \geqslant 1$, there exist 2m consecutive primes which are m couples of twin primes.*

Proof that (D₁) implies (D₁,₂). Let $f_{2i-1}(X) = X + (i-1)(2m)! + 1$ and $f_{2i}(X) = X + (i-1)(2m)! + 3$, for $i = 1, ..., m$.

I will now show that these 2m polynomials satisfy condition (*) of (D). Let $g(X) = \prod_{j=1}^{2m} f_j(X)$, so that $g(X)$ has degree 2m, leading coefficient 1. Assume that a prime number p divides $g(0), g(1), ..., g(p-1)$, thus the congruence $g(X) \equiv 0 \pmod{p}$ has p solutions. Hence, thinking of this congruence as an equation over the finite field with p elements, it has 2m solutions (in the algebraic closure of the field with p elements); thus $p \leqslant 2m$ and therefore p divides $(2m)!$. But $g(0)$ is odd, hence $p \neq 2$. From

$$g(1) = \prod_{i=1}^{m} [2+(i-1)(2m)!][4+(i-1)(2m)!] \equiv \prod_{i=1}^{m} 2 \times 4 \equiv 2^{3m} \pmod{p},$$

since p divides $g(1)$, it follows that p divides 2^{3m}, and this is impossible. Thus, condition (*) is satisfied. So, by (D₁) there exist infinitely many integers $m \geqslant 1$ such that all integers $f_j(m)$ are consecutive primes. Noting that $f_{2i}(m) - f_{2i-1}(m) = 2$, they form m couples of twin primes.   □

Another quite unexpected consequence concerns primes in arithmetic progressions. In Chapter 4, Section IV, I showed that if a, $a+d$, ..., $a + (n-1)d$ are primes and $1 < n < a$ then d is a multiple of $\prod_{p \leqslant n} p$.

Now, using (D₁), I will show:

(D$_{1,3}$) *Let* $1 < n$, *let* d *be a multiple of* $\Pi_{p \leq n} p$. *Then there exist infinitely many arithmetic progressions, with difference* d, *each consisting of* n *consecutive primes.*

Proof that (D$_1$) implies (D$_{1,3}$). Let $f_i(X) = X + id$ (for $i = 0,1, ..., n$). The polynomials $f_1(X), ..., f_n(X)$ satisfy condition (*). Indeed, if there exists a prime p dividing each of the products $\Pi_{i=1}^m f_i(j)$ ($j = 0,1, ..., p - 1$), considering the corresponding congruence modulo p, it follows that $p \leq n$. By hypothesis, p divides d. But p divides

$$\prod_{i=1}^{n} f_i(1) = (1 + d)(1 + 2d) \cdots (1 + nd)$$

and p divides d, hence p divides 1, which is absurd.

By (D$_1$), there exist infinitely many integers $m \geq 1$ such that $m+d$, $m+2d$, ..., $m+nd$ are consecutive primes. □

The reader should compare this very strong statement with what has been proved, independently of any conjecture in Chapter 4, Section IV.

Concerning Sophie Germain primes, I can prove:

(D$_2$) *For every* $m \geq 1$ *there exists infinitely many arithmetic progressions consisting of* m *Sophie Germain primes.*

Proof that (D) implies (D$_2$). Let $d = (2m+2)! \geq 4!$ and consider the $2m$ polynomials $f_1(X) = X + d$, $f_2(X) = X + 2d$, ..., $f_m(X) = X + md$, $f_{m+1}(X) = 2X + 2d + 1$, $f_{m+2}(X) = 2X + 4d + 1$, ..., $f_{2m}(X) = 2X + 2md + 1$. Thus $f_{m+i}(X) = 2f_i(X) + 1$ for $i = 1, ..., m$. These polynomials have integral coefficients, positive leading coefficient and they are irreducible. I show that they satisfy condition (*).

Let $f(X) = \Pi_{i=1}^{2m} f_i(X)$, so that $f(X)$ has degree $2m$ and leading coefficients 2^m.

Assume that a prime p divides $f(k)$ for $k = -1,0,1, ..., p - 2$. Since $f(-1) \equiv 1 \pmod{2}$, it follows that $p \neq 2$, so the congruence $f(X) \equiv 0 \pmod{p}$ has $2m$ not necessarily distinct roots (in the algebraic closure of the field with p elements). Thus, $p \leq 2m$, hence p divides $d = (2m + 2)!$. But $f(-1) \equiv 1 \pmod{3}$, so $p \neq 3$. Noting that $f(1) \equiv 3^m \pmod{p}$, then p does not divide $f(1)$, which is a contradiction.

By (D) it follows that there exist infinitely many integers k such that $f_i(k) = p_i$ and $f_{m+i}(k) = 2p_i + 1$ are primes (for $i = 1,2, ..., m$); moreover $p_1 < p_2 < \cdots < p_m$ are in an arithmetic progression with difference $d = (2m+2)!$ □

In particular, (D) implies the existence of infinitely many Sophie Germain primes, a fact which has never been proved without appealing to a conjecture. I shall return soon to a quantitative statement about the distribution of Sophie Germain primes.

Among the many statements in Schinzel & Sierpiński's paper, I quote next the following ones:

(D_3) *Let a, b, c be pairwise relatively prime non-zero integers, not all being odd. Then there exist infinitely many pairs of primes (p,q), such that $ap - bq = c$.*

This is one of the conjectures of Hardy & Littlewood in 1923 (*Partitio Numerorum*, III).

($D_{3,1}$) *Every positive rational number may be written, in infinitely many ways, in one of the forms: $(p+1)/(q+1)$, or $(p-1)/(q-1)$, where p, q are prime numbers.*

($D_{3,1,1}$) *There exist infinitely many primes of the form $\frac{1}{2} \phi(n)$ (where $n \geqslant 1$ and ϕ denotes the Euler function).*

It is not difficult to show that (D) implies (D_3), also (D_3) implies ($D_{3,1}$) and ($D_{3,1}$) implies ($D_{3,1,1}$).

Next, I consider the statement:

(D_4) *There exist infinitely many triples of consecutive integers, each being the product of two distinct primes.*

Proof that (D) implies (D_4). Let $f_1(X) = 10X + 1$, $f_2(X) = 15X + 2$, $f_3(X) = 6X + 1$. Then $f_1(0)f_2(0)f_3(0) = 2$, $f_1(1)f_2(1)f_3(1) = 11 \times 17 \times 7$; these numbers are relatively prime, and therefore condition (*) is satisfied. Thus, there exist infinitely many integers $m \geqslant 1$ such that $p = 10m + 1$, $q = 15m + 2$, $r = 6m + 1$ are primes. Then $3p + 1 = 30m + 4 = 2q$, $3p + 2 = 30m + 5 = 5r$, so $3p$, $3p + 1$, $3p + 2$ are products of two distinct primes.   ☐

The conjecture (D) is so powerful that it implies also:

(D$_5$) *There exist infinitely many composite Mersenne numbers.*

Proof that (D) implies (D$_5$). Let $f_1(X) = 4X - 1$ and $f_2(X) = 8X - 1$. Since $f_1(0)f_2(0) = 1$, condition (*) is satisfied, hence there exist infinitely many integers $m \geqslant 1$ such that $p = 4m - 1$ and $q = 8m - 1$ are primes. But $q = 2p + 1$, and $p \equiv 3 \pmod 4$, so by the Euler and Lagrange result stated in Chapter 2, Section IV, q divides $2^P - 1$. But, if $m > 1$ then the corresponding primes p, q satisfy $2^P - 1 > q = 2p + 1$, so $2^P - 1$ is composite.   ☐

I recall (see Chapter 2, Section IV) that no one has as yet succeeded in proving [without assuming the conjecture (D)] that there exist infinitely many composite Mersenne numbers. However, it is easy to prove that other sequences, similar to that of Mersenne numbers, contain infinitely composite numbers. The following result was proposed by Powell as a problem in 1981 (a solution by Israel was published in 1982):

If m, n are integers such that $m > 1$, $mn > 2$ (this excludes $m = 2$, $n = 1$), then there exist infinitely many composite numbers of the form $m^P - n$, where p is a prime number.

Proof. Let q be a prime dividing $mn - 1$, so $q \nmid m$. If p is a prime such that $p \equiv q - 2 \pmod{q - 1}$, then $m(m^P - n) \equiv m(m^{q-2} - n) \equiv 1 - mn \equiv 0 \pmod q$, hence q divides $m^P - n$. By Dirichlet's theorem on primes in arithmetic progressions, there exist infinitely many primes p, such that $p \equiv q - 2 \pmod{q - 1}$, hence there exist infinitely many composite integers $m^P - n$, where p is a prime number.   ☐

Another witness of the strength of conjecture (D) is the following:

(D$_6$) *There exist infinitely many composite integers n, such that for every a, $1 < a < n$, then $a^n \equiv a \pmod n$ (so n is an a-pseudoprime, for every such a). In particular, there exist infinitely many Carmichael numbers.*

Proof that (D) implies (D_6). Let $f_1(X) = 6X + 1$, $f_2(X) = 12X + 1$, $f_3(X) = 18X + 1$. Since $f_1(0)f_2(0)f_3(0) = 1$ then condition (*) is satisfied. By (D), there exist infinitely many integers $m \geqslant 1$ such that $p = 6m + 1$, $q = 12m + 1$, $r = 18m + 1$ are primes. By the result of Chernick (Chapter 2, Section VIII) $n = pqr$ is an a-pseudoprime, for every basis a, $1 < a < n$.   □

A famous and deep conjecture about prime numbers is due to Artin. It concerns primitive roots modulo an odd prime p.

I recall (see Chapter 2, Section II) that if p is a prime, then the $p - 1$ residue classes of integers not multiples of p form a multiplicative group. Moreover, this group is cyclic. This means that, for every odd prime p, there exists an integer g, $1 < g < p$, such that $g^{p-1} \equiv 1 \pmod{p}$, but $g^n \not\equiv 1 \pmod{p}$ for every integer n, $1 \leqslant n < p-1$. The integer g is called a primitive root modulo p. The number of primitive roots modulo p is $\phi(p - 1)$.

What happens if an integer $g \neq 0$ is given? Can it be a primitive root modulo some prime p?

Obviously, if $g = -1$, then g is primitive root only modulo 3. Also, if g is a square, it is easy to see that g is never a primitive root modulo any prime. Artin's conjecture is the following:

(A) *If g is a nonzero integer, which is neither a square nor equal to* -1, *then there exist infinitely many primes p such that g is a primitive root modulo p.*

I'll soon show that (D) implies (A). For this proof it will be necessary to use the Jacobi symbol. I ask the reader to turn back to Chapter 2, Section II if he needs to be reminded of the properties of the symbol.

Now, I shall show the following:

Proof that (D) implies (A). Let $g = a^2 b$ where b has no square factor. Since g is not a square, $b \neq 1$. Let $|b| = 2^e b_1$, where $e = 0$ or 1, and b_1 is odd.

The purpose is to find two linear polynomials $f_1(X)$, $f_2(X)$, with integer coefficients, and positive leading coefficients, satisfying (*) and also the following conditions:

(1^0)   For every $n \geqslant 1$: $\left[\dfrac{b}{f_1(n)}\right] = -1$.

(2^0) $f_1(X) - 1 = \begin{cases} 2f_2(X) & \text{if } b \neq 3 \\ 4f_2(X) & \text{if } b = 3. \end{cases}$

First let $b < 0$. Choose $f_1(X) = -4bX - 1$, $f_2(X) = -2bX - 1$. Then (2^0) is satisfied, $f_1(0)f_2(0) = 1$, so (*) is satisfied. Now I compute the Jacobi symbol $(b/f_1(n))$, noting that b, and $f_1(n)$ are relatively prime. If $b = -2b_1$, then $f_1(n) \equiv -1 \pmod 8$ so

$$\left[\dfrac{2}{f_1(n)}\right] = +1 \quad \text{and} \quad \left(\dfrac{b}{f_1(n)}\right) = \left(\dfrac{-b_1}{f_1(n)}\right),$$

which is also true if $b = -b_1$. Since $f_1(n) \equiv -1 \pmod 4$, then

$$\left[\dfrac{-1}{f_1(n)}\right] = -1,$$

so

$$\left[\dfrac{b}{f_1(n)}\right] = -\left[\dfrac{b_1}{f_1(n)}\right] = -(-1)^{\frac{b_1-1}{2} \times \frac{f_1(n)-1}{2}} \left[\dfrac{f_1(n)}{b_1}\right]$$

$$= -(-1)^{\frac{b_1-1}{2}}\left[\dfrac{f_1(n)}{b_1}\right] = -(-1)^{\frac{b_1-1}{2}}\left[\dfrac{-1}{b_1}\right] = -1.$$

Now, let $b > 0$ and assume that $b = 2b_1$. Choose $f_1(X) = 4bX + 2b-1$, $f_2(X) = 2bX + b - 1$. Then (2^0) is satisfied. For condition (*), assume that there exists a prime p dividing

$$g(0) = f_1(0)f_2(0) = (2b - 1)(b - 1),$$

$$g(1) = f_1(1)f_2(1) = (6b - 1)(3b - 1),$$

and

$$g(-1) = f_1(-1)f_2(-1) = (2b + 1)(b + 1).$$

Then p is odd, and p divides also $g(1) + g(-1) - 2g(0) = 16b^2$, so p divides b and $g(0)$, which is impossible. This proves condition (*). Now I compute the Jacobi symbol $(b/f_1(n))$, noting that b, $f_1(n)$ are relatively prime. Since that $b = 2b_1$ and b_1 is odd, then $f_1(n) \equiv 3$ (mod 8), whence $(2/f_1(n)) = -1$. It follows that

$$\left[\frac{b}{f_1(n)}\right] = \left[\frac{2}{f_1(n)}\right]\left[\frac{b_1}{f_1(n)}\right] = -\left[\frac{b_1}{f_1(n)}\right]$$

$$= -(-1)^{\frac{f_1(n)-1}{2} \times \frac{b_1-1}{2}} \left[\frac{f_1(n)}{b_1}\right] = -(-1)^{\frac{b_1-1}{2}} \left[\frac{-1}{b_1}\right] = -1.$$

Now, let $b > 0$ and assume that b is odd and $b > 3$. Choose the polynomials $f_1(X) = 4bX + m$ and $f_2(X) = 2bX + \frac{1}{2}(m-1)$, where the integer m is chosen as follows. By hypothesis $b = q_1 q_2 \cdots q_k$ with $q_1 < q_2 < \cdots < q_k$ primes ($k \geqslant 1$), $q_k \geqslant 5$. Since the number of quadratic residues modulo q_k is the same as the number of nonquadratic residues, hence at least two, there exists an integer m_0, $1 \leqslant m_0 \leqslant q_k - 1$, such that

$$\left[\frac{m_0}{q_k}\right] = -1$$

and $m_0 \not\equiv -1 \pmod{q_k}$.

By the Chinese remainder theorem, there exists an integer m such that

$$\begin{cases} m \equiv -1 \pmod{4q_1 q_2 \cdots q_{k-1}} \\ m \equiv -m_0 \pmod{q_k}. \end{cases}$$

In particular m is odd, hence the polynomials $f_1(X)$, $f_2(X)$ have integer coefficients.

I show that the condition (*) is satisfied:

$$g(0) = f_1(0)f_2(0) = \frac{1}{2} m(m-1),$$

$$g(1) = f_1(1)f_2(1) = (4b + m)(2b + \frac{1}{2}(m-1))$$

$$g(-1) = f_1(-1)f_2(-1) = (4b - m)(2b - \frac{1}{2}(m-1)).$$

Then $g(1) + g(-1) - 2g(0) = 16b^2$.

Since $(m-1)/2 \equiv -1 \pmod{2q_1 \cdots q_{k-1}}$ and q_k does not divide m and also q_k does not divide $\frac{1}{2}(m-1)$ [because $m_0 \not\equiv -1 \pmod{q_k}$], it follows that $\gcd(4b, m) = 1$, $\gcd(2b, (m-1)/2) = 1$, and therefore

$\gcd(16b^2, \frac{m(m-1)}{2}) = 1$. This means that $\gcd(g(1) + g(-1) - 2g(0), g(0)) = 1$, so there does not exist any prime p dividing the products $g(0)$, $g(1)$, $g(-1)$. In other words, condition (*) is satisfied. It is clear that condition (2^0) is satisfied. As for condition (1^0), I compute the Jacobi symbol $(b/f_1(n))$ for any integer $n \geqslant 1$. First note that $f_1(n) \equiv m \equiv -1$ (mod $4q_1 \cdots q_{k-1}$) and $f_1(n) \equiv m \not\equiv 0$ (mod q_k), so b, $f_1(n)$ are relatively prime. Then:

$$\left[\frac{b}{f_1(n)}\right] = (-1)^{\frac{f_1(n)-1}{2} \times \frac{b-1}{2}} \left[\frac{f_1(n)}{b}\right] = (-1)^{\frac{b-1}{2}} \left[\frac{f_1(n)}{b}\right]$$

$$= \left[\frac{-f_1(n)}{b}\right] = \left[\frac{-m}{q_1 q_2 \cdots q_{k-1}}\right]\left[\frac{-m}{q_k}\right]$$

$$= \left[\frac{1}{q_1 q_2 \cdots q_{k-1}}\right]\left[\frac{m_0}{q_k}\right] = -1.$$

Finally, let $b = 3$. Choose the polynomials $f_1(X) = 12X + 5$, $f_2(X) = 3X + 1$, so condition (2^0) is satisfied. Also $f_1(0)f_2(0) = 5$, $f_1(1)f_2(1) = 17 \times 4$ so condition (*) is satisfied. For every $n \geqslant 1$, 3 does not divide $f_1(n)$, and

$$\left[\frac{3}{f_1(n)}\right] = (-1)^{\frac{3-1}{2} \times \frac{f_1(n)-1}{2}} \left[\frac{f_1(n)}{3}\right] = \left[\frac{f_1(n)}{3}\right] = \left[\frac{2}{3}\right] = -1.$$

Thus, in all possible cases, there exist linear polynomials $f_1(X)$, $f_2(X)$, with integer coefficients, positive leading coefficients, satisfying condition (*) and also conditions (1^0) and (2^0).

By (D) there exist infinitely many integers $c \geqslant 1$ such that $f_1(c)$, $f_2(c)$ are prime. I'll show that for every such integer c satisfying $f_1(c) > g^4$, the order of g modulo $f_1(c)$ is $f_1(c) - 1$; that is, g is a primitive root modulo $f_1(c)$. Indeed, $f_1(c) - 1 = 2q$ or $4q$, where $q = f_2(c)$. So, if the order is smaller than $f_1(c) - 1$, then either it divides $[f_1(c)-1]/2$ or it divides 4. Thus, either

$$g^{\frac{f_1(c)-1}{2}} \equiv 1 \pmod{f_1(c)} \quad \text{or} \quad g^4 \equiv 1 \pmod{f_1(c)}.$$

The second alternative is impossible, because $g^4 < f_1(c)$. As for the first alternative, since $g = a^2 b$ then

$$1 \equiv g^{\frac{f_1(c)-1}{2}} \equiv \left(\frac{g}{f_1(c)} \right) \equiv \left(\frac{b}{f_1(c)} \right) \equiv -1 \ (\mathrm{mod} \ f_1(c))$$

hence $f_1(c) = 2$, an absurdity. This concludes the proof. $\square$

Lest the reader gets a false impression, I hasten to describe some of the work that has been done in attempting to prove Artin's conjecture.

To tell the truth, the statement of the conjecture was supplemented by a quantitative assertion, which I will now explain. As before, let g be a nonzero integer which is neither -1 nor a square. For every $x > 1$ let $\mathrm{prim}_g(x)$ denote the number of primes $p \leqslant x$ such that g is a primitive root modulo p.

The quantitative version of Artin's conjecture is the following statement (see Hooley, 1967):

(A') *Let $k \geqslant 1$ be the largest integer such that g is a kth power (so k is odd), let g_1 be the square-free kernel of g. Then the following limit exists and its value is*

$$\lim_{x \to \infty} \frac{\mathrm{prim}_g(x)}{\pi(x)} = \prod_{p|k} \frac{1 - \dfrac{1}{p-1}}{1 - \dfrac{1}{p(p-1)}} \cdot \prod_{p \geqslant 2} \left[1 - \frac{1}{p(p-1)} \right] \cdot D(g),$$

where

$$D(g) = \begin{cases} 1 \ \text{ if } \ g_1 \not\equiv 1 \ (\mathrm{mod} \ 4) \\ 1 - \mu(|g_1|) \displaystyle\prod_{\substack{p \,|\, k \\ p \,|\, g_1}} \frac{1}{p-2} \displaystyle\prod_{\substack{p \,|\, k \\ p \,|\, g_1}} \frac{1}{p^2 - p - 1}. \end{cases}$$

Note that the infinite product above is absolutely convergent since

$$\sum_{p \geqslant 2} \frac{1}{p(p-1)} \leqslant \sum_{n=2}^{\infty} \frac{1}{n(n-1)} = \sum_{n=2}^{\infty} \left(\frac{1}{n-1} - \frac{1}{n} \right) = 1.$$

The constant $A = \prod_{p \geqslant 2} (1 - \frac{1}{p(p-1)})$ is called Artin's constant. It is instructive to see how to compute A.

Let $\alpha, \beta = (1 \pm \sqrt{5})/2$ be the roots of $X^2 - X - 1$, so that $p^2 - p - 1 = (p - \alpha)(p - \beta)$. Then

$$A = \prod_p \frac{p^2 - p - 1}{p(p-1)} = \prod_p \frac{(p-\alpha)(p-\beta)}{p(p-1)} = \prod_p \frac{\left(1 - \dfrac{\alpha}{p}\right)\left(1 - \dfrac{\beta}{p}\right)}{1 - \dfrac{1}{p}}$$

hence

$$\log A = \sum_p \log\left(1 - \frac{\alpha}{p}\right) + \sum_p \log\left(1 - \frac{\beta}{p}\right) - \sum_p \log\left(1 - \frac{1}{p}\right)$$

$$= \left[-\frac{\alpha}{p} - \frac{1}{2} \times \frac{\alpha^2}{p^2} - \frac{1}{3} \times \frac{\alpha^3}{p^3} - \frac{1}{4} \times \frac{\alpha^4}{p^4} - \cdots \right]$$

$$+ \left[-\frac{\beta}{p} - \frac{1}{2} \times \frac{\beta^2}{p^2} - \frac{1}{3} \times \frac{\beta^3}{p^3} - \frac{1}{4} \times \frac{\beta^4}{p^4} - \cdots \right]$$

$$- \left[-\frac{1}{p} - \frac{1}{2} \times \frac{1}{p^2} - \frac{1}{3} \times \frac{1}{p^3} - \frac{1}{4} \times \frac{1}{p^4} - \cdots \right].$$

The sums $\alpha + \beta = a_1$, $\alpha^2 + \beta^2 = a_2$, $\alpha^3 + \beta^3 = a_3$, ..., $\alpha^k + \beta^k = a_k$, ..., may be computed by Newton's formulas and satisfy the following recurrence: $a_1 = 1$, $a_2 = 3$ and for $k \geqslant 2$, $a_k = a_{k-1} + a_{k-2}$. Therefore

$$\log A = -\frac{2}{2} \sum_p \frac{1}{p^2} - \frac{3}{3} \sum_p \frac{1}{p^3} - \frac{6}{4} \sum_p \frac{1}{p^4} - \cdots - \frac{b_k}{k} \sum_p \frac{1}{p^k} - \cdots$$

where $b_k = a_k - 1$.

There exist tables of $\sum_p 1/p^k$ with 50 decimals for $2 \leqslant k \leqslant 167$ (by Liénard, 1948). In this way, Wrench gave, in 1961, the following value of A up to 25 decimals:

$A = 0.37395\ 58136\ 19202\ 28805\ 47280\ 54346\ 41641\ 51116\ 29249.$

In 1967, Hooley was able to show that Artin's conjecture, even in its quantitative form, is true, under the assumption of an "extended analogue of Riemann's hypothesis." (I have to remain vague on this condition, since its formulation requires concepts beyond those introduced here.)

Another positive indication is provided by results of Bilharz (1937) and Weil (1948). Now it is a question of function fields of one variable over a finite field and a statement which may be regarded somewhat as a kind of Artin's conjecture. It is well known that Weil proved for such fields, that an analogue of Riemann's hypothesis is true, and from this, he deduced that the analogue of Artin's conjecture is true.

To know that Artin's conjecture follows from this or that conjecture, or that an analogue statement is actually true, is better than nothing. But this is not yet entirely satisfactory.

It is only quite recently that a substantial progess on Artin's conjecture was achieved. Firstly, there was the path-breaking paper by Gupta & Ram Murty (1984), and then, the crowning result of Heath-Brown, who was kind enough to communicate his result to me:

Dear Professor Ribenboim: 16.11.85

My paper is entitled "Artin's Conjecture for Primitive Roots" and is accepted by the *Quart. J. Math. Oxford*. The main theorem is:

Let q, r, s be nonzero integers which are multiplicatively independent ($q^e r^f s^g = 1$, e, f, g integers, imply $e = f = g = 0$). Suppose none of q, r, s, $-3qr$, $-3qs$, $-3rs$, qrs is a square. Then the number $N(x)$ of primes $p \leqslant x$ for which at least one of q, r, s is a primitive root, satisfies

$$N(x) \geqslant c \, \frac{x}{(\log x)^2}$$

(for some positive constant c).

Corollary 1. If q, r, s are as above, then Artin's conjecture holds for at least one of them.

Corollary 2. There are at most two primes and at most three positive square-free integers for which Artin's conjecture fails.

Corollary 3. Let S be the set of nonzero integers (different from -1, and not squares) for which Artin's conjecture fails. Then, $\#\{n \in S \mid |n| \leqslant x\} \leqslant c(\log x)^2$ (for some positive constant c).

These results improve on those of Gupta & Ram Murty (*Invent. Math.*, **78**, 1984, 127-130) and also those in a paper I was shown by Ram Murty, which is as yet unpublished,* I think. However, the key idea is contained in the paper by Gupta & Ram Murty, and my improvement stems largely from using sharper sieve results.

With best wishes,

(signed Roger Heath-Brown).

Well done, and this is all, for the moment about Artin's conjecture.

---

*This paper, co-authored with Srinivasan, has now appeared (1987).

II. Prime Values of Polynomials of Arbitrary Degree

Now I turn to polynomials which may be nonlinear.

Historically, the first conjecture was by Bouniakowsky, in 1857, and it concerns one polynomial of degree at least two:

(B) *Let $f(X)$ be an irreducible polynomial, with integral coefficients, positive leading coefficient and degree at least 2. Assume that the following condition is satisfied:*

(*) *there does not exist any integer $n > 1$ dividing all the values $f(k)$, for every integer k.*

Then there exist infinitely many natural numbers m such that $f(m)$ is a prime.

Just as for the conjectures (D) and (D$_0$), the following conjecture is equivalent to (B):

(B$_0$) *If $f(X)$ satisfies the same hypothesis as in* (B) *there exists a natural number m such that $f(m)$ is a prime.*

Before I discuss conjecture (B), let it be made clear that there are, in fact, very few results about prime values of polynomials. If $f(X)$ has degree $d \geqslant 2$ and integral coefficients, for every $x \geqslant 1$ recall the notation

$$\pi_{f(X)}(x) = \{n \geqslant 1 \mid |f(n)| \leqslant x \text{ and } |f(n)| \text{ is a prime}\}.$$

In 1922, Nagell showed that $\lim_{x \to \infty} \pi_{f(X)}(x)/x = 0$, so there are few prime values. In 1931, Heilbronn proved the more precise statement: There exists a positive constant C (depending on $f(X)$), such that

$$\pi_{f(X)}(x) \leqslant C \frac{x}{\log x}, \quad \text{for every } x \geqslant 1.$$

Not much more seems to be known for arbitrary polynomials. However, for special types of polynomials, there are interesting conjectures and extensive computations. I'll soon discuss this in more detail.

If one is happy with theorems concerning almost-primes as values of polynomials, then I note that sieve methods have been used to obtain rather good results. Let $f(X)$ be a polynomial with integral coefficients and degree $d \geqslant 1$; assume also that the condition (*) of conjecture (B) is satisfied. Richert showed in 1969 that there exist infinitely many natural numbers m such that $f(m)$ is a $(d+1)$-almost-prime. As a matter of fact, Richert gave a quantitative version under the following hypothesis: for every prime p, the number $\rho(p)$ of solutions of the congruence $f(X) \equiv 0 \pmod{p}$ is less than p [this implies condition (*)]. Then, for every sufficiently large $x > x_0 = x_0(f)$,

$$\#\{m \mid 1 \leqslant m \leqslant x,\ f(m) \in P_{d+1}\} \geqslant \frac{2}{3} \prod_p \frac{1 - \dfrac{\rho(p)}{p}}{1 - \dfrac{1}{p}} \times \frac{s}{\log x}$$

[here P_{d+1} denotes the set of $(d+1)$-almost-primes].

The special case when $d = 1$ had been obtained by Ricci (1937), while the general result was indicated without proof by Buchstab (see more references in Halberstam & Richert's book) for the polynomial $X^2 - 2$. Ankeny & Onishi showed in 1964 that there exist infinitely many integers n such that $n^2 - 2 \in P_3$.

The question of existence of infinitely many primes p such that $f(p)$ is a prime, is even more difficult. In particular, as I have already said, it is not known that there exist infinitely many primes for which the polynomials $f(X) = X + 2$, respectively $f(X) = 2X + 1$, assume prime values (existence of infinitely many twin primes, or infinitely many Sophie Germain primes). However, if happiness comes from almost-primes, then there are good reasons to be contented, once more with sieve methods and the unavoidable book of Halberstam & Richert. The latter author proved in 1969:

Let $f(X)$ be a polynomial with integral coefficients, positive leading coefficient, degree $d \geqslant 1$ (and different from X). Assume that for every prime p, the number $\rho(p)$ of solutions of $f(X) \equiv 0 \pmod{p}$ is less than p; moreover if $p \leqslant d + 1$ and p does not divide $f(0)$ assume also that $\rho(p) < p - 1$. Then, for every sufficiently large $x \geqslant x_0 = x_0(f)$,

$\#\{p \text{ prime} \mid p \leqslant x, \ f(p) \in P_{2d+1}\}$

$$\geqslant \frac{4}{3} \prod_{p \mid f(0)} \frac{1 - \dfrac{\rho(p)}{p - 1}}{1 - \dfrac{1}{p}} \times \prod_{p \mid f(0)} \frac{1 - \dfrac{\rho(p) - 1}{p - 1}}{1 - \dfrac{1}{p}} \times \frac{x}{(\log x)^2} \ .$$

Thus, there exist infinitely many primes p such that $f(p) \in P_{2d+1}$.

The following particular case was proved by Rieger (1969): there exist infinitely many primes p such that $p^2 - 2 \in P_5$.

Now, back to Bouniakowsky's hypothesis!

Bouniakowsky drew no consequences from his conjecture. Once more, this was the task of Schinzel & Sierpiński, who studied a more general hypothesis, stated one century later, but independently.

The following proposition, which has never been proved, follows easily if the conjecture (B) is assumed to be true:

(B$_1$) *Let a, b, c be relatively prime integers, such that $a \geqslant 1$ and $a+b$ and c are not simultaneously even.*

If $b^2 - 4ac$ is not a square, then there exist infinitely many natural numbers m such that $am^2 + bm + c$ is a prime number.

Proof that (B) implies (B$_1$). Let $f(X) = aX^2 + bX + c$; since $b^2 - 4ac$ is not a square, then $f(X)$ is irreducible. I show that condition (*) is satisfied: $f(0) = c$, $f(1) = a + b + c$, $f(-1) = a - b + c$. If there exists a prime p dividing $f(0)$, $f(1)$, $f(-1)$, then p divides $2b$ and $2(a+c)$; also p divides $2c$ so p divides $2a$; since $\gcd(a,b,c) = 1$ then $p = 2$ so c and $a + b$ would be even, against the hypothesis.

By (B), there exist infinitely many natural numbers m such that $am^2 + bm + c$ is a prime number. □

In particular, this result refers to primes of the form $m^2 + m + 1$, or also of the form $(m^2+1)/2$, with m odd; if $m = 2n + 1$, these are therefore primes of the form $2n^2 + 2n + 1$. For this kind, Shanks showed in 1964, that there exist infinitely many primes of the form $2n^2 + 2n + 1$ if and only if the power series

$$\left[X + X^3 + X^6 + \cdots + X^{k(k+1)/2} + \cdots \right]^2$$
$$- 3 \left[X + X^3 + X^6 + \cdots + X^{k(k+1)/2} \right]$$

has infinitely many negative coefficients.

(B_1) in turn implies:

(B_2) *If k is an integer such that −k is not a square, there exist in-finitely many natural numbers m such that $m^2 + k$ is a prime number.*

Proof that (B_1) implies (B_2). Take $a = 1$, $b = 0$, $c = k$, so $b_2 - 4ac = -4k$ is not a square. Since all the hypotheses are satisfied, by (B_1) there exist infinitely many natural numbers m such that $m_2 + k$ is a prime number.

□

In particular, (B) implies that there exist infinitely many primes of the form $m^2 + 1$.

The "game of primes of the form $m^2 + 1$" is not just as innocent as one would inadvertently think at first. It is deeply related with the class number of real quadratic fields.

I recall the well known difficult classical question, which originated with Gauss and remains open: are there infinitely many real quadratic fields $Q(\sqrt{d})$, with $d > 0$, d square-free, with class number $h(d) = 1$?

(For imaginary quadratic fields, this was discussed in Chapter 3, Section II.)

And how about the same question for real quadratic fields with any given class number $h(d) = h_0 > 1$?

For $d = p$, a prime of the form $m^2 + 1$, Chowla conjectured (see Chowla & Friedlander, 1976):

If $p = m^2 + 1$ is a prime and $m > 26$, then $h(p) > 1$.

This conjecture clearly represents a first step towards the main problem. There has been interesting work on this matter. For example, Felthi showed in 1980 that if $d = 16m^2 + 1$ (d square-free, but not necessarily prime), then $h(d) > 1$ for d sufficiently large.

In a series of three papers, one in 1987, the others to appear hopefully in 1987 (two of which in collaboration with Williams), Mollin placed himself in a more embracing situation, analogous to the one considered in Chapter 3, Section II, in regard to Euler's polynomials $X^2 + X + q$.

Let d be a square-free positive integer, and let

$$
\alpha = \begin{cases}
\dfrac{\sqrt{d-1}}{2} & \text{when } d \equiv 1 \pmod 4 \\[2mm]
\sqrt{d} & \text{when } d \not\equiv 1 \pmod 4
\end{cases}
$$

and

$$
f_d(X) = \begin{cases}
-X^2 + X + \dfrac{d-1}{4} & \text{when } d \equiv 1 \pmod 4 \\[2mm]
-X^2 + d & \text{when } d \not\equiv 1 \pmod 4.
\end{cases}
$$

The purpose was to study the relationship between, among others, the following conditions:

1) d is one of the nineteen numbers 2, 3, 5, 6, 7, 11, 13, 17, 21, 29, 37, 53, 77, 101, 173, 197, 293, 437, 677
2) $f_d(x)$ is a prime for every integer x, $2 \leqslant x < \alpha$
3) $h(d) = 1$.

Clearly (1) implies (2) and (3). But there are other discriminants d, even odd ones (what a phrase!!!), for which $h(d) = 1$, and d is not in the list (1). That (2) implies (3) was shown using a previous result by Kutsuna (1980). A more interesting implication was first shown by Mollin with no extraneous assumption. Namely, if $d = 4m^2 + 1 \geqslant 5$, then (3) implies (2), that is, $-x^2 + x + m$ is prime for $x = 2, 3, ..., m-1$.

In particular, Felthi's result is a special case: if $d = 16m^2 + 1 \geqslant 65$ then $2 < -4 + 2 + (2m)^2$, hence $h(d) > 1$.

The implication (2) → (1) was established by Mollin & Williams under the expedient assumption that the generalized Riemann hypothesis holds for the zeta function of the number field $Q(\sqrt{d})$ (I do not enter into more explanations). Theoretical estimates gave the result for $d > 10^{13}$ and numerical computations took care of the smaller values of d.

This is the analogue of Rabinovitch result of Chapter 3, Section II − a beautiful relationship.

Mollin conjectures also that if $q > 2$ is a prime and $d = 4q^2 + 1$ is also a prime, then $-x^2 + x + q^2$ is a prime for every $x = 2, 3, ..., q-1$ if and only if $q \leqslant 13$.

Of course, all this does not settle the original conjecture of Chowla, which remains open when $m = 2q$ and $p = 4q^2 + 1$ is a prime. But, in their recent paper, Mollin & Williams used the generalized Riemann hypothesis for $Q(\sqrt{p})$ to establish Chowla's conjecture.

The following statement was also conjectured by Hardy & Littlewood, in 1923:

(B$_3$) *Let d be an odd integer, $d > 1$; let k be an integer which is not a eth power of an integer, for any factor $e > 1$ of d. Then there exist infinitely many natural numbers m such that $m^d + k$ is a prime.*

Proof that (B) implies (B$_3$). Consider the polynomial $f(X) = X^d + k$. Anticipating a further application, k may be allowed to be any rational number, not necessarily an integer.

An old theorem of Capelli (1898) quoted in Schinzel's book (1982), states:

$f(X)$ is reducible if and only if: either $-k$ is a eth power of a rational number (for some e dividing d, $1 < e$), or 4 divides d and $k = 4h^4$ (for some rational number h).

In the present situation, $f(X)$ is irreducible. Since $f(0) = k$, $f(1) = k + 1$, then the condition (*) is satisfied. So, by (B) there exist infinitely many natural numbers m such that $m^d + k$ is a prime. □

In his joint paper with Sierpiński, Schinzel proposed the following conjectures:

(H) *Let $s \geqslant 1$, let $f_1(X), ..., f_s(X)$ be irreducible polynomials, with integral coefficients and positive leading coefficient. Assume that the following condition holds:*

(*) there does not exist any integer $n > 1$ dividing all the products $f_1(k)f_2(k) ... f_s(k)$, for every integer k.

Then there exist infinitely many natural numbers m such that all numbers $f_1(m), f_2(m), ..., f_s(m)$ are primes.

(H$_0$) *Under the same assumptions for $f_1(X), ..., f_s(X)$, there exists a natural number m such that the numbers $f_1(m), ..., f_s(m)$ are primes.*

Once again, (H) and (H_0) are equivalent. If all the polynomials $f_1(X)$, ..., $f_s(X)$ have degree 1, these conjectures are Dickson's (D), (D_0). If $s = 1$, they are Bouniakowsky's conjectures (B), (B_0).

Some support to the conjecture (H) is provided by Ricci (1937). It may be inferred, from Ricci's theorem 1, without assuming the truth of any conjecture:

If the polynomials $f_1(X)$, ..., $f_s(X)$ satisfy the conditions indicated in (H), there exists a constant $k \geqslant 1$, depending on the given polynomials, such that there exist infinitely many natural numbers m for which each of the integers $f_1(m)$, ..., $f_s(m)$ is a k-almost-prime.

A more precise quantitative formulation is theorem 10.4 in Halberstam & Richert's book.

This is, of course, weaker than what is asserted in (H).

Here are some consequences of (H), which, seemingly, would not be obtainable either from (D) or from (B).

(H_1) *Let a, b, c, d be integers such that $a > 0$, $d > 0$ and $b^2 - 4ac$ is not a square. Assume also that there exist integers x_0, y_0 such that $\gcd(x_0 y_0, 6ad) = 1$ and $ax_0^2 + bx_0 + c = dy_0$. Then there exist infinitely many primes p, q such that $ap^2 + bp + c = dq$.*

Proof that (H) implies (H_1). Let $f_1(X) = dX + x_0$, $f_2(X) = ad X^2 + (2ax_0 + b)X + y_0$. Since $(2ax_0 + b)^2 - 4ady_0 = (2ax_0 + b)^2 - 4a(ax_0^2 + bx_0 + c) = b^2 - 4ac$ is not a square, the polynomial $f_2(X)$ is irreducible, just like $f_1(X)$.

I verify condition (*). Let $g(X) = f_1(X)f_2(X)$; this is polynomial of degree 3, with leading coefficient ad^2. If there exists a prime p dividing $g(m)$ for every integer m, then p divides $g(m) - g(m-1) = \Delta g(m)$, $g(m-1) - g(m-2) = \Delta g(m-1)$, $g(m-2) - g(m-3) = \Delta g(m-2)$ which are the first difference of $g(X)$, evaluated at m, $m - 1$, $m - 2$. By the same token, p divides $\Delta^2 g(m) = \Delta g(m) - \Delta g(m-1)$, $\Delta^2 g(m-1) = \Delta g(m-1) - \Delta g(m-2)$, and again p divides $\Delta^3 g(m) = \Delta^2 g(m) - \Delta^2 g(m-1)$. But $\Delta^3 g(X) = 6ad^2$. If p divides also $g(0) = x_0 y_0$ then p divides $\gcd(6ad, x_0 y_0) = 1$, which is an absurdity. This shows that condition (*) is satisfied. By (H) there exist infinitely many natural numbers m such that $f_1(X) = p$, $f_2(X) = q$ are primes. Since $af_1(X)^2 + bf_1(X) + c = df_2(X)$, then $ap^2 + bp + c = dq$. $\square$

$(H_{1,1})$ *Every rational number* $r > 1$ *may be written in infinitely many ways in the form* $r = (p^2-1)/(q-1)$, *where* p, q *are prime numbers.*

Proof that (H_1) implies $(H_{1,1})$. Let $r = d/a$ with $d > a > 0$. Take $b = 0$, $c = d - a$ in the statement of (H_1). Since $b^2 - 4ac = -4a(d-a) < 0$, then it is not a square. Let $x_0 = y_0 = 1$, then the hypotheses of (H_1) are satisfied, hence there exist infinitely many primes p, q such that $ap^2 + d - a = dq$, therefore

$$\frac{d}{a} = \frac{d}{a}q - p^2 + 1, \quad \text{hence} \quad r = \frac{p^2 - 1}{q - 1}. \quad \square$$

$(H_{1,1,1})$ *There exist infinitely many right-angled triangles, with sizes measured by integers, two of which are prime numbers.*

Proof that $(H_{1,1})$ implies $(H_{1,1,1})$. Let $r = 2$, so there exist infinitely many primes p, q such that $2 = (p^2-1)/(q-1)$, hence $p^2 = 2q - 1$ and therefore $p^2 + (q-1)^2 = q^2$. $\quad \square$

Further consequences of (H_1) are:

$(H_{1,2})$ *There exist infinitely many primes* p *such that* $\frac{p(p+1)}{2} + 1$ *is also a prime.*

$(H_{1,3})$ *There exist infinitely many primes* p *such that* $\frac{p(p+1)}{2} - 2$ *is also a prime.*

Here is another consequence of (H) similar to (B_3):

(H_2) *Let* d *be an odd integer,* $d > 1$. *Let* k *be an even integer which is not an eth power of an integer, for any factor* $e > 1$ *of* d. *Then there exist infinitely many primes* p *such that* $p^d + k$ *is a prime.*

Proof that (H) implies (H_2). Consider the polynomials $f_1(X) = X$ and $f_2(X) = X^d + k$ [which is irreducible, as indicated in the proof of (B_3)]. Now, $f_1(1)f_2(1) = 1 + k$, $f_1(-1)f_2(-1) = 1 - k$, hence $\gcd(1+k, 1-k) = 1$, because k is even. It follows that condition (*) is satisfied, and so there exist infinitely many primes p such that $p^d + k$ is also a prime. $\quad \square$

The next consequence allows one to produce prime, as well as composite values of polynomials:

(H_3) *Let $s \geq 1$ and let $f_1(X)$, ..., $f_s(X)$ be polynomials satisfying the assumptions indicated in (H). Let $t \geq 1$, $g_1(X)$, ..., $g_t(X)$ be irreducible polynomials, with integral coefficients and positive leading coefficient. If each $g_i(X)$ is different from the polynomials $f_j(X)$ (for $j = 1,2, ..., s$), there exist infinitely many natural numbers m such that $f_1(m)$, $f_2(m)$, ..., $f_s(m)$ are primes and $g_1(m)$, $g_2(m)$, ..., $g_t(m)$ are composite integers.*

Now I shall present a result of Schinzel in connection with Carmichael's conjecture. Recall from Chapter 2, Section II, that Carmichael has conjectured: for every $m \geq 1$ there exists $n \geq 1$, $n \neq m$, such that $\phi(n) = \phi(m)$. The notation $N_\phi(k) = s$ (with $s \geq 1$) means that there exist exactly s integers n_1, ..., n_s, such that $\phi(n_i) = k$ ($i = 1, ..., s$).

Using (H_3), Schinzel proved in 1961:

($H_{3,1}$) *For every $s > 1$ there exist infinitely many integers $k > 1$ such that $N_\phi(k) = s$.*

Proof that (H_3) implies ($H_{3,1}$). There are two cases. First, let $s = 2h$ be even.

For $i = 1,2, ..., 2h$ let $f_i(X) = 2X^{2i-1} + 1$ and $f_{2h+1}(X) = X$. By Capelli's theorem [quoted in the proof of (B_3)], each polynomial $f_i(X) = 2(X^{2i-1} + \frac{1}{2})$ is irreducible. Also, their leading coefficients are positive. Since $f_1(-1)f_2(-1) \cdots f_{2h}(-1)f_{2h+1}(-1) = (-1)^{2h+1} = -1$, then condition (*) is satisfied. By (H) there exist infinitely many primes p such that $f_i(p)$ are primes (for $i = 1,2, ..., 2h$). Let $p > 5$ be such a prime. Then $k = 4p^{4h}$ is not a multiple of 8. Thus, if m is a natural number such that $\phi(m) = 4p^{4h}$, then m can have at most two odd prime factors; thus $m = q^a$, or $2q^a$, or $4q^a$, or $q^a r^b$, or $2q^a r^b$, where q, r are distinct odd primes, $a \geq 1$, $b \geq 1$.

If $a > 1$, then $q(q-1)$ divides $\phi(m) = 4p^{4h}$ and therefore $q = p$ and $p - 1$ divides $4p^{4h-1}$, with $p > 5$ — which is impossible.

So $a = 1$, and the same argument shows that $b = 1$.

Now, $m = q$ or $2q$ is impossible, because this implies that $q - 1 = \phi(m) = 4p^{4h} \equiv 4 \equiv -1 \pmod 5$, so that 5 divides q, that is $q = 5$, a contradiction.

If $m = 4q$, then $2(q-1) = \phi(m) = 4p^{4h} \equiv 1 \pmod 3$ so $q \equiv 0 \pmod 3$, which is absurd.

If $m = qr$ or $2qr$, then $(q-1)(r-1) = \phi(m) = 4p^{4p}$, hence $q - 1 = 2p^n$, $r - 1 = 2p^{4h-n}$ with $0 \leqslant n \leqslant 4h$. If n is even, then $q = 2p^n + 1 \equiv 0 \pmod 3$, which is impossible. Thus $n = 2i - 1$ (with $1 \leqslant i \leqslant h$ by symmetry of roles of q and r), and $m = (2p^{2i-1}+1)(2p^{4h-2i+1}+1) = f_i(p)f_{2h-i+1}(p)$ or $m = 2f_i(p)f_{2h-i+1}(p)$. But the numbers $f_i(p)$ are all primes, so the $s = 2h$ values of m given above are exactly those such that $\phi(m) = k = 4p^{4h}$. Hence if s is even, then $N_\phi(k) = s$, for infinitely many integers k.

Now, let s be odd, $s = 2h + 3$ (with $h \geqslant 0$, since $s \geqslant 3$). Let $f_i(X) = 2X^{6i-3} + 1$, $f_{h+i}(X) = 6X^{6i-1} + 1$ (for $i = 1,2, ..., h$), $f_{2h+1}(X) = X$, $f_{2h+2}(X) = 6X^{6h+2} + 1$ and let $g_j(X) = 2X^{6j-5} + 1$, $g_{h+j}(X) = 2X^{6j-1} + 1$ (for $j = 1,2, ..., h$), $g_{2h+1}(X) = 2X^{6h+1} + 1$, $g_{2h+2}(X) = 12X^{6h+2} + 1$. All the above polynomials are irreducible (as follows from Capelli's criterion already indicated), with integral coefficients and positive leading coefficients. Moreover,

$$\prod_{i=1}^{2h+2} f_i(-1) = -5^h \times 7, \qquad \prod_{i=1}^{2h+2} f_i(1) = 3^h \times 7^{h+1},$$

$$\prod_{i=1}^{2h+2} f_i(2) \not\equiv 0 \pmod 7,$$

hence condition (*) is satisfied by the polynomials $f_1(X), ..., f_{2h+2}(X)$.

By (H_3) there exist infinitely many primes $p > 3$ such that $f_i(p) = 2p^{6i-3} + 1$, $f_{h+i}(p) = 6p^{6i-1} + 1$, $f_{2h+2}(p) = 6p^{6h+2} + 1$ are primes (for $i = 1, ..., h$), while $g_j(p) = 2p^{6j-5} + 1$, $g_{h+j}(p) = 2p^{6j-1} + 1$, $g_{2h+1}(p) = 2p^{6h+1} + 1$, $g_{2h+2}(p) = 12p^{6h+2} + 1$ are composite integers (for $j = 1, ..., h$). Moreover, if n is even, $0 < n \leqslant 6h + 2$, then $2p^n + 1 \equiv 0 \pmod 3$, so $2p^n + 1$ is also composite.

By hypothesis, if n is odd, $n \neq 6i - 3$, then $2p^n + 1$ is also composite.

Let $k = 12p^{6h+2}$ and assume that m is a natural number such that $\phi(m) = 12p^{6h+2}$. As in the first case, $m = q$, or $2q$, or $4q$, or qr, or $2qr$, where q, r are distinct odd primes. Indeed, if $m = 9q$, or $18q$, then $6(q-1) = \phi(m) = 12p^{6h+2}$, hence $q = 2p^{6h+2} + 1$ – but this number is composite, which is a contradiction.

If $m = q$ or $2q$, then $q = 12p^{6h+2} + 1$, but this number is composite.

If $m = 4q$, then $2(q-1) = \phi(m) = 12p^{6h+2}$, hence $q = 6p^{6h+2} + 1$ and $m = 4(6p^{6h+2} + 1)$.

If $m = qr$ or $2qr$, then $(q-1)(r-1) = \phi(m) = 12p^{6h+2}$, hence

$$\begin{cases} q - 1 = 2p^n \\ r - 1 = 6p^{6h+1-n} \end{cases} \quad \text{(or vice versa)},$$

so $q = 2p^n + 1$, $r = 6p^{6h+2-n} + 1$. But for n even, $0 < n \leqslant 6h + 2$, or n odd and $n \neq 6i - 3$, the integer $2p^n + 1$ is composite. So, the possible cases are $n = 0$ and $n = 6i - 3 \leqslant 6h + 2$. Thus

$$\begin{cases} q = 3, \\ r = 6p^{6h+2} + 1 = f_{2h+2}(p), \end{cases}$$

hence $m = 3f_{2h+2}(p)$ or $6f_{2h+2}(p)$, or

$$\begin{cases} q = 2p^{6i-3} + 1 = f_i(p), \\ r = 6p^{6(h-i)+5} + 1 = f_{2h-i+1}(p), \end{cases}$$

hence $m = f_i(p)f_{2h-i+1}(p)$ or $2f_i(p)f_{2h-i+1}(p)$.

Since the numbers $f_i(p)$ are all primes ($i = 1,2, ..., h$) this yields exactly $2h + 2 + 1 = 2h + 3 = s$ solutions, that is, $N_\phi(k) = s$, for infinitely many integers k.   □

For every $x \geqslant 1$, let

$$p(x) = \lim_{y \to \infty} \sup (\pi(y + x) - \pi(y)).$$

I have already mentioned in Chapter 4, Section II, Hardy & Littlewood's conjecture (1923): if $x \geqslant 2$, then $p(x) \leqslant \pi(x)$.

As I indicated, this inequality may well not be always true.

Using the conjecture (H), Schinzel was able to give another equivalent formulation to the above conjecture of Hardy & Littlewood.

Schinzel & Sierpiński showed that (H) implies:

(H$_4$) $p(X) \leqslant \pi(x)$ for every $x \geqslant 2$ if and only if $\pi(x+y) \leqslant \pi(x) + \pi(y)$ for every $x,y \geqslant 2$.

It is interesting to make the following observation. If there exists an $x \geqslant 2$ such that $\rho(x) \geqslant 2$, then

$$\liminf_{k \to \infty} (p_{k+1} - p_k) < x.$$

Indeed, by hypothesis, if $\epsilon > 0$ there exists an infinite sequence $y_1 < y_2 < \cdots$ such that $\pi(y_i+x) - \pi(y_i) \geqslant 2 - \epsilon$, hence necessarily $\pi(y_i+x) - \pi(y_i) \geqslant 2$ (since the left-hand side is an integer); without loss of generality, it may be assumed that $y_1 < y_1 + x < y_2 < y_2 + x < y_3 < \cdots$; so, there exist prime numbers p_i such that $y_1 < p_1 < p_2 \leqslant y_1 + x < y_2 < p_3 < p_4 \leqslant y_2 + x < \cdots$. Hence

$$\liminf_{k \to \infty} (p_{k+1} - p_k) < x.$$

This implies at once that there exists some even number $2m < x$ which is equal in infinitely many ways to the difference of two consecutive primes — a fact which has never been proved.

Thus, if there exist integers $x, y \geqslant 2$ such that $\pi(x+y) > \pi(x) + \pi(y)$ and if the conjecture (H) is assumed to be true then there exists $x \geqslant 2$ such that $\rho(x) > \pi(x) \geqslant 1$; thus $\rho(x) \geqslant 2$ and therefore there exists an even number $2m$ representable as the difference of consecutive primes, in infinitely many ways.

Subtle relationship! ...

Still in the same paper, there is the following conjecture by Sierpiński:

(S) *For every integer $n > 1$, let the n^2 integers $1,2, ..., n^2$ be written in an array with n rows, each with n integers, like an $n \times n$ matrix:*

1	2	$\cdots$	n
$n + 1$	$n + 2$	$\cdots$	$2n$
$2n + 1$	$2n + 2$	$\cdots$	$3n$
. . . . .	. . . . .	$\cdots$	. .
$(n-1)n+1$	$(n-1)n+2$	$\cdots$	n^2 .

Then, there exists a prime number in each row.

Of course, 2 is in the first row. By the theorem of Bertrand and Tschebycheff, there is a prime number in the second row.

More can be said about the first few rows using a strengthening of Bertrand and Tschebycheff's theorem. For examples, in 1932, Breusch showed that if $n \geqslant 48$, then there exists a prime between n and $(9/8)n$. Thus, if $0 < k \leqslant 7$ and $n \geqslant 48$, then there exists a prime p such that $kn + 1 \leqslant p \leqslant (9/8)(kn+1) \leqslant (k+1)n$. By direct calculation, the same result holds also for $n \geqslant 9$. So, there is a prime in each of the first 8 rows.

Using the explicit result of Rohrbach & Weis (1964), quoted in Chapter 4, Section II, if $n \geqslant 13$, then each of the first 13 rows of the array contains a prime.

Similarly, by the prime number theorem, for every $h \geqslant 1$ there exists $n_0 = n_0(h) > h$ such that, if $n \geqslant n_0$, then there exists a prime p such that $n < p < (1 + \frac{1}{h})n$.

And from this, it follows that, if $n \geqslant n_0$, then each of the first h rows of the array contains a prime. Indeed, this is clearly true for the first row. If $1 \leqslant k \leqslant h - 1$, then there exists a prime p such that $n_0 \leqslant kn + 1 < p < (1 + \frac{1}{h})(kn + 1)$. But $(1 + \frac{1}{h})(kn + 1) \leqslant (k+1)n$ because $1 + \frac{1}{h} \leqslant n(1 - \frac{k}{h})$, noting that $(h+1)/(h-k) \leqslant h + 1 \leqslant n_0 \leqslant n$.

I mention here that Ricci proved in the paper already quoted, the following proposition (independent of any conjecture):

There exists $k \geqslant 1$ such that for every $n \geqslant 1$, every row of the array of n^2 integers $1, 2, ..., n^2$ contains at least one k-almost-prime. With the present sieve methods, it is possible to take $k = 2$, for $n > n_0$.

Just like the preceding conjectures, (S) has also several interesting consequences.

(S_1) *For every $n \geqslant 2$ there exist at least two primes p, p' such that* $(n-1)^2 < p < p' < n^2$.

Proof that (S) implies (S_1). The last two rows of the array of n^2 integers are

$$(n-1)^2 = (n-2)n + 1, \quad (n-2)n + 2, ..., \quad (n-2)n + n = (n - 1)n$$

$$(n-1)n + 1 \qquad\qquad (n-1)n + 2, ..., \quad (n-1)n + n = n^2.$$

So, by (S) there exists a prime in each of the last two rows of the array.   □

$(S_{1,1})$ *For every $n \geqslant 1$ there exist at least four primes p, p', p'', p''' such that $n^3 < p < p' < p'' < p''' < (n+1)^3$.*

Proof that (S_1) implies $(S_{1,1})$. The statement is true for $n = 1,2,3$. Let $n \geqslant 4$. I show that, if $m \geqslant 1$, then $m^2 \leqslant n^3 < (n+1)^3 \leqslant (m+3)^2$ is impossible.

Indeed, if the above inequalities hold then

$$n^3 + 3n^2 + 3n + 1 \leqslant m^2 + 6m + 9 \leqslant n^3 + 6m + 9,$$

so $3n^2 + 3n < 6m + k < 6n^{3/2} + 8$, hence $3n(n^{1/2} - 1)^2 < 8$, which is impossible when $n \geqslant 4$. So, if m^2 is the largest square such that $m^2 \leqslant n^3$, then

$$m^2 \leqslant n^3 < (m+1)^2 < (m+2)^2 < (m+3)^2 \leqslant (n+1)^3.$$

By (S_1) there exist primes p, p', p'', p''' such that

$$n^3 < (m+1)^2 < p < p' < (m+2)^2 < p'' < p''' < (m+3)^2 \leqslant (n+1)^3. \quad □$$

It should be noted that both statements (S_1), $(S_{1,1})$ have not yet been proved without appealing to the conjecture (S). However, it is easy to show, that for every sufficiently large n, there exists a prime p between n^3 and $(n+1)^3$; this is done using Ingham's result that $d_n = O(p_n^{(5/8)+\epsilon})$ for every $\epsilon > 0$.

Schinzel stated the following "transposed" form of Sierpiński's conjecture:

(S') *For every integer $n > 1$, let the n^2 integers $1,2, ..., n^2$ be written in an array with n rows, each with n integers [just like in (S)]. If $1 \leqslant k \leqslant n$ and $\gcd(k,n) = 1$, then the kth column contains at least one prime number.*

This time, Schinzel & Sierpiński drew no consequence from the conjecture (S'). I suppose it was Sunday evening and they were tired. However, in 1963, this conjecture was spelled out once more by Kanold, as I have said in Chapter 4, Section IV.

I conclude with the translation of the following comment in

Schinzel & Sierpiński's paper:

> We do not know what will be the fate of our hypotheses, however we think that, even if they are refuted, this will not be without profit for number theory.

It is my impression that none of the conventional present day methods of number theory will lead to a proof of any of the conjectures (D), (B), (H), (S), (S'). Perhaps, it will be the role of logicians, investigating the inner structure of arithmetic, to decide whether such statements are, or are not, provable from the Peano axioms.

III. Some Probabilistic Estimates

A. Partitio Numerorum

It is instructive to browse through the paper *Partitio Numerorum, III: On the expression of a number as a sum of primes*, by Hardy & Littlewood (1923). There one finds a conscious systematic attempt — for the first time in such a scale — to derive heuristic formulas for the distribution of primes satisfying various additional conditions.

I shall present here a selection of the probabilistic conjectures, extracted from Hardy & Littlewood's paper (I will keep their labeling, since it is classical). The first conjecture concerns Goldbach's problem:

Conjecture A: *Every sufficiently large even number 2n is the sum of two primes. The asymptotic formula for the number of representations is*

$$r_2(2n) \sim C_2 \frac{2n}{(\log 2n)^2} \prod_{\substack{p>2 \\ p \mid n}} \frac{p-1}{p-2}$$

with

$$C_2 = \prod_{p>2} \left[1 - \frac{1}{(p-1)^2} \right] = 0.66016 \ldots .$$

Note that C_2 is the same as the twin primes constant (see Chapter 4, Section III).

The following conjecture deals with (not necessarily consecutive) primes with given difference $2k$, in particular, with twin primes:

Conjecture B: *For every even integer* $2k \geqslant 2$ *there exist infinitely many primes* p *such that* $p + 2k$ *is also a prime. For* $x \geqslant 1$, *let*

$$\pi_{X,X+2k}(x+2k) = \#\{p \text{ prime} \mid p+2k \text{ is prime and } p \leqslant x\}.$$

Then

$$\pi_{X,X+2k}(x + 2k) \sim 2C_2 \frac{x}{(\log x)^2} \prod_{\substack{p>2 \\ p|k}} \frac{p-1}{p-2},$$

where C_2 *is the twin primes constant.*

In particular, if $2k = 2$ this gives the asymptotic estimate for the twin primes counting function:

$$\pi_2(x) = \pi_{X,X+2}(x + 2) \sim 2C_2 \frac{x}{(\log x)^2}.$$

Soon, I'll show how conjecture B for twin primes and other similar conjectures may be "proved" — all in the framework of a quantitative version of the work of Schinzel & Sierpiński.

The conjecture E is about primes of the form $m^2 + 1$:

Conjecture E: *There exist infinitely many primes of the form* $m^2 + 1$. *For* $x > 1$ *let*

$$\pi_{X^2+1}(x) = \#\{p \text{ prime} \mid p \leqslant x \text{ and } p \text{ is of the form } p = m^2 + 1\}.$$

Then

$$\pi_{X^2+1}(x) \sim C \frac{\sqrt{x}}{\log x}$$

where

$$C = \prod_{p \geqslant 3} \left[1 - \frac{(-1/p)}{p-1} \right] = 1.37281346 \ldots$$

(here

$$(-1/p) = (-1)^{(p-1)/2}$$

is the Legendre symbol).

The values of $\pi_{X^2+1}(x)$ predicted by the conjecture are in significant agreement with the actual values indicated below (the last two
were computed by Wunderlich in 1973):

x	$\pi_{X^2+1}(x)$
10^6	112
10^8	841
10^{10}	6656
10^{12}	54110
10^{14}	456362

Here is the place to mention almost-primes as values of $X^2 + 1$ (a
special case of theorem of Richert, stated at the beginning of this
section). Already in 1954, Kuhn proved that there exist infinitely
many natural numbers m such that $m^2 + 1 \in P_3$. But the best result
up to now is due to Iwaniec (1978), where, instead, of almost-3-
primes, infinitely many numbers $m^2 + 1$ are in P_2.

One more step, but a giant one, is still needed!

More generally:

Conjecture F: *Let* $a > 0$, b, c *be integers such that* $\gcd(a,b,c) = 1$,
$b^2 - 4ac$ *is not a square and* $a + b$, c *are not both even. Then there are
infinitely many primes of the form* $am^2 + bm + c$ *[this was statement*
(B$_1$)] and the number $\pi_{aX^2+bX+c}(x)$ *of primes* $am^2 + bm + c$ *which are
less than* x *is given asymptotically by*

$$\pi_{aX^2+bX+c}(x) \sim \frac{\epsilon C}{\sqrt{a}} \frac{\sqrt{x}}{\log x} \prod_{\substack{p>2 \\ p|\gcd(a,b)}} \frac{p}{p-1}$$

where

$$\epsilon = \begin{cases} 1 & if \ a + b \ is \ odd \\ 2 & if \ a + b \ is \ even \end{cases}$$

$$C = \prod_{\substack{p>2 \\ p \mid a}} \left[1 - \frac{\left[\dfrac{b^2 - 4ac}{p}\right]}{p - 1} \right]$$

and $\left(\dfrac{b^2 - 4ac}{p}\right)$ denotes the Legendre symbol.

In particular, the conjecture is applicable to primes of the form $m^2 + k$, where $-k$ is not a square.

Another conjecture:

Conjecture H: *Every sufficiently large integer n is either a square or a sum of a square and a prime. The number $N(n)$ of representations is given asymptotically by*

$$N(n) \sim \frac{\sqrt{n}}{\log n} \prod_{p>2} \left[1 - \frac{(n/p)}{p - 1} \right],$$

where (n/p) is the Legendre symbol.

And still another one:

Conjecture I: *Every sufficiently large odd integer n is the sum of a prime and twice a square. The number $N(n)$ of representations is given asymptotically by*

$$N(n) \sim \frac{\sqrt{2n}}{\log n} \prod_{p>2} \left[1 - \frac{(2n/p)}{p - 1} \right]$$

where $(2n/p)$ is the Legendre symbol.

Conjecture I was spelled out in response to numerical calculations begun by Stern already in 1856. The numbers 5777, 5993 are the only odd ones less than 9000, not of the form $p + 2m^2$. Of course, this does not contradict the conjecture, since it should only hold for all large odd integer, and no indication is provided about this size.

Now, cubes:

Conjecture K: *Let k be an integer which is not a cube. Then there exist infinitely many primes of the form $m^3 + k$. The number $\pi_{x^3+k}(x)$ of such primes less than x is given asymptotically by*

$$\pi_{X^3+k}(x) \sim \frac{x^{1/3}}{\log x} \prod_{\substack{p \equiv 1(\text{mod } 3) \\ p \mid k}} \left[1 - \frac{2(-k)_p}{p-1} \right]$$

where

$$(-k)_p = \begin{cases} 1 & \text{when } -k \text{ is a cubic residue modulo } p, \\ -\tfrac{1}{2} & \text{when } -k \text{ is not a cubic residue modulo } p. \end{cases}$$

Conjecture L: *Every sufficiently large integer n is either a cube or the sum of a positive cube and a prime. The number N(n) of representations is given asymptotically by*

$$N(n) \sim \frac{n^{1/3}}{\log n} \prod_{p \equiv 1(\text{mod } 3)} \left[1 - \frac{(2n)_p}{p-1} \right]$$

where

$$(n)_p = \begin{cases} 1 & \text{when } n \text{ is a cubic residue modulo } p, \\ -\tfrac{1}{2} & \text{when } n \text{ is not a cubic residue modulo } p. \end{cases}$$

The next conjecture concerns expressions of primes as sums of three positive cubes. As a preliminary explanation, if p is any prime such that $p \equiv 1$ (mod 3), then p decomposes as the product of two prime elements in the cyclotomic field of cubic roots of unity; say $p = \alpha_p \, \alpha_p'$; this decomposition is only unique up to units. It is possible to choose α_p of the following form: $\alpha_p = a_p + b_p \zeta_3$ where

$$\zeta_3 = \cos \frac{2\pi}{3} + i \sin \frac{2\pi}{3},$$

and a_p, b_p are integers, such that $a_p \equiv -1$ (mod 3), $b_p \equiv 0$ (mod 3). Let $A_p = 2a_p - b_p$.

Conjecture N: *There exist infinitely many prime p of the form* $p = k^3 + \ell^3 + m^3$, *where k, ℓ, m are positive integers. The number* $\pi_{X^3+Y^3+Z^3}(x)$ *of triples (k,ℓ,m) with* $p \leqslant x$, *is given asymptotically by*

$$\pi_{X^3+Y^3+Z^3}(x) \sim (\Gamma(4/3))^3 \frac{x}{\log x} \prod_{p \equiv 1(\text{mod } 3)} \left[1 - \frac{A_p}{p^2} \right];$$

where Γ denotes the gamma function.

Conjecture P: *There exist infinitely many pairs of primes of the form (m^2+1, m^2+3). The number of such primes $m^2 + 3$ less than n is given asymptotically by*

$$Q(n) \sim \frac{3\sqrt{n}}{(\log n)^2} \prod_{p>3} \frac{p(p - v)}{(p - 1)^2}$$

where

$$v = \begin{cases} 0 & when \ (-1/p) = (-3/p) = -1, \\ 2 & when \ (-1/p)(-3/p) = -1, \\ 4 & when \ (-1/p) = (-3/p) = 1. \end{cases}$$

These are the most important conjectures in the paper of Hardy & Littlewood.

As it should be expected, the conjectures of Hardy & Littlewood inspired a considerable amount of computation, intended to determine accurately the constants involved in the formulas, and to verify that the predictions fitted well with the observation. As the constants were often given by slowly convergent infinite products, it was imperative to modify these expressions, to make them more easily computable. According to the case, this led to calculations with Riemann or Epstein zeta functions, L-series, etc. The methods and difficulties encountered are amply illustrated in numerous papers of Shanks (especially those of 1960, 1975; see also his book, third edition, 1985).

For example, concerning conjecture F relative to the famous Euler polynomial, Shanks showed (1975):

$$\pi_{X^2+X+41}(N) \sim 3.31977317741 \int_2^N \frac{dt}{\log t}.$$

In relation to conjecture K, Shanks proved also in 1975 that the constant in Hardy & Littlewood's formula for $\pi_{X^3+2}(N)$ is $1.298539557557843 \ ... \ $.

In 1962, Bateman & Horn indicated a quantitative form of the conjectures of Schinzel & Sierpiński.

Let $s \geqslant 1$, let $f_1(X)$, ..., $f_s(X)$ be irreducible polynomials, with integral coefficients and positive leading coefficients. Assume that the following condition [already considered in (H)] holds:

(*) There does not exist any integer $n > 1$ dividing all the products $f_1(k)f_2(k)\cdots f_s(k)$, for every integer k.

Denote by $d_i \geqslant 1$ the degree of $f_i(X)$.

For every integer $N > 1$ let $Q(N) = Q_{f_1,\ldots,f_s}(N)$ denote the number of integers n, $1 \leqslant n \leqslant N$, such that $f_1(n), f_2(n)$, ..., $f_s(n)$ are primes. Then:

$$Q(N) \sim C_{f_1,\ldots,f_s} \frac{1}{d_1\cdots d_s} \sum_{n=2}^{N} \frac{1}{(\log n)^s}.$$

"**Proof.**" If p is an arbitrary prime, let u_p denote the chance that none of the integers of a random s-tuple is divisible by p. Then

$$u_p = \left(\frac{p-1}{p}\right)^s = \left(1 - \frac{1}{p}\right)^s.$$

Similarly, let t_p denote the chance that none of the integers $f_1(n)$, $f_2(n)$, ..., $f_s(n)$ is divisible by p. If $w(p)$ denotes the number of solutions x, $0 \leqslant x \leqslant p-1$, of the congruence

$$f_1(X)f_2(X)\cdots f_s(X) \equiv 0 \pmod{p},$$

then

$$t_p = \frac{p - w(p)}{p} = 1 - \frac{w(p)}{p}.$$

It may be shown that the infinite product $\Pi_p\, t_p/u_p$ is convergent, say to a limit $C = C_{f_1,\ldots,f_s}$.

By the prime number theorem, the probability that a large number m be a prime is

$$\frac{\pi(m)}{m} \sim \frac{1}{\log m}.$$

Thus, the probability that the numbers $f_1(n)$, ..., $f_s(n)$ are simultaneously prime is — if these events were independent —

$$\frac{1}{\log f_1(n) \cdots \log f_s(n)}.$$

However, the s-tuples $(f_1(n), ..., f_s(n))$ are not random. The constant C obtained above should be viewed as measuring the extent to which the above s-tuples are not random. Thus, it is reasonable to state that the probability that $f_1(n), ..., f_s(n)$ be simultaneously prime is

$$C \frac{1}{\log f_1(n) \cdots \log f_s(n)} \sim C \frac{1}{d_1 \cdots d_s} \times \frac{1}{(\log n)^s},$$

because $\log f_i(n) \sim d_i \log n$. Then

$$Q(N) \sim C \frac{1}{d_1 \cdots d_s} \sum_{n=2}^{N} \frac{1}{(\log n)^s}. \qquad \square$$

This is essentially the same as

$$Q(N) \sim C \frac{1}{d_1 \cdots d_2} \int_2^N \frac{dt}{(\log t)^s}.$$

Another reasonable asymptotic formula is also

$$Q(N) \sim C \frac{1}{d_1 \cdots d_s} \times \frac{N}{(\log N)^s}.$$

Special cases are, for example:

(a) $f_1(X) = X$, $f_2(X) = X + 2k$ $(k \geqslant 1)$
(b) $f_1(X) = X$, $f_2(X) = 2X + 1$.

Case (a) corresponds to pairs of primes with difference $2k$, in particular to twin primes; case (b) corresponds to Sophie Germain primes.

In both cases, $w(2) = 1$, $w(p) = 2$ if $p > 2$ and a simple calculation shows that the constant C_{f_1,f_2} is equal to twice the twin-primes constant: $2C_2 = 1.32032$.

In 1965, Bateman & Horn also investigated the prime values of $X^2 + X + 1$ and other polynomials, like $X^3 - 2$, $X^3 + 2$, $2X^3 - 1$, $2X^3 + 1$, which have non-abelian groups over the field of rational numbers. In each instance, the actual calculations confirmed the predictions indicated heuristically.

B. Polynomials with Many Successive Composite Values

I want now to report on the new results of McCurley, which I find very interesting, despite their noneffective aspects. According to Bouniakowsky's conjecture, if $f(X)$ is an irreducible polynomial with integral coefficients and satisfying condition (*), then there exists a smallest integer $m \geqslant 1$ such that $f(m)$ is a prime. Denote it by $p(f)$.

If $f(X) = dX + a$, with $d \geqslant 2$, $1 \leqslant a \leqslant d - 1$, $\gcd(a,d) = 1$, then of course, $p(f)$ exists. With the notation of Chapter 4, Section IV,

$$p(dX + a) = \frac{p(d,a) - a}{d} \ .$$

Recall that Prachar and Schinzel gave lower bounds for $p(d,a)$.

The work of McCurley provides an extension of the above results for polynomials $f(X)$ of arbitrary degree.

An essential tool for McCurley was the recent result of Odlyzko (included in the paper by Adleman, Pomerance & Rumely, 1983, which was quoted in Chapter 2):

There exists an absolute constant $C > 0$ and infinitely many integers $d > 1$ having at least $e^{C[\log d/(\log \log d)]}$ factors of the form $p - 1$, where p is an odd prime.

In 1984, McCurley proved:

For every $\epsilon > 0$ there exist infinitely many irreducible polynomials $f_i(X) = X^{d_i} + k_i$ (with $2 \leqslant d_1 < d_2 < \cdots$, with each $k_i > 1$), such that if

$$m < M_i = e^{(C-\epsilon)(\log d_i)/(\log \log d_i)} ,$$

then $f_i(m)$ is composite.

This may be rewritten as $p(f_i) \geqslant M_i$.

The degree of the above polynomials belong to the very thin set indicated by Odlyzko.

Since M_i becomes very large, these polynomials likely assume very few prime values. It should be noted that the method of proof does not give any such polynomial explicitly. However, in a computer search, the following polynomials were discovered (the first example is due to Shanks, 1971):

$f(X)$	$f(m)$ is composite for all m up to
$X^6 + 1091$	3905
$X^6 + 82991$	7979
$X^{12} + 4094$	170624
$X^{12} + 488669$	616979

The smallest prime value of the last polynomial has no less than 70 digits.

With another method, McCurley showed in 1966:

For every $d \geq 1$ there exists an irreducible polynomial $f(X)$, of degree d, satisfying the condition (*) and such that if

$$|m| < e^{C\sqrt{L(f)/(\log L(f))}},$$

then $|f(m)|$ is composite.

In the above statement, $L(f)$ is the length of $f(X) = \sum_{k=0}^{d} a_k X^k$ which is defined by $L(f) = \sum_{k=0}^{d} \|a_k\|$ and $\|a_k\|$ is the number of digits in the binary expansion of $|a_k|$, with $\|0\| = 1$.

Note that this last result is applicable to polynomials of any degree, and also the proof yields explicit polynomials with the desired property.

This result should be compared with the recent conjecture of Adleman & Odlyzko (1983):

There exists $a > 0$, $\ell_0 \geq 1$ and $C > 0$, such that, if $f(X)$ is an irreducible polynomial satisfying (*) and with $L(f) \geq \ell_0$, then

$$p(f) \leq Ce^{(L(f))^a}.$$

According to McCurley's result, if the conjecture is true then $a \geq \frac{1}{2}$.

For binomials $f(X) = X^d + k$, with $d \geqslant 1$, McCurley (1986) generalized Rankin's method (used in the estimate of gaps between primes):

For every $d \geqslant 1$ there exists a constant $C(d) > 0$ and infinitely many integers $k \geqslant 1$ such that $X^d + k$ is irreducible, satisfies (*) and

$$p(X^d + k) > C(d) \frac{\log k}{\log_3 k} \left[\frac{\log_2 k \, \log_4 k}{\log_3 k} \right]^{\tau(d)}$$

where $\tau(d)$ denotes the number of divisors of d.

The proof gives

$$C(1) = e^\gamma - \epsilon, \quad C(2) = \frac{2e^{2\gamma}}{\pi^2} - \epsilon$$

(good for all sufficiently large k).

In particular, taking $d = 1$, if k is such that it lies between the primes p_n, p_{n+1}, then replacing k by p_n and using the prime number theorem, Rankin's result follows:

There exist infinitely many n such that

$$\frac{p_{n+1} - p_n}{\log p_n} \times \frac{(\log_3 p_n)^2}{(\log_2 p_n)(\log_4 p_n)} > e^\gamma - \epsilon$$

(see Chapter 4, Section II).

Here is another one of McCurley's results. Given $d \geqslant 2$, $u \geqslant 2$, let $K(d,u)$ be the set of all integers $k \geqslant 1$ such that $X^d + k$ is irreducible, satisfies (*) and $p(X^d + k) > u$. Let $k(d,u)$ be the smallest integer in $K(d,u)$. Then:

There exists $C > 0$ such that, if $d \geqslant 2$, $u \geqslant 2$, then

$$k(d,u) < e^{e^M}, \quad M = u^{1/\tau(d)} d^C.$$

McCurley determined $p(X^d + k)$ for several polynomials. From his tables, I note:

d	$k \leqslant m$	max $p(X^d + k)$
2	10^6	$p(X^2 + 576239) = 402$
3	10^6	$p(X^3 + 382108) = 297$
4	150000	$p(X^4 + 72254) = 2505$
5	10^5	$p(X^5 + 89750) = 339$

C. Distribution of Mersenne Primes

The problem of the distribution of Mersenne primes is quite different. It is not a question of distribution of prime values of polynomials, but of functions $2^q - 1$, where q itself is a prime.

There are different heuristic theories, more or less in a happy
agreement with reality. In 1983, Wagstaff modified an earlier
conjecture by Gillies (1964) and stated his own conjecture:

(a)   The number of Mersenne primes less than x is about

$$\frac{e^{\gamma}}{\log 2} \log \log x = (2.5695...) \log \log x.$$

(b)   The expected number of Mersenne primes M_q with $x < q < 2x$ is
      about $e^{\gamma} = 1.7806 \ldots$.

(c)   The probability that M_q is a prime is about

$$\frac{e^{\gamma}}{\log 2} \times \frac{\log aq}{\log 2} = (2.5695...) \times \frac{\log aq}{q},$$

where
$$a = \begin{cases} 2 & \text{if } q \equiv 3 \pmod 4, \\ 6 & \text{if } q \equiv 1 \pmod 4. \end{cases}$$

Wagstaff's analysis is based on various considerations:

(a)   The divisors of M_q are known to be of the form $2kq + 1$, and
      moreover they are congruent to ± 1 modulo 8.
(b)   Use of Bateman & Horn's count of the number of primes $q \leqslant x$
      such that $2kq + 1$ is also a prime.
(c)   A fine study by Shanks & Kravitz (1967) of the probability for
      $2kq + 1$ to divide M_q.
(d)   A modification of Gillies argument, as suggested by Lenstra
      (1980).

In his article of 1983 in *The Mathematical Intelligencer*, and in his
book of 1984, Schroeder refers also to Eberhart's conjecture, mentioned explicitly in Slowinski's paper of 1978/9. According to this
conjecture, if q_n is the nth prime such that M_{q_n} is a Mersenne prime,

then q_n is approximately equal to $(3/2)^n$ (at least it should be, if n is large). With Wagstaff's approach, Eberhart's conjecture ought to be phrased rather as

$$q_n \sim (2^{e^{-\gamma}})^n \quad \text{(note that } 2^{e^{-\gamma}} = 1.4759 \ldots).$$

But, really, there is no serious reason supporting this conjecture — the theory is weak, the numerical evidence is minimal, and worst of all, despite all my efforts, I could never locate Eberhart to discuss the conjecture with him.

In spite of my negative comments, there is some accurate prediction in Schroeder's article. Indeed, if the Mersenne primes M_q, with $q = 132049$ and 216091, found by Slowinski turn out to be the 29th and 30th Mersenne primes, in size this will be in a very satisfactory agreement with the forecast for q to be around 130000, resp. 215000.

D. The log log Philosophy

I want to thank Serre for calling my attention to the following heuristic method, based on the formula:

$$\sum_{p \leqslant x} \frac{1}{p} = \log \log x + O(1).$$

For every prime p, let $f(p)$ be a natural number and for every $x > 1$ let

$$N(x) = \#\{p \text{ prime} \mid p \leqslant x, f(p) \equiv 0 \bmod p\}.$$

If the function $f(p) \bmod p$ is "random," then heuristically $N(x)$ is about $\log \log x$, so it grows very slowly to infinity.

Thus, for example, if $a \geqslant 2$ and if the remainder modulo p of

$$q_p(a) = \frac{a^{p-1} - 1}{p} \quad \text{(for } p \mid a)$$

is assumed to be random, then it would follow that the number of primes p such that $q_p(a) \equiv 0 \pmod{p}$ is infinite; in particular, the number of Wieferich primes (see Chapter 5, Section III) is infinite. However, these should be very scarce, hence very difficult to detect in actual computation.

It is also expedient to assume, for lack of reasons to the contrary, that the events $q_p(\ell) \equiv 0 \pmod{p}$, for distinct prime bases $\ell = 2,3,5, \ldots,$ are pairwise independent.   This may be used to estimate heuristically the probability that the first case of Fermat's last theorem be false for an exponent p.   Recall that (First FLT) is true when $p \leqslant 7 \times 10^{14}$ and also, if (First FLT) is false for p, then $q_p(\ell) \equiv 0 \pmod{p}$, for all the 24 primes $\ell \leqslant 89$ (this was shown by Granville & Monagan, see Chapter 5, Section III).   Thus, if $x > 7 \times 10^{14}$ the probability that (First FLT) is false for $p < x$ is

$$\sum_{7\times10^{14}<p\leqslant x} \frac{1}{p^{24}} < \sum_{p} \frac{1}{p^{24}} - \sum_{p\leqslant 7\times10^{14}} \frac{1}{p^{24}} \leqslant \frac{1}{10^{350}} \, .$$

Even though such arguments are fuzzy and cannot be rigorously justified, they constitute a good guide to suggest what should be true.

IV. The Density of the Set of Regular Primes

In Chapter 5, Section I, I have discussed the regular primes p, namely those not dividing the class number of the p-cyclotomic field $Q(\zeta_p)$,

$$\zeta_p = \cos \frac{2\pi}{p} + i \sin \frac{2\pi}{p} \, .$$

Kummer characterized the regular primes p by the following condition: p does not divide the numerators of the Bernoulli numbers $B_{2k} = N_{2k}/D_{2k}$ (N_{2k}, D_{2k} are relatively prime integers, $D_{2k} > 0$) for $2k = 2, 4, \ldots, p - 3$.

As I said, Jensen proved that there exist infinitely many irregular primes. However, it is unknown whether there exist infinitely many regular primes.

My aim in this section is to show that, heuristically, there must exist infinitely many regular primes, and, more precisely, the set of regular primes has density $1/\sqrt{e}$ in the set of all primes.

Here is exactly the statement proved by Siegel in 1964:

Assume that, modulo any prime number, the residue classes of the numerators of the Bernoulli numbers are randomly distributed. Then

$$\lim_{n \to \infty} \frac{\#\{p_k \mid 1 \leqslant k \leqslant n, \ p_k \text{ is a regular prime}\}}{n} = \frac{1}{\sqrt{e}} \sim 0.61.$$

The assumption is unverifiable, but once it is accepted, the conclusion follows rigorously.

As I'm arriving at the end of this book, it is harmless to devote a number of pages to this proof. A few readers who have arrived this far, may wish to go directly to the conclusion — they will miss an argument of a very different nature, which in fact, pleases me very much (the argument ...).

Before entering into the details of the proof, I shall indicate the main steps.

As usual, let $p_1 = 2$, $p_2 = 3$, $p_3 = 5$, ..., p_k, ... be the increasing sequence of primes. Let $q_k = (p_k - 3)/2$ for $k \geqslant 3$.

Define also

$$\rho_k = 1 - \left[1 - \frac{1}{p_k}\right]^{q_k},$$

for $k \geqslant 3$, $\rho = 1 - 1/\sqrt{e}$, thus $\rho = \lim_{k \to \infty} \rho_k$.

Let S be the set of sequences $x = (x_i)_{i \geqslant 1}$ of positive integers.

For every $\epsilon > 0$, $n \geqslant 3$, it is convenient to say that the sequence x satisfies the condition $(C_{\epsilon,n})$ whenever

$$\left| \frac{1}{n-2} \#\{p_k \mid 3 \leqslant k \leqslant n, \ p_k \text{ divides } x_1 \cdots x_{q_k}\} - \rho \right| < \epsilon.$$

Let $S_{\epsilon,n}$ be the set of all sequences $x \in S$ such that x satisfies condition $(C_{\epsilon,n})$.

If $\tau = (T_n)_{n \geqslant 3}$ is any increasing sequence of positive integers, let $S_{\epsilon,n,\tau}$ be the set of all $x \in S_{\epsilon,n}$ such that $1 \leqslant x_i \leqslant T_n$ for $i = 1, ..., q_n$.

Clearly, if τ, τ' are sequences as above, if $T_n \leqslant T'_n$, then $S_{\epsilon,n,\tau} \subseteq S_{\epsilon,n,\tau'}$ and thus $S_{\epsilon,n} = \cup_\tau S_{\epsilon,n,\tau}$ is the union of an increasing sequence of sets.

Let $Y_{\epsilon,n} \colon S_{\epsilon,n} \to \{0,1\}$ be defined by

$$Y_{\epsilon,n}(x) = \begin{cases} 1 & \text{if } x \in S_{\epsilon,n}, \\ 0 & \text{if } x \notin S_{\epsilon,n}. \end{cases}$$

$Y_{\epsilon,n}$ will be thought of as a random variable defined on the probability space $S_{\epsilon,n}$: in other words, it will be defined $P(x \in S_{\epsilon,n}) = P(Y_{\epsilon,n}(x) = 1)$, the probability for x to be in $S_{\epsilon,n}$.

It will be shown that $\lim_{n\to\infty} P(Y_{\epsilon,n}(x) = 1) = 1$.

Since the sequence $\nu = ((N_2),(N_4), ...)$ is assumed to be random with respect to divisibility, then the probable value of $Y_{\epsilon,n}$ is $Y_{\epsilon,n}(\nu)$; thus $P(Y_{\epsilon,n}(x) = 1) = Y_{\epsilon,n}(\nu)$ and $\lim_{n\to\infty} Y_{\epsilon,n}(\nu) = 1$. So for every $\delta > 0$ there exists n_0 such that, if $n > n_0$, then $Y_{\epsilon,n}(\nu) > 1 - 1/\delta$; therefore $Y_{\epsilon,n}(\nu) = 1$, that is $\nu \in S_{\epsilon,n}$, so condition $(C_{\epsilon,n})$ holds for ν. This is rephrased as

$$\lim_{n\to\infty} \frac{1}{n} \#\{p_k \mid 3 \leqslant k \leqslant n, p_k \text{ divides } N_2 ... N_{p_k-3}\} = \rho.$$

Thus, the objective is to define $P(Y_{\epsilon,n}(x) = 1)$ and show that $\lim_{n\to\infty} P(Y_{\epsilon,n}(x) = 1) = 1$.

For this purpose, let $\tau = (T_n)_{n\geqslant3}$, $\epsilon > 0$, $n \geqslant 3$ be given let $M(\epsilon,n,\tau) = M(\epsilon,n,T_n)$ be the set of all q_n-tuples of integers $(y_1, ..., y_{qn})$ where $1 \leqslant y_i \leqslant T_n$ and

$$\left| \frac{1}{n-2} \#\{p_k \mid 3 \leqslant k \leqslant n, p_k \text{ divides } y_1, ..., y_{q_k}\} - \rho \right| < \epsilon.$$

The condition for x to belong to $S_{\epsilon,n,\tau}$ depends only on the first q_n coordinates of x.

Thus it is natural to define the probability for x to be in $S_{\epsilon,n,\tau}$ as

$$P(x \in S_{\epsilon,n,\tau}) = \frac{\#M(\epsilon,n,\tau)}{T_n^{q_n}}$$

and $P(x \in S_{\epsilon,n}) = \lim_\tau P(x \in S_{\epsilon,n,\tau})$.

It will be shown that, if τ is any sequence increasing sufficiently fast (this will be made precise in the course of the proof), then $\lim_{n\to\infty} P(x \in S_{\epsilon,n,\tau}) = 1$, hence

$$\lim_{n\to\infty} P(x \in S_{\epsilon,n}) = \lim_{n\to\infty} \lim_{\tau} P(x \in S_{\epsilon,n,\tau}) = \lim_{\tau} \lim_{n\to\infty} P(x \in S_{\epsilon,n,\tau}) = 1.$$

In summary, it is required to prove:

(1^0) If τ, τ' are sequences as above, if $n \geqslant 3$ and $s_n = p_3 p_4 \cdots p_n$ divides T_n and also T_n', then

$$\frac{\#M(\epsilon,n,\tau)}{T_n^{q_n}} = \frac{\#M(\epsilon,n,\tau')}{T_n'^{q_n}}.$$

Call this ratio $\mu(\epsilon,n)$.

(2^0) If $\tau = (T_n)_{n\geqslant 3}$ is increasing fast, say $T_n \geqslant T_n'$, T_n' is a multiple of ρ_n, and $T_n' \geqslant \max\{|N_2|, ..., |N_{p_n-3}|\}$ for every $n \geqslant 3$,

$$\lim_{n\to\infty} \frac{\#M(\epsilon,n,\tau)}{T_n^{q_n}} = \lim_{n\to\infty} \mu(\epsilon,n).$$

(3^0) $\lim_{n\to\infty} \mu(\epsilon,n) = 1$.

First, I shall establish (1^0). For simplicity, let $n \geqslant 3$, $s = s_n$, $p = p_n$, $q = q_n$, $T = T_n$, $T' = T_n'$.

Lemma A. *If s divides T and T', then*

$$\frac{\#M(\epsilon,n,\tau)}{T^q} = \frac{\#M(\epsilon,n,\tau)}{T'^q}.$$

Proof. If $3 \leqslant k \leqslant n$, let

$$W_k = \{(\alpha_1, ..., \alpha_q) \in (Z/Zp_k)^q \mid \alpha_1\alpha_2 \cdots \alpha_{q_k} = 0\}$$

$$V_k = \{(\alpha_1, ..., \alpha_q) \in (Z/Zp_k)^q \mid \alpha_1\alpha_2 \cdots \alpha_{q_k} \neq 0\}.$$

Let $w_k = \#(W_k)$, $v_k = \#(V_k)$. Then

$$v_k = (p_k - 1)^{q_k} p_k^{q-q_k} = p_k^q \left[1 - \frac{1}{p_k}\right]^{q_k},$$

$$w_k = p_k^q - v_k = p_k^q \left[1 - \left[1 - \frac{1}{p_k}\right]^{q_k}\right].$$

Let $\Phi_0: Z/Z_s \to \Pi_{k=3}^m Z/Zp_k$ be the ring-isomorphism provided by the Chinese remainder theorem:

$$\Phi_0(a \bmod s) = (a \bmod p_3, \ldots, a \bmod p_n).$$

Let

$$\Phi: (Z/Zs)^q \to \prod_{k=3}^n (Z/Zp_k)^q$$

be the ring-isomorphism defined by

$$\Phi(a_1 \bmod s, \ldots, a_q \bmod s) = (\beta^{(3)}, \ldots, \beta^{(n)})$$

where

$$\beta^{(k)} = (a_1 \bmod p_k, \ldots, a_q \bmod p_k) \in (Z/Zp_k)^q.$$

Let $1 \leqslant r \leqslant n$,

$$K_r = \{(k_1, \ldots, k_r) \mid 3 \leqslant k_1 < k_2 < \cdots < k_r \leqslant n\}$$

and define the set

$$W_{(k_1, \ldots, k_r)} : \text{ set of all } (a_1 \bmod s, \ldots, a_q \bmod s) \in (Z/Zs)^q$$

such that if

$$\Phi(a_1 \bmod s, \ldots, a_q \bmod s) = (\beta^{(3)}, \ldots, \beta^{(n)}),$$

then $\beta^{(k_i)} \in W_{k_i}$ for $i = 1, \ldots, r$, while $\beta^{(j)} \in V_j$ for every

$j \neq k_1, \ldots, k_r$.

Let $w_{(k_1, \ldots, k_r)} = \#W_{(k_1, \ldots, k_r)}$. So

$$w_{(k_1, \ldots, k_r)} = \prod_{i=1}^r w_{k_i} \prod_{j \neq k_i} v_j.$$

Note that if $(k_1, \ldots, k_r) \neq (k_1', \ldots, k_r')$ then

$$W_{(k_1, \ldots, k_r)} \cap W_{(k_1', \ldots k_r')} = \emptyset.$$

Let

$$W_r = \bigcup_{(k_1,\ldots,k_r)\in K_r} W_{(k_1,\ldots,k_r)} \subseteq (\mathbf{Z}/\mathbf{Z}s)^q,$$

and let $w_r = \#W_r$. Then

$$w_r = \sum_{(k_1,\ldots,k_r)\in K_r} w_{(k_1,\ldots,k_r)} .$$

Let

$$\overline{W}_{(k_1,\ldots,k_r),T} = \{(x_1, \ldots, x_q) \mid 1 \leqslant x_i \leqslant T \text{ for } i = 1, \ldots, q$$

and $\quad (x_1 \bmod s, \ldots, x_q \bmod s) \in W_{(k_1,\ldots,k)}\}$,

and let

$$\overline{w}_{(k_1,\ldots,k_r),T} = \#\overline{W}_{(k_1,\ldots,k_r),T} .$$

Since s divides T, then

$$\overline{w}_{(k_1,\ldots,k_r),T} = w_{(k_1,\ldots,k_r)} \left[\frac{T}{s}\right]^q .$$

Again, if $(k_1, \ldots, k_r) \neq (k_1', \ldots, k_r')$, then

$$\overline{W}_{(k_1,\ldots,k_r),T} \cap \overline{W}_{(k_1',\ldots,k_r'),T} = \emptyset .$$

Let

$$\overline{W}_{r,T} = \bigcup_{(k_1,\ldots,k_r)\in K_r} \overline{W}_{(k_1,\ldots,k_r),T}$$

and let $w_{r,T} = \#(\overline{W}_{r,T})$. Then

$$\overline{w}_{r,T} = \sum_{(k_1,\ldots,k_r)\in K_r} \overline{w}_{(k_1,\ldots,k_r),T}$$

$$= \sum_{(k_1,\ldots,k_r)\in K_r} w_{(k_1,\ldots,k_r)} \left[\frac{T}{s}\right]^q = w_r \left[\frac{T}{s}\right]^q .$$

Now,

$$M(\epsilon,n,T) = \bigcup_{\left|\frac{r}{n-2} - \rho\right|} \overline{W}_{r,T} .$$

Indeed, if $(x_1, ..., x_q) \in M(\epsilon,n,T)$, if r is the number of primes p_k, $3 \leqslant k \leqslant n$, such that p_k divides $x_1 \cdots x_{q_k}$, then $(x_1, ..., x_q) \in \overline{W}_{r,T}$, and conversely.

If $3 \leqslant r, r' \leqslant n, r \neq r'$, then $\overline{W}_{r,T} \cap \overline{W}_{r',T} = \emptyset$ hence

Thus
$$\#M(\epsilon,n,T) = \sum_{\left|\frac{r}{n-2} - \rho\right| < \epsilon} \overline{w}_{r,T} = \sum_{\left|\frac{r}{n-2} - \rho\right| < \epsilon} w_r \left(\frac{T}{s}\right)^q.$$

$$\frac{\#M(\epsilon,n,T)}{T^q} = \frac{1}{s^q} \sum_{\left|\frac{r}{n-2} - \rho\right| < \epsilon} w_r,$$

and this number is independent of T, concluding the proof of the lemma.   □

For the proof of (2^0), it will be necessary to recall some facts about Bernoulli numbers and establish three lemmas.

Writing $B_{2k} = N_{2k}/D_{2k}$, then the fundamental theorem of von Staudt & Clausen says that $D_{2k} = \Pi_{(p-1)|2k}p$. This describes the denominator completely, once the divisors d of $2k$ and the primes $d+1$ are determined.

Recall, from Chapter 4, Section I, that

$$|B_{2k}| = \frac{|N_{2k}|}{D_{2k}} \sim 4\sqrt{\pi k} \left(\frac{k}{\pi e}\right)^{2k}.$$

The above results are classical and very well explained, for example, in Chapter I of Rademacher's book (1973).

Lemma B.

$$\log |N_{2k}| = 2k \log k + O(k).$$

Proof.

$$0 < \log D_{2k} = \sum_{(p-1)|2k} \log p \leqslant \sum_{p \leqslant 2k+1} \log p < C'(2k + 1) < Ck$$

by Tschebycheff's result (where C, C' are appropriate constants).

If δ is such that $0 < \delta < \frac{1}{2}$, there exists $k_0 \geqslant 3$ such that, if $k \geqslant k_0$, then

$$1 - \delta < \frac{|N_{2k}|}{D_{2k}4\ \sqrt{\pi k}(k/\pi e)^{2k}} < 1 + \delta.$$

Then

$$-\log 2 - 2k \log(\pi e) < \log(1-\delta) + \log D_{2k} + \log(4\sqrt{\pi})$$

$$+ \tfrac{1}{2} \log k - 2k \log(\pi e) < \log|N_{2k}| - 2k \log k$$

$$< \log D_{2k} + \log(4\sqrt{\pi}) + \tfrac{1}{2} \log k - 2k \log(\pi e) + \log(1+\delta)$$

$$< Ck + \log(4\sqrt{\pi}) + \tfrac{1}{2} \log k + \log (\tfrac{3}{2}).$$

Therefore

$$|\log|N_{2k}| - 2k \log k| < \max\{\log 2 + 2k \log(\pi e),$$

$$Ck + \log(4\sqrt{\pi}) + \tfrac{1}{2} \log k + \log \tfrac{3}{2} \}.$$

Hence $\log |N_{2k}| = 2k \log k + O(k)$. □

For $n \geqslant 3$, let $s_n = p_3 p_4 \cdots p_n$.

Lemma C. *If $U_n \geqslant \max\{|N_2|, |N_4|, ..., |N_{p_n-3}|\}$ for $n \geqslant 3$, then*

(1) $\quad \displaystyle \liminf \frac{\log U_n}{p_n} > 1.$

(2) $\quad p_n s_n = o(U_n).$

Proof. (1) Write $q_n = (p_n-3)/2$. By Lemma B, given δ, $0 < \delta < 1$, there exists $C > 0$ such that for n sufficiently large:

$$\log U_n \geqslant \log |N_{2q_n}| = 2q_n \log q_n - Cq_n$$

$$> (p_n-3) \log(p_n-3) - C\, \frac{p_n - 3}{2} > 2p_n > p_n(1+\delta).$$

(2) By the prime number theorem,

$$\log s_n = \sum_{k=3}^{n} \log p_k = \theta(p_n) - \log 2 - \log 3 \sim p_n .$$

Thus, given $\delta > 0$, for n sufficiently large,

$$0 < \frac{\log p_n}{p_n} < \frac{\delta}{4}, \quad 1 - \frac{\delta}{4} < \frac{\log s_n}{p_n} < 1 + \frac{\delta}{4} .$$

From (1), $\log U_n > (1+\delta)p_n$, hence

$$\log U_n - \log p_n - \log s_n > \frac{\delta}{2} p_n ,$$

therefore

$$\lim_{n \to \infty} \frac{U_n}{p_n s_n} = \infty . \qquad \square$$

Here is another technical lemma, which is easy:

Lemma D. *Let U_n, s_n be as before. Assume that s_n divides U_n and $T_n \geq U_n$ (for every $n \geq 3$). Then*

$$(U_n + s_n)^{q_n} - U_n^{q_n} = o(T_n^{q_n}).$$

Proof. Write $U_n = s_n u_n$. Then

$$\frac{(U_n+s_n)^{q_n} - U_n^{q_n}}{T_n^{q_n}} \leq \frac{(U_n+s_n)^{q_n} - U_n^{q_n}}{U_n^{q_n}} = \frac{(u_n+1)^{q_n} - u_n^{q_n}}{u_n^{q_n}}$$

$$= \frac{1}{u_n^{q_n}} \left[\binom{q_n}{1} u_n^{q_n-1} + \binom{q_n}{2} u_n^{q_n-2} + \cdots + 1 \right]$$

$$\leq \frac{p_n}{u_n} + \frac{p_n^2}{u_n^2} + \cdots + \frac{p_n^{q_n}}{u_n^{q_n}} < \sum_{i=1}^{\infty} \left(\frac{p_n}{u_n} \right)^i = \frac{p_n}{u_n} \times \frac{1}{1 - \frac{p_n}{u_n}}$$

$$= \frac{p_n s_n}{U_n} \times \frac{1}{1 - \frac{p_n s_n}{U_n}} .$$

By Lemma C, the last expression tends to 0 as n tends to infinity.

$\square$

Now, I can prove (2^0).

Lemma E. *Let* $T_n \geq T_n'$, *where* T_n' *is the smallest multiple of* s_n *such that* $T_n' \geq \max\{|N_2|, ..., |N_{p_n-3}|\}$ *for every* $n \geq 3$. *Then*

$$\lim_{n \to \infty} \frac{\#M(\epsilon,n,T)}{T_n^{q_n}} = \lim_{n \to \infty} \mu(\epsilon,n).$$

Proof. Choose U_n such that s_n divides U_n and $U_n \leq T_n < U_n + s_n$, so $T_n' \leq U_n$. By Lemma D,

$$\lim_{n \to \infty} \frac{(U_n+s_n)^{q_n} - U_n^{q_n}}{T_n^{q_n}} = 0$$

hence also

$$\lim_{n \to \infty} \frac{T_n^{q_n} - U_n^{q_n}}{T_n^{q_n}} = 0,$$

so

$$\lim_{n \to \infty} \left[\frac{U_n}{T_n}\right]^{q_n} = 1.$$

Then, by Lemma A

$$\frac{\#M(\epsilon,u,U_n)}{U_n^{q_n}} \leq \frac{\#M(\epsilon,n,T_n)}{T_n^{q_n}} \times \left[\frac{T_n}{U_n}\right]^{q_n}$$

$$\leq \frac{\#M(\epsilon,n,U_n+s_n)}{(U_n+s_n)^{q_n}} \times \left[\frac{U_n+s_n}{U_n}\right]^{q_n} = \frac{\#M(\epsilon,n,U_n)}{U_n^{q_n}} \times \left[\frac{U_n+s_n}{U_m}\right]^{q_n},$$

and finally, by Lemma C,

$$\lim_{n \to \infty} \frac{\#M(\epsilon,n,T)}{T_n^{q_n}} = \lim_{n \to \infty} M(\epsilon,n). \qquad \square$$

It remains to prove (3^0). If $T \geq 1$, $k \geq 3$, let $S_{k,T}$ be the set of all sequences $x = (x_i)_{i \geq 1}$ with $1 \leq x_i \leq T$ for $i = 1, ..., q_k$. Note that $S_{3,T} \supseteq S_{4,T} \supseteq \cdots$.

Let $X_{k,T}: S_{k,T} \to \{0,1\}$ be defined by

$$X_{k,T}(x) = \begin{cases} 1 & \text{if } p_k \text{ divides } x_1 \cdots x_{q_k}, \\ 0 & \text{otherwise} . \end{cases}$$

Thus the value of $X_{k,T}$ depends only on the first q_k coordinates of x. Let $I_{k,T}$ be the set of all $(y_1, ..., y_{q_k})$ such that $1 \leqslant y_i \leqslant T$ for $i = 1, ..., q_k$ and p_k divides $y_1 \cdots y_{q_k}$.

Then the probability for $X_{k,T}(x)$ to be 1 is defined as

$$P(X_{k,T}(x) = 1) = \frac{\#(I_{k,T})}{T^{q_k}} .$$

Lemma F. *Assume that* $k \geqslant 3$ *and* p_k *divides* T. *Then*

$$P(X_{k,T}(x) = 1) = 1 - \left[1 - \frac{1}{p_k} \right]^{q_k} = \rho_k ;$$

so it is independent of T, *provided* p_k *divides* T.

Proof. For simplicity, write $p = p_k$, $q = q_k$.

If $p|T$ there are $(T - \frac{T}{p})^q$ q-tuples $(x_1, ..., x_q)$ with $1 \leqslant x_i \leqslant T$ (for $i = 1, ..., q$), such that p does not divide $x_1 x_2 \cdots x_q$. Hence

$$\#(I_{q,T}) = T^q - \left[T - \frac{T}{p} \right]^q = T^q \left[1 - \left[1 - \frac{1}{p} \right]^q \right]$$

and so

$$P(X_{k,T}(x) = 1) = 1 - \left[1 - \frac{1}{p} \right]^q . \qquad \square$$

Now let $\tau = (T_n)_{n \geqslant 3}$ be a sequence of positive integers, such that s_n divides T_n for every $n \geqslant 3$. Let

$$A_{n,\tau} = \frac{1}{n-2} \sum_{k=3}^{n} X_{k,T_n},$$

so $A_{n,\tau}$ is a mapping from S_{n,T_n} to

$$\left\{ 0, \frac{1}{n-2}, \frac{2}{n-2}, ..., 1 \right\}.$$

It is the mean of the random variables X_{k,T_n} ($3 \leqslant k \leqslant n$). Since the expectation of each X_{k,T_n} is equal to ρ_k, then the expectation of A_{n,T_n} is

$$\frac{1}{n-2} \sum_{k=3}^{n} \rho_k \ ;$$

note that, as n tends to infinity, it converges to ρ.

Lemma G. *For every $\epsilon > 0$, $\lim_{n \to \infty} \mu(\epsilon, n) = 1$.*

Proof. With above notations, from the law of large numbers, the arithmetic mean A_{n,T_n} of the random variables X_{k,T_n} ($3 \leqslant k \leqslant n$), converges in probability to ρ.

In other words, given $\epsilon > 0$,

$$\lim_{n \to \infty} P(|A_{n,T_n}(x) - \rho| < \epsilon) = 1.$$

Now

$$A_{n,T_n}(x) = \frac{1}{n-2} \, \#\{p_k \mid 3 \leqslant k \leqslant n, \ p_k \text{ divides } x_1 \cdots x_{q_k}\}.$$

So, $P(|A_{n,T_n}(x) - \rho| < \epsilon)$ is equal to

$$\frac{\#M(\epsilon, n, T_n)}{T_n^{q_n}} = \mu(\epsilon, n)$$

by Lemma A, under the present assumption that s_n divides T_n for every $n \geqslant 3$. This concludes the proof of the lemma. $\square$

From the previous steps, it follows the heuristic statement of Siegel on the distribution of irregular primes.

I add here that the numerical computations of Wagstaff strongly support Siegel's heuristic theorem.

CONCLUSION

Dear Reader:

If you are not like certain persons, who turn to the last page of a thrilling book to find out who the criminal was, then I may assume that you have dutifully read this book, and I am almost tempted to write "Dear friend," or better "Dear friend of prime numbers."

You may now be thinking about what you learned, perhaps already trying some of the problems and putting to work your personal computer, as well as your brain.

I hope that this presentation of various topics of the theory of prime numbers conveys an approximate picture of the main problems being investigated.

On the one hand, there are many efforts which may be easily explained to students and even to laymen. They may be compared to athletic performances, involving patience, endurance, a fine technique, and leading to records. Not surprisingly, to surpass previous records, quite often an entirely new method needs to be developed. So, despite the futility of the exercise, interesting problems may have to be solved, and a whole new understanding evolves. Names of record holders are, in a few instances, the object of a certain attention, but are, alas, soon led into oblivion, replaced by other names. "Sic transit gloria" ... as wisely exclaims Yates, borrowing from Thomas a Kempis.

The other side of the study is profound. A full understanding of the intricacies of the distribution of prime numbers is one of the

greatest challenges for mathematicians. In the past, as well as today, some of the most powerful minds have tried to grasp, analyze,
count, foresee, and discover amazing facts and relationships between
prime numbers. It is comforting to know that much has yet to be
revealed.

I suppose also that you may be wondering about my decision to
include certain topics and not others, to write in detail some proofs
totally omitting others of more important results. I owe you an explanation, if not an apology.

In writing this book I wanted to produce a work of synthesis, to
develop the theory of prime numbers as a discipline where the
natural questions are systematically studied.

In the Introduction, I believe, I made clear the reasons for dividing the book into its various parts, which are devoted to the answer
of what I consider the imperative questions. This organization is
intended to prevent any young student from having the same impression as I had in my early days (it was, alas, long ago ...) that number
theory dealt with multiple unrelated problems.

This justifies the general plan but not the choice of details, especially the ones which are missing (we always cry for the dear absent
ones ...). It is evident that a choice had to be made. I wanted this
book to be small and lightweight, so the hand could hold it − not a
bulky brick, which cannot be carried in a pocket, nor read while
waiting or riding in a train (in Canada, we both wait for and ride
in trains).

The numerous ways to show the existence of infinitely many
primes are intended to stress different aspects of the distribution of
primes. The inclusion of other proofs at odd spots in the text, often
of minor results, was aimed at breaking the rhythm and providing
young readers with some detailed work which they should be able to
follow easily, thus encouraging them to continue. I took care to
choose pleasant and clever proofs, and let myself be guided by my
momentary whims − in French, one would say *par mon caprice*. True
enough, by the end, in Chapter 6, there are some proofs which are
rather long and technical, but by then the battle has taken place,
and there is no harm for the reader in skipping the harder proofs
altogether. I admit readily that the proofs of all important theorems

are absent, but they are already written in detail in many good books. Besides, I do include ample references.

It should be apparent that Chapter 4 concerns the most striking aspects of the theory of prime numbers – the uniformity at large, coupled with an intrinsic unpredictability. The prime number theorem is a centerpiece.

However, the point of view and level of the present text was not conducive to an extensive technical discussion of the prime number theorem. So, in guise of epilogue, I wish to quote Bateman's interesting

Citations for Some Possible Prizes for Work on the Prime Number Theorem

1. To P. L. Tschebycheff for his two papers *Sur la fonction qui détermine la totalité des nombres premiers inférieurs à une limite donnée*, Journal de Math., 17(1), 1852, 341-365 and *Mémoire sur les nombres premiers*, Journal de Math., 17(1), 1852, 366-390.

2. To Bernhard Riemann for his paper *Über die Anzahl der Primzahlen unter einer gegebenen Grösse*, Monatsberichte der Königlichen Preussischen Akademie der Wissenschaften zu Berlin aus dem Jahre 1859 (1860), 671-680.

3. To Jacques Hadamard for his two papers *Étude sur les propriétés des fonctions entières et en particulier d'une fonction considérée par Riemann*, Journal de Math., 9(4), 1893, 171-215 and *Sur la distribution des zéros de la fonction $\zeta(s)$ et ses conséquences arithmétiques*, Bulletin de la Soc. Math. de France, 24, 1896, 199-220.

4. To C.-J. de la Vallée Poussin for his two papers *Recherches analytiques sur la théorie des nombres; Première partie: La fonction $\zeta(s)$ de Riemann et les nombres premiers en général*, Annales de la Soc. Scientifique de Bruxelles, 20, 1896, 183-256 and *Sur la fonction $\zeta(s)$ de Riemann et le nombre des nombres premiers inférieurs à une limite donnée*, Mémoires couronnés et autres memoires publiés par l'Acad. Roy. des Sciences, des Lettres et des Beaux-Arts de Belgique, 59, No. 1, 1899-1900, 74 pages.

5. To Edmund Landau for his paper *Neuer Beweis des Primzahl-satzes und Beweis des Primidealsatzes*, Mathematische Annalen, **56**, 1903, 645-670 in his book *Handbuch der Lehre von der Verteilung der Primzahlen*, Teubner, Leipzig, 1909.

6. To G. H. Hardy and J. E. Littlewood for their many contributions to analysis and number theory, in particular their two papers *New proofs of the prime-number theorem and similar theorems*, Quarterly Journal of Math., **46**, 1915, 215-219 and *Contributions to the theory of the Riemann zeta function and the theory of the distribution of primes*, Acta Math., **41**, 1918, 119-196.

7. To Norbert Wiener for his memoir *Tauberian theorems*, Annals of Mathematics, 33(2), 1932, 1-100.

8. To I. M. Vinogradov for his papers on exponential sums, particularly the three papers: *On Weyl's sums*, Mat. *Sbornik*, **42**, 1935, 521-530; *A new method of resolving certain general questions of the theory of numbers*, Mat. Sbornik, **43**(1), 1936, 9-19; *A new method of estimation of trigonometrical sums*, Mat. Sbornik, **43**(1), 1936, 175-188.

9. To Arne Beurling for his paper *Analyse de la loi asymptotique de la distribution des nombres premiers généralisés*, *I*, Acta Math., **68**, 1937, 255-291.

10. To Atle Selberg for his many contributions to the theory of the zeta function and the distribution of primes and in particular for his paper, *An elementary proof of the prime number theorem*, Annals of Mathematics, 50(2), 1949, 305-313.

11. To Paul Erdös for his many papers in the theory of numbers, and in particular for his paper, *On a new method in elementary number theory which leads to an elementary proof of the prime number theorem*, Proceedings of the National Academy of Sciences, **35**, 1949, 374-385.

To the reader whose curiosity is still alive and whose mind is not intimidated by deep mathematical ideas and sophisticated techniques, the way is open to further his studies along the various directions indicated here.

GOOD LUCK!

Whatever shortcomings there are in my treatment of the subject, it is my hope that the beauty found here and there will compensate for what was expected, but not found, at various other places.

And now, fancying that you will reopen the book once more at the beginning, it is the moment to explain the reference to Bashō. He was a classical Japanese poet, who lived from 1644 to 1694. Son of a samurai, in the service of the Lord of Ueno Castle, Bashō learned the art of haiku and later became the most important representative of this form of poetry. You certainly know that these are microscopic poems of seventeen syllables, like a seashell containing an ocean of thoughts. It is perhaps amusing to tell that Bashō was a *nom de plume*, which means banana-tree. This plant, exotic and new in Japan, was given to him by his disciples and tended by the poet in his little garden with the greatest care.

Narrow Road to a Far Province is the title of his famous travel diary, where Bashō tells in prose or haikai verses of his impressions while visiting the places in Japan celebrated by the most famous poets. This journey took him through mountains into the Northern provinces and eventually by the West Coast, up to Lake Biwa. It was a trip where he found much beauty, sometimes revealed in an unexpected way.

I cannot resist quoting some of his verses, for your enjoyment.

At his departure, middle of May:

> *Loath to let spring go,*
> *Birds cry, and even fishes',*
> *Eyes are wet with tears.*

Of a little girl, composed by his travelling companion Sora:

> *Were she a flower,*
> *She would be a wild, fring'd pink,*
> *Petals manifold.*

On clogs, walking their way:

> *In the hills, 'tis May*
> *Bless us, holy shoes, as we*
> *Go upon our way.*

Where the heroes were buried:

> *A mound of summer grass:*
> *Are warriors' heroic deeds*
> *Only dreams that pass?*

Fruits in autumn:

> *How cool the autumn air!*
> *I'll peel them and enjoy them —*
> *The melon and the pear.*

Adieu:

> *Sadly, I part from you,*
> *Like a clam torn from its shell.*
> *I go, and autumn too.*

Should I be explicit and say why Bashō is on the title page of my book? I do not have to add that beauty comes in different forms, and it is our good fortune if we know how to discern it.

Jean Pierre Serre has been, probably unknowingly, a good guide for me. His mathematical pursuit and taste deserve to be imitated.

But many other mathematicians have also greatly helped me, encouraging my project with the right words, and taking their time to write many pages of detailed suggestions. I derived much comfort from their support. In alphabetical order:

J. Brillhart,

J. B. Friedlander,

E. Grosswald,

R. K. Guy,

D. R. Heath-Brown,

K. Inkeri,

J. P. Jones,

M. Jutila,

W. Keller,

A. M. Odlyzko,

C. Pomerance,

A. Schinzel,

D. Shanks,

S. S. Wagstaff, Jr.,

M. Waldschmidt,

H. C. Williams.

BIBLIOGRAPHY

A. General References

The books listed below are highly recommended for the quality of contents and presentation. This list is, of course, not exhaustive.

1909   LANDAU, E.   *Handbuch der Lehre der Verteilung der Primzahlen.*
Teubner, Leipzig, 1909.
Reprinted by Chelsea, Bronx, N.Y., 1974.

1927   LANDAU, E.   *Vorlesungen über Zahlentheorie* (in 3 volumes).
S. Hirzel, Leipzig, 1927.
Reprinted by Chelsea, Bronx, N.Y., 1969.

1938   HARDY, G. H. & WRIGHT, E. M.   *An Introduction to the Theory of Numbers.*
Clarendon Press, Oxford, 1938 (5th edition, 1979).

1952   DAVENPORT, H.   *The Higher Arithmetic.*
                                Hutchinson, London, 1952.
                                Reprinted by Dover, New
                                York, 1983.

1953   TROST, E.   *Primzahlen.*
                                Birkhäuser, Basel, 1953.

1957   PRACHAR, K.   *Primzahlverteilung.*
                                Springer-Verlag, Berlin,
                                1957 (2nd edition 1978).

1962   SHANKS, D.   *Solved and Unsolved Problems
                                in Number Theory.*
                                Spartan, Washington, 1962.
                                3rd edition by Chelsea, Bronx, N.Y.,
                                1985.

1963   AYOUB, R.G.   *An Introduction to the Theory
                                of Numbers.*
                                Amer. Math. Soc.,
                                Providence, R.I., 1963.

1974   HALBERSTAM, H. &   *Sieve Methods.*
       RICHERT, H.E.   Academic Press, New York, 1974.

1975   ELLISON, W.J. &   *Les Nombres Premiers.*
       MENDÈS-FRANCE, M.   Hermann, Paris, 1975.

1976   ADAMS, W.W. &   *Introduction to Number Theory.*
       GOLDSTEIN, L.J.   Prentice-Hall, Englewood Cliffs,
                                N.J., 1976.

1981   GUY, R.K.   *Unsolved Problems in Number
                                Theory.*
                                Springer-Verlag, New York, 1981.

1982   HUA, L.K.   *Introduction to Number Theory.*
                                Springer-Verlag, New York, 1982.

B. Specific References

Chapter 1

1878 KUMMER, E.E.   Neuer elementarer Beweis, dass die
                            Anzahl aller Primzahlen eine
                            unendliche ist.
                            Monatsber. Akad. d. Wiss., Berlin,
                            1878/9, 777-778.

1890 STIELTJES, T.J.   Sur la théorie des nombres.
                            Étude bibliographique.
                            An. Fac. Sci. Toulouse, 4, 1890,
                            1-103.

1897 THUE, A.   Mindre meddelelser II. Et
                            bevis for at primtallenes
                            antal er unendeligt.
                            Arch. f. Math. og Naturv.,
                            Kristiania, 1897, 19, No. 4,
                            3-5.
                            Reprinted in *Selected
                            Mathematical Papers* (edited by
                            T. Nagell, A. Selberg & S. Selberg),
                            31-32.
                            Universitetsforlaget, Oslo, 1977.

1924 PÓLYA, G. &   *Aufgaben und Lehrsätze
      SZEGÖ, G.   aus der Analysis*, 2 vols.
                            Springer-Verlag, Berlin,
                            1924 (4th edition, 1970).

1955 FÜRSTENBERG, H.   On the infinitude of primes.
                            Amer. Math. Monthly, 62, 1955,
                            p. 353.

1959   GOLOMB, S.W.   A connected topology for the
                                 integers.
                                 Amer. Math. Monthly, 66, 1959,
                                 663-665.

1963   MULLIN, A.A.   Recursive function theory (A modern
                                 look on an Euclidean idea).
                                 Bull. Amer. Math. Soc., 69,
                                 1963, p. 737.

1964   EDWARDS, A.W.F.   Infinite coprime sequences.
                                 Math. Gazette, 48, 1964, 416-422.

1967   SAMUEL, P.   *Théorie Algébrique des
                                 Nombres.*
                                 Hermann, Paris, 1967.
                                 English translation published by
                                 Houghton-Mifflin, Boston, 1970.

1968   COX, C.D. &   On a sequence of prime numbers.
       VAN DER   J. Austr. Math. Soc., 1968, 8,
       POORTEN, A.J.   571-574.

1972   BORNING, A.   Some results for $k! \pm 1$ and
                                 $2 \cdot 3 \cdot 5 \cdots p + 1$.
                                 Math. Comp., 26, 1972, 567-570.

1980   TEMPLER, M.   On the primality of $k! + 1$ and
                                 $2*3* \cdots *p + 1$.
                                 Math. Comp., 34, 1980, 303-304.

1980   WASHINGTON, L.C.   The infinitude of primes
                                 via commutative algebra.
                                 Unpublished manuscript.

1982   BUHLER, J.P.,   Primes of the form $n! \pm 1$
       CRANDALL, R.E. &   and $2 \cdot 3 \cdot 5 \cdots p \pm 1$.
       PENK, M.A.   Math. of Comp., 38, 1982,
                                 639-643.

1985 DUBNER, H. &   The development of a powerful
      DUBNER, H.   low-cost computer for number
                        theory applications.
                        J. Recr. Math., 18 (2), 1985/6,
                        92-96.

1987 DUBNER, H.   Factorial and primorial primes.
                        J. Recr. Math., 19, 1987, 197-203.

Chapter 2

1801 GAUSS, C.F.   *Disquisitiones Arithmeticae.*
                        G. Fleischer, Leipzig, 1801.
                        English translation by A.A. Clarke.
                        Yale Univ. Press, New Haven, 1966.
                        Revised English translation by
                        W.C. Waterhouse.
                        Springer-Verlag, New York, 1986.

1844 EISENSTEIN, F.G. Aufgaben.
                        Journal f.d. reine u. angew. Math., 27,
                        1844, p. 87.
                        Reprinted in *Mathematische Werke*,
                        Vol. I, p. 112.
                        Chelsea, Bronx, N.Y., 1975.

1852 KUMMER, E.E.   Über die Erganzungssätze zu den
                        allgemeinem Reciprocitätsgesetzen.
                        Journal f.d. reine u. angew. Math.,
                        44, 1852, 93-146.
                        Reprinted in *Collected Papers*,
                        Vol. I, (edited by A. Weil, 485-538).
                        Springer-Verlag, New York, 1975.

1876 LUCAS, E.   Sur la recherche des grands
                        nombres premiers.
                        Assoc. Française p. l'Avanc. des
                        Sciences, 5, 1876, 61-68.

1877 PEPIN, T.   Sur la formule $2^{2^n} + 1$.
                               C.R. Acad. Sci. Paris, 85, 1877,
                               329-331.

1878 LUCAS, E.   Théorie des fonctions numériques
                               simplement périodiques.
                               Amer. J. Math., 1, 1878, 184-240 and
                               289-321.

1878 PROTH, F.   Théorèmes sur les nombres
                               premiers.
                               C.R. Acad. Sci. Paris, 85, 1877,
                               329-331.

1886 BANG, A.S.   Taltheoretiske Undersøgelser.
                               Tidskrift f. Math., ser. 5,4,
                               1886, 70-80 and 130-137.

1891 LUCAS, E.   *Théorie des Nombres.*
                               Gauthier-Villars, Paris, 1891.
                               Reprinted by A. Blanchard,
                               Paris, 1961.

1892 ZSIGMONDY, K.   Zur Theorie der Potenzreste.
                               Monatsh. f. Math., 3, 1892,
                               265-284.

1903 MALO, E.   Nombres qui, sans être premiers,
                               vérifient exceptionnellement une
                               congruence de Fermat.
                               L'Interm. des Math., 10, 1903, p. 88.

1904 BIRKHOFF, G.D. &   On the integral divisors of $a^n - b^n$.
      VANDIVER, H.S.   Annals of Math., (2), 5, 1904,
                               173-180.

1904 CIPOLLA, M.   Sui numeri composti P, che
                             verificano la congruenza di Fermat
                             $a^{P-1} \equiv 1 \pmod{P}$.
                             Annali di Matematica, (3), 9, 1904,
                             139-160.

1912 CARMICHAEL, R.D.   On composite numbers P which
                             satisfy the Fermat congruence
                             $a^{P-1} \equiv 1 \pmod{P}$.
                             Amer. Math. Monthly, 19, 1912,
                             22-27.

1913 CARMICHAEL, R.D.   On the numerical factors of the
                             arithmetic forms $\alpha^n \pm \beta^n$.
                             Annals of Math., 15, 1913, 30-70.

1913 DICKSON, L.E.   Finiteness of odd perfect and
                             primitive abundant numbers with n
                             distinct prime factors.
                             Amer. J. Math., 35, 1913, 413-422.
                             Reprinted in *The Collected
                             Mathematical Papers*
                             (edited by A.A. Albert),
                             Vol. I, 349-358.
                             Chelsea, Bronx, N.Y., 1975.

1914 POCKLINGTON, H.C.   The determination of the prime or
                             composite nature of large numbers
                             by Fermat's theorem.
                             Proc. Cambridge Phil. Soc., 18,
                             1914/6, 29-30.

1921 PIRANDELLO, L.   *Fu Mattia Pascal.*
                             Bemporad & Figlio, Firenze, 1921.

1922 CARMICHAEL, R.D.   Note on Euler's ϕ-function.
                             Bull. Amer. Math. Soc., 28, 1922,
                             109-110.

1925   CUNNINGHAM, A.J.C.   *Factorization* of $y^n \pm 1$,
       & WOODALL, H.J. $y = 2, 3, 5, 6, 7, 10, 11, 12$
                                Up to High Powers (n).
                                Hodgson, London, 1925.

1929   PILLAI, S.S.   On some functions connected with
                                $\phi(n)$.
                                Bull. Amer. Math. Soc., 35, 1929,
                                832-836.

1930   LEHMER, D.H.   An extended theory of Lucas'
                                functions.
                                Annals of Math., 31, 1930, 419-448.
                                Reprinted in *Selected Papers*
                                (edited by D. McCarthy),
                                Vol. I, 11-48.
                                Ch. Babbage Res. Centre, St.
                                Pierre, Manitoba, Canada, 1981.

1932   LEHMER, D.H.   On Euler's totient function.
                                Bull. Amer. Math. Soc., 38, 1932,
                                745-757.
                                Reprinted in *Selected Papers*
                                (edited by D. McCarthy),
                                Vol. I, 319-325.
                                Ch. Babbage Res. Centre, St. Pierre,
                                Manitoba, Canada, 1981.

1932   WESTERN, A.E.   On Lucas' and Pepin's tests for
                                the primeness of Mersenne's
                                numbers.
                                J. London Math. Soc., 7, 1932,
                                130-137.

1935   ARCHIBALD, R.C.   Mersenne's numbers.
                                Scripta Math., 3, 1935,
                                112-119.

1935   ERDÖS, P.   On the normal number of prime factors
                                    of $p-1$ and some related problems
                                    concerning Euler's ϕ-function.
                                    Quart. J. Math. Oxford, 6, 1935,
                                    205-213.

1935   LEHMER, D.H.   On Lucas' test for the primality
                                    of Mersenne numbers.
                                    J. London Math. Soc., 10, 1935, 162-165.
                                    Reprinted in *Selected Papers*
                                    (edited by D. McCarthy),
                                    Vol. I, 86-89.
                                    Ch. Babbage Res. Centre, St. Pierre,
                                    Manitoba, Canada, 1981.

1936   LEHMER, D.H.   On the converse of Fermat's theorem.
                                    Amer. Math. Monthly, 43, 1936,
                                    Reprinted in *Selected Papers*
                                    (edited by D. McCarthy),
                                    Vol. I, 90-95.
                                    Ch. Babbage Res. Centre, St. Pierre,
                                    Manitoba, Canada, 1981.

1939   CHERNICK, J.   On Fermat's simple theorem.
                                    Bull. Amer. Math. Soc., 45, 1939,
                                    269-274.

1941   KANOLD, H.J.   Unterschungen über ungerade
                                    volkommene Zahlen.
                                    Journal f. d. reine u. angew.
                                    Math., 183, 1941, 98-109.

1942   KANOLD, H.J.   Verschärfung einer notwendigen
                                    Bedingung für die Existenz
                                    einer ungeraden vollkommenen
                                    Zahl.
                                    Journal f. d. reine u. angew.
                                    Math., 184, 1942, 116-123.

1943   BRAUER, A.   On the non-existence of odd perfect
                               numbers of the form $p^{\alpha}q_1^2q_2^2\cdots q_{t-1}^2q_t^4$.
                               Bull. Amer. Math. Soc., 49, 1943,
                               712-718, and page 937.

1944   KANOLD, H.J.   Folgerungen aus dem Vorkommen
                               einer Gauss' schen Primzahl in
                               der Primfaktorenzerlegung einer
                               ungeraden vollkommenen Zahl.
                               Journal f. d. reine u. angew.
                               Math., 186, 1944, 25-29.

1944   PILLAI, S.S.   On the smallest primitive root of a
                               prime.
                               J. Indian Math. Soc., (N.S.), 8,
                               1944, 14-17.

1944   SCHUH, F.   Can $n-1$ be divisible by $\phi(n)$ when
                               n is composite? (in Dutch).
                               Mathematica, Zutphen, B, 12, 1944,
                               102-107.

1945   KAPLANSKY, I.   Lucas' tests for Mersenne numbers.
                               Amer. Math. Monthly, 52, 1945,
                               188-190.

1947   KLEE, V.L.   On a conjecture of Carmichael.
                               Bull. Amer. Math. Soc., 53, 1947,
                               1183-1186.

1947   LEHMER, D.H.   On the factors of $2^n \pm 1$.
                               Bull. Amer. Math. Soc., 53,
                               1947, 164-167.
                               Reprinted in *Selected Papers*
                               (edited by D. McCarthy),
                               Vol. III, 1081-1084.
                               Ch. Baggage Res. Centre, St. Pierre,
                               Manitoba, Canada, 1981.

1947 LEVIT, R.J.   The non-existence of a certain
type of odd perfect number.
Bull. Amer. Math. Soc., 53,
1947, 392-396.

1948 ORE, O.   On the averages of the divisors of a
number.
Amer. Math. Monthly, 55, 1948,
615-619.

1948 STEUERWALD, R.   Über die Kongruenz $a^{n-1} \equiv 1$
(mod n).
Sitzungsber. math.-naturw. Kl.
Bayer. Akad. Wiss. München, 1948,
69-70.

1949 BERNHARD, H.A.   On the least possible odd perfect
number.
Amer. Math. Monthly, 56, 1949,
628-629.

1949 FRIDLENDER, V.R.   On the least nth power
non-residue (in Russian).
Doklady Akad. Nauk SSSR
(N.S.), 66, 1949, 351-352.

1949 ERDÖS, P.   On the converse of Fermat's
theorem.
Amer. Math. Monthly, 56,
1949, 623-624.

1949 KÜHNEL, U.   Verschärfung der notwendigen
Bedingungen für die Existenz von
ungeraden vollkommenen Zahlen.
Math. Z., 52, 1949, 202-211.

1949 SHAPIRO, H.N.   Note on a theorem of Dickson.
Bull. Amer. Math. Soc., 55, 1949,
450-452.

1950   BEEGER, N.G.W.H.   On composite numbers n for which
                                 $a^{n-1} \equiv 1 \pmod{n}$, for every a,
                                 prime to n.
                                 Scripta Math., 16, 1950, 133-135.

1950   GIUGA, G.   Su una presumibile proprietà
                                 caratteristica dei numeri primi.
                                 Ist. Lombardo Sci. Lett. Rend. Cl.
                                 Sci. Mat. Nat., (3), 14 (83), 1950,
                                 511-528.

1950   GUPTA, H.   On a problem of Erdös.
                                 Amer. Math. Monthly, 57, 1950,
                                 326-329.

1950   KANOLD, H.J.   Sätze über Kreisteilungspolynome
                                 und ihre Anwendungen auf einige
                                 zahlentheoretische Probleme, II.
                                 Journal f.d. reine u. angew. Math.,
                                 187, 1950, 355-366.

1950   SALIÉ, H.   Über den kleinsten positiven
                                 quadratischen Nichtrest einer
                                 Primzahl.
                                 Math. Nachr., 3, 1949, 7-8.

1950   SOMAYAJULU, B.S.K.R.   On Euler's totient function
                                 $\phi(n)$.
                                 Math. Student, 18, 1950, 31-32.

1951   BEEGER, N.G.W.H.   On even numbers m dividing
                                 $2^m - 2$.
                                 Amer. Math. Monthly, 58,
                                 1951, 553-555.

1951   MORROW, D.C.   Some properties of D numbers.
                                 Amer. Math. Monthly, 58, 1951,
                                 329-330.

1951   WEBBER, G.C.   Nonexistence of odd perfect
                             numbers of the form
                             $3^{2\beta} \cdot p^{\alpha} \cdot s_1^{2\beta_1} s_2^{2\beta_2} s_3^{2\beta_3}$.
                             Duke Math. J., 18, 1951,
                             741-749.

1952   DUPARC, H.J.A.   On Carmichael numbers.
                             Simon Stevin, 29, 1952, 21-24.

1952   GRÜN, O.   Über ungerade vollkommene Zahlen.
                             Math. Zeits., 55, 1952, 353-354.

1953   KANOLD, H.J.   Einige neuere Bedingungen für
                             die Existenz ungerader
                             vollkommener Zahlen.
                             Journal f. d. reine u. angew.
                             Math., 192, 1953, 24-34.

1953   KNÖDEL, W.   Carmichaelsche Zahlen.
                             Math. Nachr., 9, 1953, 343-350.

1953   TOUCHARD, J.   On prime numbers and perfect numbers.
                             Scripta Math., 19, 1953, 35-39.

1954   KANOLD, H.J.   Über die Dichten der Mengen der
                             vollkommenen und der befreundeten
                             Zahlen.
                             Math. Zeits., 61, 1954, 180-185.

1954   ROBINSON, R.M.   Mersenne and Fermat numbers.
                             Proc. Amer. Math. Soc., 5
                             1954, 842-846.

1954   SCHINZEL, A.   Quelques théorèmes sur les
                             fonctions $\phi(n)$ et $\sigma(n)$.
                             Bull. Acad. Polon. Sci., Cl. III, 2,
                             1954, 467-469.

1954 SCHINZEL, A.   Generalization of a theorem of
                                    B.S.K.R. Somayajulu on the Euler's
                                    function $\phi(n)$.
                                    Ganita, 5, 1954, 123-128.

1954 SCHINZEL, A. &   Sur quelques propriétés des
      SIERPIŃSKI, W.   fonctions $\phi(n)$ et $\sigma(n)$.
                                    Bull. Acad. Polon. Sci., Cl. III, 2,
                                    1954, 463-466.

1955 ARTIN, E.   The orders of the linear groups.
                                    Comm. Pure & Appl. Math., 8, 1955,
                                    355-365.
                                    Reprinted in *Collected Papers*
                                    (edited by S. Lang & J.T. Tate),
                                    387-397.
                                    Addison-Wesley, Reading, Mass., 1965.

1955 HORNFECK, B.   Zur Dichte der Menge der
                                    vollkommenen Zahlen.
                                    Arch. Math., 6, 1955, 442-443.

1955 LABORDE, P.   A note on the even perfect numbers.
                                    Amer. Math. Monthly, 62, 1955,
                                    348-349.

1955 WARD, M.   The intrinsic divisors of Lehmer
                                    numbers.
                                    Annals of Math., (2), 62, 1955,
                                    230-236.

1956 HORNFECK, B.   Bemerkung zu meiner Note über
                                    vollkommene Zahlen.
                                    Arch. Math., 7, 1956, p. 273.

1956 KANOLD, H.J.   Über einen Satz von L.E. Dickson, II.
                                    Math. Ann., 132, 1956, 246-255.

1956   ORE, O.   *Number Theory and its History.*
                              McGraw-Hill, New York, 2nd edition,
                              1956.

1956   SCHINZEL, A.   Sur l'équation $\phi(x) = m$.
                              Elem. Math., 11, 1956, 75-78.

1956   SCHINZEL, A.   Sur un problème concernant la
                              fonction $\phi(n)$.
                              Czechoslovak Math. J., 6, (81), 1956,
                              164-165.

1956   SIERPIŃSKI, W.   Sur une propriété de la fonction
                              $\phi(n)$.
                              Publ. Math. Debrecen, 4, 1956, 184-185.

1957   HORNFECK, B &   Über die Haufigkeit vollkommener
       WIRSING, E.   Zahlen.
                              Math. Ann., 133, 1957, 431-438.

1957   KANOLD, H.J.   Über die Verteilung der vollkommenen
                              Zahlen und allgemeinerer
                              Zahlenmengen.
                              Math. Ann., 132, 1957, 442-450.

1957   KANOLD, H.J.   Über mehrfach vollkommene Zahlen, II.
                              Journal f. d. reine u. angew. Math.,
                              197, 1957, 82-96.

1957   McCARTHY, P.J.   Odd perfect numbers.
                              Scripta Math., 23, 1957, 43-47.

1957   ROBINSON, R.M.   The converse of Fermat's
                              theorem.
                              Amer. Math. Monthly, 64,
                              1957, 703-710.

1958   ERDÖS, P.   Some remarks on Euler's ϕ-function.
                              Acta Arithm., 4, 1958, 10-19.

1958 JARDEN, D.   *Recurring Sequences.*
                                Riveon Lematematika, Jerusalem, 1958
                                (3rd edition, 1973).

1958 PERISASTRI, M.   A note on odd perfect numbers.
                                Math. Student, 26, 1958, 179-181.

1958 ROBINSON, R.M.   A report on primes of the form
                                $k \cdot 2^n + 1$ and on factors of
                                Fermat numbers.
                                Proc. Amer. Math. Soc., 9,
                                1958, 673-681.

1958 SCHINZEL, A.   Sur l'équation $\phi(x + k) = \phi(x)$.
                                Acta Arithm., 4, 1958, 181-184.

1958 SIERPIŃSKI, W.   Sur les nombres premiers de la
                                forme $n^n + 1$.
                                L'Enseign. Math., (2), 4, 1958,
                                211-212.

1959 DURST, L.K.   Exceptional real Lehmer sequences.
                                Pacific J. Math., 9, 1959,
                                437-441.

1959 ROTKIEWICZ, A.   Sur les nombres composés qui
                                divisent $a^{n-1} - b^{n-1}$.
                                Rend. Circ. Mat. Palermo, (2), 8,
                                1959, 115-116.

1959 ROTKIEWICZ, A.   Sur les nombres pairs a pour lesquels
                                les nombres $a^n b - ab^n$, respectivement
                                $a^{n-1} - b^{n-1}$ sont divisibles par n.
                                Rend. Circ. Mat. Palermo, (2), 8,
                                1959, 341-342.

1959 SATYANARAYANA,   Odd perfect numbers
      M.   Math. Student, 27, 1959, 17-18.

1959   SCHINZEL, A.   Sur les nombres composés n qui
                                 divisent $a^n - a$.
                                 Rend. Circ. Mat. Palermo, (2), 7,
                                 1958, 1-5.

1959   SCHINZEL, A. &   Sur l'équation $\phi(x+k) = \phi(x)$, II.
       WAKULICZ, A.   Acta Arithm., 5, 1959, 425-426.

1959   WARD, M.   Tests for primality based on
                                 Sylvester's cyclotomic numbers.
                                 Pacific J. Math., 9, 1959,
                                 1269-1272.

1959   WIRSING, E.   Bemerkung zu der Arbeit über
                                 vollkommene Zahlen.
                                 Math. Ann., 137, 1959, 316-318.

1960   INKERI, K.   Tests for primality.
                                 Annales Acad. Sci. Fennicae,
                                 Ser. A, I, 279, Helsinki, 1960,
                                 19 pages.

1961   ERDÖS, P.   Remarks in number theory, I (in
                                 Hungarian, with English summary).
                                 Mat. Lapok, 12, 1961, 10-17.

1961   NORTON, K.K.   Remarks on the number of factors of
                                 an odd perfect number.
                                 Acta Arithm., 6, 1961, 365-374.

1961   WARD, M.   The prime divisors of Fibonacci
                                 numbers.
                                 Pacific J. Math., 11, 1961,
                                 379-389.

1962   BURGESS, D.A.   On character sums and L-series.
                              Proc. London Math. Soc., (3), 12
                              1962, 193-206.

1962   CROCKER, R.   A theorem on pseudo-primes.
                              Amer. Math. Monthly, 69,
                              1962, p. 540.

1962   MĄKOWSKI, A.   Generalization of Morrow's D numbers.
                              Simon Stevin, 36, 1962, p. 71.

1962   SCHINZEL, A.   The intrinsic divisors of Lehmer
                              numbers in the case of negative
                              discriminant.
                              Arkiv. för Mat., 4, 1962, 413-416.

1962   SCHINZEL, A.   On primitive prime factors of $a^n - b^n$.
                              Proc. Cambridge Phil. Soc., 58, 1962,
                              555-562.

1962   SHANKS, D.   *Solved and Unsolved Problems in*
                              Number Theory.
                              Spartan, Washington, 1962.
                              3rd edition by Chelsea, Bronx, N.Y.,
                              1985.

1963   SCHINZEL, A.   On primitive prime factors of Lehmer
                              numbers, I.
                              Acta Arithm., 8, 1963, 211-223.

1963   SCHINZEL, A.   On primitive prime factors of Lehmer
                              numbers, II.
                              Acta Arithm., 8, 1963, 251-257.

1963   SURYANARAYANA,   On odd perfect numbers, II.
       D.   Proc. Amer. Math., 14, 1963,
                              896-904.

1964   LEHMER, E.   On the infinitude of Fibonacci
                             pseudo-primes.
                             Fibonacci Quart., 2, 1964, 229-230.

1965   ERDÖS, P.   Some recent advances and current
                             problems in number theory, in
                             Lectures in Modern Mathematics,
                             Vol. III (edited by T.L. Saaty),
                             196-244.
                             Wiley, New York, 1965.

1965   ROTKIEWICZ, A.   Sur les nombres de Mersenne
                             dépourvus de facteurs
                             carrés et sur les nombres
                             naturels n tels que
                             $n^2 \mid 2^n - 2$.
                             Matem. Vesnik (Beograd), 2,
                             (17), 1965, 78-80.

1966   GROSSWALD, E.   *Topics from the Theory of Numbers.*
                             Macmillan, New York, 1966;
                             2nd edition Birkhäuser, Boston,
                             1984.

1966   MUSKAT, J.B.   On divisors of odd perfect numbers.
                             Math. Comp., 20, 1966, 141-144.

1967   MOZZOCHI, C.J.   A simple proof of the Chinese
                             remainder theorem.
                             Amer. Math. Monthly, 74, 1967,
                             p. 998.

1967   TUCKERMAN, B.   Odd perfect numbers: A search
                             procedure, and a new lower
                             bound of 10^{36}.
                             IBM Res. Rep., RC-1925, 1967,
                             and Notices Amer. Math. Soc.,
                             15, 1968, p. 226.

1968   SCHINZEL, A.   On primitive prime factors of
                               Lehmer numbers, III.
                               Acta Arithm., 15, 1968, 49-69.

1969   LEHMER, D.H.   Computer technology applied to the
                               theory of numbers.
                               In *Studies in Number Theory*
                               (edited by W.J. Le Veque), 117-151.
                               Math. Assoc. of America,
                               Washington, 1969.

1970   ELLIOTT, P.D.T.A.   On the mean value of $f(p)$.
                               Proc. London Math. Soc., (3), 21,
                               1970, 28-96.

1970   LIEUWENS, E.   Do there exist composite numbers
                               for which $k\phi(M) = M - 1$ holds?
                               Nieuw. Arch. Wisk., (3), 18, 1970,
                               165-169.

1970   PARBERRY, E.A.   On primes and pseudo-primes related
                               to the Fibonacci sequence.
                               Fibonacci Quart., 8, 1970, 49-60.

1970   SURYANARAYANA,   A theorem concerning odd
       D. & HAGIS, Jr., P.   perfect numbers.
                               Fibonnaci Quart., 8, 1970,
                               337-346 and 374.

1971   LIEUWENS, E.   *Fermat Pseudo-Primes.*
                               Ph.D. Thesis, Delft, 1971.

1971   MORRISON, M.A. &   The factorization of F_7
       BRILLHART, J.   Bull. Amer. Math. Soc., 77,
                               1971, page 264.

1972   HAGIS, Jr., P. &   A new result concerning the structure
       McDANIEL, W.L.   of odd perfect numbers.
                                 Proc. Amer. Math. Soc., 32, 1972,
                                 13-15.

1972   MANN, H.B. &   A necessary and sufficient condition
       SHANKS, D.   for primality and its source.
                                 J. Comb. Th., ser. A, 13, 1972,
                                 131-134.

1972   MILLS, W.H.   On a conjecture of Ore.
                                 Proc. 1972 Number Th. Conf. in
                                 Boulder, p. 142-146.

1972   RIBENBOIM, P.   *Algebraic Numbers.*
                                 Wiley-Interscience, New York,
                                 1972.

1972   ROBBINS, N.   The nonexistence of odd perfect
                                 numbers with less than seven
                                 distinct prime factors.
                                 Notices Amer. Math. Soc., 19, 1972,
                                 page A52.

1972   ROTKIEWICZ, A.   *Pseudoprime Numbers and their*
                                 Generalizations.
                                 Stud. Assoc. Fac. Sci. Univ.
                                 Novi Sad, 1972.

1972   ROTKIEWICZ, A.   Un problème sur les nombres pseudo-
                                 premiers.
                                 Indag. Math., 34, 1972, 86-91.

1972   WILLIAMS, H.C.   The primality of $N = 2A3^n - 1$.
                                 Can. Math. Bull., (4), 15,
                                 1972, 585-589.

1973   HAGIS, Jr., P.   A lower bound for the set of odd
                                   perfect numbers.
                                   Math. Comp., 27, 1973, 951-953.

1973   HONSBERGER, R.   *Mathematical Gems*, I.
                                   Math. Assoc. of America,
                                   Washington, 1973.

1973   ROTKIEWICZ, A.   On pseudoprimes with respect to the
                                   Lucas sequences.
                                   Bull. Acad. Polon. Sci., 21, 1973,
                                   793-797.

1974   LIGH, S. &   A note on Mersenne numbers.
       NEAL, L.   Math. Mag., 47, 1974, 231-233.

1974   POMERANCE, C.   On Carmichael's conjecture.
                                   Proc. Amer. Math. Soc., 43, 1974,
                                   297-298.

1974   SCHINZEL, A.   Primitive divisors of the expression
                                   $A^n - B^n$ in algebraic number fields.
                                   Journal f. d. reine u. angew. Math.,
                                   268/269, 1974, 27-33.

1974   SINHA, T.N.   Note on perfect numbers.
                                   Math. Student, 42, 1974, p. 336.

1975   BRILLHART, J.,   New primality criteria and
       LEHMER, D.H. &   factorizations of $2^m \pm 1$.
       SELFRIDGE, J.L.   Math. Comp., 29, 1975, 620-647.

1975   GUY, R.K.   How to factor a number.
                                   Proc. Fifth Manitoba Conf.
                                   Numerical Math., 1975, 49-89
                                   (Congressus Numerantium, XVI,
                                   Winnipeg, Manitoba, 1976).

1975 HAGIS, Jr., P. &   On the largest prime divisor of an
McDANIEL, W.L.   odd perfect number, II.
    Math. Comp., 29, 1975, 922-924.

1975 MORRISON, M.A.   A note on primality testing using
    Lucas sequences.
    Math. Comp., 29, 1975, 181-182.

1975 MORRISON, M.A. &   A method of factoring and the
BRILLHART, J.   factorization of F_7.
    Math. Comp., 29, 1975, 183-208.

1975 POMERANCE, C.   The second largest prime factor of
    an odd perfect number.
    Math. Comp., 29, 1975, 914-921.

1975 PRATT, V.R.   Every prime has a succinct
    certificate.
    SIAM J. Comput., 4, 1975,
    214-220.

1975 STEWART, C.L.   The greatest prime factor of
    $a^n - b^n$.
    Acta Arithm., 26, 1975, 427-433.

1976 BUXTON, M. &   An extension of lower bounds for odd
ELMORE, S.   perfect numbers.
    Notices Amer. Math. Soc., 23, 1976,
    page A55.

1976 DIFFIE, W. &   New directions in cryptography.
HELLMAN, M.E.   IEEE Trans. on Inf. Th., IT-22,

1976 ERDÖS, P. &   On the greatest prime factor of
SHOREY, T.N. $2^p - 1$ for a prime p, and
    other expressions.
    Acta Arithm., 30, 1976, 257-265.

1976   LEHMER, D.H.   Strong Carmichael numbers.
                                       J. Austral. Math. Soc., A, 21,
                                       1976, 508-510.
                                       Reprinted in *Selected Papers*
                                       (edited by D. McCarthy),
                                       Vol. I, 140-142.
                                       Ch. Babbage Res. Centre,
                                       St. Pierre, Manitoba, Canada,
                                       1981.

1976   MENDELSOHN, N.S.   The equation $\phi(x) = k$.
                                       Math. Mag., 49, 1976, 37-39.

1976   MILLER, G.L.   Riemann's hypothesis and tests for
                                       primality.
                                       J. Comp. Syst. Sci., 13, 1976,
                                       300-317.

1976   RABIN, M.O.   Probabilistic algorithms.
                                       In *Algorithms and Complexity*
                                       (edited by J.F. Traub), 21-39.
                                       Academic Press, New York, 1976.

1976   RECAMÁN SANTOS, B.   Twelve and its totatives.
                                       Math. Mag., 49, 1970, 239-240.

1976   TRAUB, J.F.   *Algorithms and Complexity.*
       (editor)   Academic Press, New York, 1976.

1976   YORINAGA, M.   On a congruential property of
                                       Fibonacci numbers. Numerical
                                       experiments. Considerations and
                                       Remarks.
                                       Math. J. Okayama Univ., 19, 1976,
                                       5-10; 11-17.

1977   AIGNER, A.   Das quadratfreie Kern der
                             Eulerschen ϕ-Funktion.
                             Monatsh. Math., 83, 1977, 89-91.

1977   KISHORE, M.   Odd perfect numbers not divisible by
                             3 are divisible by at least ten distinct
                             primes.
                             Math. Comp. 31, 1977, 274-279.

1977   KISHORE, M.   *The Number of Distinct Prime Factors
                             for Which $\sigma(N) = 2N$,*
                             $\sigma(N) = 2N \pm 1$ *and* $\phi(N) \mid N - 1$.
                             Ph.D. Thesis, University of Toledo,
                             Ohio, 1977, 39 pages.

1977   MALM, D.E.G.   On Monte-Carlo primality tests.
                             Notices Amer. Math. Soc., 24,
                             1977, A-529, abstract 77T-A22.

1977   POMERANCE, C.   On composite n for which
                             $\phi(n) \mid n - 1$. II
                             Pacific J. Math., 69, 1977,
                             177-186.

1977   POMERANCE, C.   Multiply perfect numbers, Mersenne
                             primes and effective computability.
                             Math. Ann., 226, 1977, 195-206.

1977   SOLOVAY, R. &   A fast Monte-Carlo test for
       STRASSEN, V.   primality.
                             SIAM J. Comput., 6, 1977, 84-85.

1977   STEWART, C.L.   On divisors of Fermat, Fibonacci,
                             Lucas, and Lehmer numbers.
                             Proc. London Math. Soc., (3), 35,
                             1977, 425-447.

1977 STEWART, C.L.   Primitive divisors of Lucas and
                             Lehmer numbers.
                             In *Transcendence Theory: Advances
                             and Applications* (edited by A. Baker &
                             D.W. Masser), 79-92.
                             Academic Press, London, 1977.

1977 WILLIAMS, H.C.   On numbers analogous to the
                             Carmichael numbers.
                             Can. Math. Bull., 20, 1977, 133-143.

1978 COHEN, G.L.   On odd perfect numbers.
                             Fibonacci Quart., 16, 1978, 523-527.

1978 HAUSMAN, M. &   Adding totatives.
      SHAPIRO, H.N.   Math. Mag., 51, 1978, 284-288.

1978 KISS, P. &   On a function concerning second
      PHONG, B.M.   order recurrences.
                             Ann. Univ. Sci. Budapest, 21, 1978,
                             119-122.

1978 RIVEST, R.L.   Remarks on a proposed cryptanalytic
                             attack on the M.I.T. public-key
                             cryptosystem.
                             Cryptologia, 2, 1978.

1978 RIVEST, R.L.,   A method for obtaining digital
      SHAMIR, A. &   signatures and public-key
      ADLEMAN, L.M.   cryptosystems.
                             Comm. ACM, 21, 1978, 120-126.

1978 WILLIAMS, H.C.   Primality testing on a computer.
                             Ars Comb., 5, 1978, 127-185.

1978 YORINAGA, M.   Numerical computation of Carmichael
                             numbers.
                             Math. J. Okayama Univ., 20, 1978,
                             151-163.

1979   CHEIN, E.Z.   Non-existence of odd perfect numbers
                              of the form $q_1^{a_1} q_2^{a_2} \cdots q_6^{a_6}$ and
                              $5^{a_1} q_2^{a_2} \cdots q_9^{a_9}$.
                              Ph.D. Thesis, Pennsylvania State
                              Univ., 1979.

1980   BAILLIE, R. &   Lucas pseudoprimes.
       WAGSTAFF, Jr., S.S.   Math. Comp., 35, 1980, 1391-1417.

1980   COHEN, G.L. &   On the number of prime factors of
       HAGIS, Jr., P. n if $\phi(n) \mid (n-1)$.
                             Nieuw Arch. Wisk., (3), 28, 1980,
                             177-185.

1980   HAGIS, Jr., P.   Outline of a proof that every odd
                             perfect number has at least eight
                             prime factors.
                             Math. Comp., 35, 1980, 1027-1032.

1980   MONIER, L.   Evaluation and comparison of two
                             efficient probabilistic primality
                             testing algorithms.
                             Theoret. Comput. Sci., 12, 1980,
                             97-108.

1980   POMERANCE, C.,   The pseudoprimes to 25.10^9.
       SELFRIDGE, J.L. &   Math. Comp., 35, 1980, 1003-1026.
       WAGSTAFF, Jr., S.S.

1980   RABIN, M.O.   Probabilistic algorithm for
                             testing primality.
                             J. Nb. Th., 12, 1980, 128-138.

1980   WAGSTAFF, Jr., S.S.   Large Carmichael numbers
                             Math. J. Okayama Univ., 22,
                             1980, 33-41.

1980   WALL, D.W.   Conditions for $\phi(N)$ to properly
                               divide $N - 1$.
                               In *A Collection of Manuscripts
                               Related to the Fibonacci Sequence*
                               (edited by V.E. Hoggatt &
                               M. Bicknell-Johnson), 205-208.
                               18th Anniv. Vol., Fibonacci Assoc.,
                               San Jose, 1980.

1981   BRENT, R.P. &   Factorization of the eighth
       POLLARD, J.M.   Fermat number.
                               Math. Comp., 36, 1981, 627-630.

1981   DIXON, J.D.   Asymptotically fast factorization of
                               integers.
                               Math. Comp., 36, 1981, 255-260.

1981   GROSSWALD, E.   On Burgess' bound for primitive
                               roots modulo primes and an
                               application to $\Gamma(p)$.
                               Amer. J. Math., 103, 1981,
                               1171-1183.

1981   KISHORE, M.   On odd perfect, quasi perfect and odd
                               almost perfect numbers.
                               Math. Comp., 36, 1981, 583-586.

1981   KNUTH, D.E.   *The Art of Computer Programming*,
                               Vol. 2, 2nd edition.
                               Addison-Wesley, Reading, Mass.,
                               1981.

1981   KONHEIM, A.G.   *Cryptography: A Primer.*
                               Wiley-Interscience, New York, 1981.

1981   LENSTRA, Jr., H.W.   Primality testing algorithms (after
                                 Adleman, Rumely and Williams).
                                 In Séminaire Bourbaki,
                                 exposé No. 576.
                                 Lecture Notes in Math.,
                                 #901, 243-257.
                                 Springer-Verlag, Berlin, 1981.

1981   LÜNEBURG, H.   Ein einfacher Beweis für den Satz
                                 von Zsigmondy über primitive
                                 Primteiler von $A^N - 1$.
                                 In *Geometries and Groups* (edited by
                                 M. Aigner & D. Jungnickel).
                                 Lecture Notes in Math., #893,
                                 219-222.
                                 Springer-Verlag, New York, 1981.

1981   POMERANCE, C.   Recent developments in primality
                                 testing.
                                 Math. Intelligencer, 3, No. 3,
                                 1981, 97-105.

1981   SHOREY, T.N. &   On divisors of Fermat, Fibonacci,
       STEWART, C.L.   Lucas and Lehmer numbers, II.
                                 J. London Math. Soc., (2), 23,
                                 1981, 17-23.

1982   BRENT, R.P.   Succinct proofs of primality for the
                                 factors of some Fermat numbers.
                                 Math. Comp., 38, 1982, 253-255.

1982   HOOGENDOORN, P.J.   On a secure public-key
                                 cryptosystem.
                                 In *Computational Methods in
                                 Number Theory* (edited by
                                 H.W. Lenstra, Jr. & R. Tijdeman),
                                 Part I, 159-168.
                                 Math. Centre Tracts, #154,
                                 Amsterdam, 1982.

1982   LENSTRA, Jr., H.W.   Primality testing.
                                  In *Computational Methods in
                                  Number Theory* (edited by
                                  H.W. Lenstra, Jr. & R. Tijdeman),
                                  Part I, 55-77.
                                  Math. Centre Tracts, #154,
                                  Amsterdam, 1982.

1982   MASAI, P. &   A lower bound for a counterexample
       VALETTE, A.   in Carmichael's conjecture.
                                  Boll. Un. Mat. Ital., (6), 1-A,
                                  1982, 313-316.

1982   POMERANCE, C.   Analysis and comparison of some
                                  integer factoring algorithms.
                                  In *Computational Methods in
                                  Number Theory* (edited by
                                  H.W. Lenstra, Jr. & R. Tijdeman),
                                  Part I, 89-139.
                                  Math. Centre Tracts, #154, Amsterdam,
                                  1982.

1982   VAN EMDE BOAS, P.   Machine models, computational
                                  complexity and number theory.
                                  In *Computational Methods in
                                  Number Theory* (edited by
                                  H.W. Lenstra, Jr. & R. Tijdeman),
                                  Part I, 7-42.
                                  Math. Centre Tracts, #154,
                                  Amsterdam, 1982.

1982   VORHOEVE, M.   Factorizations algorithms of
                             exponential order.
                             In *Computational Methods in*
                             Number Theory (edited by
                             H.W. Lenstra, Jr. & R. Tijdeman),
                             Part I, 79-87.
                             Math. Centre Tracts, #154,
                             Amsterdam, 1982.

1982   WILLIAMS, H.C.   A class of primality tests for
                             trinomials which includes the
                             Lucas-Lehmer test.
                             Pacific J. Math., 98, 1982,
                             477-494.

1982   WILLIAMS, H.C.   The influence of computers in the
                             development of number theory.
                             Comp. & Maths. with Appl., 8,
                             1982, 75-93.

1982   WOODS, D. &   Larger Carmichael numbers.
       HUENEMANN, J.   Comput. Math. & Appl., 8, 1982,
                             215-216.

1983   ADLEMAN, L.M.,   On distinguishing prime numbers from
       POMERANCE, C. &   composite numbers.
       RUMELY, R.S.   Annals of Math., (2), 117, 1983,
                             173-206.

1983   BRILLHART, J.,   *Factorizations of $b^n \pm 1$,*
       LEHMER, D.H., $b = 2, 3, 5, 6, 7, 10, 11, 12$
       SELFRIDGE, J.L.,   *up to High Powers.*
       TUCKERMAN, B. &   Contemporary Math., Vol. 22,
       WAGSTAFF, Jr., S.S.   Amer. Math. Soc.,
                             Providence, R.I., 1983.

1983   HAGIS, Jr., P.   Sketch of a proof that an odd perfect
                                         number relatively prime to 3 has at
                                         least eleven prime factors.
                                         Math. Comp., 40, 1983, 399-404.

1983   KELLER, W.   Factors of Fermat numbers and large
                                         primes of the form $k \cdot 2^n + 1$.
                                         Math. Comp., 41, 1983, 661-673.

1983   KISHORE, M.   Odd perfect numbers not divisible by
                                         3, II.
                                         Math. Comp., 40, 1983, 405-411.

1983   POMERANCE, C. &   Implementation of the continued
       WAGSTAFF, Jr., S.S.   fraction integer factoring algorithm.
                                         Congressus Numerantium, 37, 1983,
                                         99-118.

1983   POWELL, B.   Problem 6420 (On primitive roots).
                                         Amer. Math. Monthly, 90, 1983, p. 60.

1983   RUMELY, R.S.   Recent advances in primality testing.
                                         Notices Amer. Math. Soc., 30, No. 5,
                                         1983, 475-477.

1983   SINGMASTER, D.   Some Lucas pseudoprimes.
                                         Abstracts Amer. Math. Soc., 4,
                                         1983, No. 83T-10-146, p. 197.

1983   YATES, S.   Titanic primes.
                                         J. Recr. Math., 16, 1983/4, 250-260.

1984   COHEN, H. &   Primality testing and Jacobi sums.
       LENSTRA, Jr., H.W.   Math. Comp., 42, 1984, 297-330.

1984   DIXON, J.D.   Factorization and primality tests.
                                         Amer. Math. Monthly, 91, 1984,
                                         333-352.

1984   KEARNES, K.   Solution of problem 6420.
                               Amer. Math. Monthly, 91, 1984, p. 521.

1984   NICOLAS, J.L.   Tests de primalité.
                               Expo. Math., 2, 1984, 223-234.

1984   POMERANCE, C.   *Lecture Notes on Primality Testing
                               and Factoring*
                               (Notes by G.M. Gagola Jr.).
                               Math. Assoc. America, Notes, No. 4,
                               1984, 34 pages.

1984   ROTKIEWICZ, A.   On the congruence $2^{n-2} \equiv 1 \pmod{n}$.
                               Math. Comp., 43, 1984, 271-272.

1984   WILLIAMS, H.C.   An overview of factoring.
                               In *Advances in Cryptology* (edited by
                               D. Chaum), 71-80.
                               Plenum, New York, 1984.

1984   WILLIAMS, H.C.   Factoring on a computer.
                               Math. Intelligencer, 6, 1984,
                               No. 3, 29-36.

1984   YATES, S.   Sinkers of the Titanics.
                               J. Recr. Math., 17, 1984/5,
                               268-274.

1985   BACH, E.   *Analytic Methods in the Analysis and
                               Design of Number-Theoretic Algorithms.*
                               A.C.M. Distinguished Dissertations.
                               M.I.T. Press, Cambridge, 1985.

1985   BEDOCCHI, E.   Note on a conjecture on prime
                               numbers.
                               Rev. Mat. Univ. Parma (4), 11,
                               1985, 229-236.

1985   BOSMA, W.   Primality testing using elliptic
                              curves.
                              Math. Inst., Univ. Amsterdam,
                              Report 85-12, 54 pages, 1985.

1985   DUBNER, H. &   The development of a powerful
       DUBNER, R.   low-cost computer for number theory
                              applications.
                              J. Recr. Math., 18, 1985/6, 92-96.

1985   KELLER, W.   The 17th prime of the form $5 \cdot 2^n + 1$.
                              Abstract Amer. Math. Soc., 6,
                              No. 1, 1985, p. 121.

1985   ONDREJKA, R.   Ten extraordinary primes.
                              J. Recr. Math., 18, 1985/86, 87-92.

1985   RIESEL, H.   *Prime Numbers and Computer Methods*
                              for Factorization.
                              Birkhäuser, Boston, 1985.

1985   WAGON, S.   Perfect numbers.
                              Math. Intelligencer, 7, No. 2, 1985,
                              66-68.

1986   ADLEMAN, L.M. &   Recognizing primes in random
       HUANG, M.-D.A.   polynomial time.
                              Preprint, 1986.

1986   ERDÖS, P. &   On the number of false witnesses
       POMERANCE, C.   for a composite number.
                              Math. Comp., 46, 1986, 259-279.

1986   KISS, P.,   On Lucas pesudoprimes which are
       PHONG, B.M. &   products of s primes.
       LIEUWENS, E.   In *Fibonacci Numbers and their*
                              Applications (edited by
                              A.N. Philippou, G.E. Bergum &
                              A.F. Horadam), 131-139.
                              Reidel, Dordrecht, 1986.

1986   LENSTRA, Jr., H.W.   Elliptic curves and number-
                                  theoretic algorithms.
                                  Math. Inst., Univ. Amsterdam,
                                  Report 86-19, 18 pages, 1986.

1986   LENSTRA, Jr., H.W.   Factoring integers with elliptic
                                  curves.
                                  Math. Inst., Univ. Amsterdam.
                                  Report 86-18, 19 pages, 1986.

1986   SHEN, M.K.   On the congruence $2^{n-k} \equiv 1 \pmod{n}$.
                                  Math. Comp., 46, 1986, 715-716.

1986   WAGON, S.   Carmichael's "Empirical Theorem".
                                  Math. Intelligencer, 8, No. 2, 1986,
                                  61-62.

1986   WAGON, S.   Primality testing.
                                  Math. Intelligencer, 8, No. 3, 1986,
                                  58-61.

1987   ADLEMAN, L.M. &   Open problems in number theoretic
       McCURLEY, K.S.   complexity, in *Discrete Algorithms and
                                  Complexity*, 237-262.
                                  Academic Press, New York, 1987.

1987   BRILLHART, J.,   Tables of Fibonacci and Lucas
       MONTGOMERY, P.L.   factorizations.
       & SILVERMAN, R.D.   To appear.

1987   COHEN, H. &   Implementation of a new primality test
       LENSTRA, A.K.   (and Supplement).
                                  Math. Comp., 48, 1987, 103-121,
                                  and S1-S4.

1987   KISS, P. &   On a problem of A. Rotkiewicz.
       PHONG, B.M.   Math. Comp., 48, 1987, 751-755.

1987 McDANIEL, W.L.   The generalized pseudoprime
                             congruence $a^{n-k} \equiv b^{n-k} \pmod{n}$.
                             C.R. Math. Rep. Acad. Sci. Canada,
                             9, 1987, 143-148.

1987 POMERANCE, C.   Very short primality proofs.
                             Math. Comp., 48, 1987, 315-322.

1987 YOUNG, J. &   The twentieth Fermat number is
      BUELL, D.A.   composite.
                             Math. Comp. (to appear).

1988 McDANIEL, W.L.   The existence of solutions of the
                             generalized pseudoprime congruence
                             $a^{cn-k} \equiv b^{cn-k} \pmod{n}$.
                             Preprint, to appear in 1988(?)

1988 PHONG, B.M.   Generalized solution of Rotkiewicz's
                             problem.
                             Mat. Lapok, to appear in 1988(?)

Chapter 3

1839/ DIRICHLET, G.L.   Recherches sur diverses
1840   applications de l'analyse infinitésimale
                            à la théorie des nombres.
                            Journal f. d. reine u. angew.
                            Math., 19, 1839, 342-369 and
                            21, 1840, 1-12 and 134-135.
                            Reprinted in *Werke* (edited by
                            L. Kronecker), Vol. I, 411-496.
                            G. Reimer, Berlin, 1889.
                            Reprinted by Chelsea, Bronx, N.Y., 1969.

1852 KUMMER, E.E.   Über die Erganzungssätze zu den
                            allgemeinem Reciprocitätsgestzen.
                            Journal f. d. reine u. angew. Math.,
                            44, 1852, 93-146.
                            Reprinted in *Collected Papers* (edited
                            by A. Weil), Vol. I, 485-538.
                            Springer-Verlag, New York, 1975.

1912   FROBENIUS, F.G.   Über quadratische Formen, die
                              viele Primzahlen darstellen.
                              Sitzungsber. d. Königl. Akad. d.
                              Wiss. zu Berlin, 1912, 966-980.
                              Reprinted in *Gesammelte Abhandlungen*,
                              vol. III, 573-587.
                              Springer-Verlag, Berlin, 1968.

1912   RABINOVITCH, G.   Eindeutigkeit der Zerlegung in
                              Primzahlfaktoren in quadratischen
                              Zahlkörpern.
                              Proc. Fifth Intern. Congress Math.,
                              Cambridge, Vol. 1, 1912, 418-421.

1918   LANDAU, E.   Über die Klassenzahl imaginär
                              quadratischer Zahlkörper.
                              Göttinger Nachr., 1918,
                              285-295.

1933   DEURING, M.   Imaginär-quadratische
                              Zahlkörper mit Klassenzahl Eins.
                              Invent. Math., 5, 1968, 169-179.

1933   LEHMER, D.H.   On imaginary quadratic fields whose
                              class number is unity.
                              Bull. Amer. Math. Soc., 39, 1933,
                              p. 360.

1934   HEILBRONN, H.   On the class number in imaginary
                              quadratic fields.
                              Quart. J. Pure & Appl. Math., Oxford,
                              Ser. 2, 5, 1934, 150-160.

1934   HEILBRONN, H. &   On the imaginary quadratic corpora of
       LINFOOT, E.H.   class-number one.
                              Quart. J. Pure & Appl. Math., Oxford,
                              Ser. 2, 5, 1934, 293-301.

1934 MORDELL, L.J.   On the Riemann hypothesis and
                             imaginary quadratic fields with a
                             given class number.
                             J. London Math. Soc., 9, 1934,
                             289-298.

1935 SIEGEL, C.L.   Über die Classenzahl quadratischer
                             Zahlkörper.
                             Acta Arithm., 1, 1935, 83-86.
                             Reprinted in *Gesammelte Abhandlungen*
                             (edited by K. Chandrasekharan &
                             H. Maaß), Vol. I, 406-409.
                             Springer-Verlag, Berlin, 1966.

1936 LEHMER, D.H.   On the function $x^2 + x + A$.
                             Sphinx, 6, 1936, 212-214.

1938 SKOLEM, T.   *Diophantische Gleichungen.*
                             Springer-Verlag, Berlin, 1938.

1943 REINER, I.   Functions not formulas for primes.
                             Amer. Math. Monthly, 50, 1943,
                             619-621.

1946 BUCK, R.C.   Prime-representing functions.
                             Amer. Math. Monthly, 53, 1946,
                             p. 265.

1947 MILLS, W.H.   A prime-representing function.
                             Bull. Amer. Math. Soc., 53, p. 604.

1951 WRIGHT, E.M.   A prime-representing function.
                             Amer. Math. Monthly, 58, 1951,
                             616-618.

1952 HEEGNER, K.   Diophantische Analysis und
                             Modulfunktionen.
                             Math. Zeits., 56, 1952, 227-253.

1952 SIERPIŃSKI, W.   Sur une formule donnant tous
                              les nombres premiers.
                              C.R. Acad. Sci. Paris, 235,
                              1952, 1078-1079.

1960 PUTNAM, H.   An unsolvable problem in number
                              theory.
                              J. Symb. Logic, 1960, 220-232.

1962 COHN, H.   *Advanced Number Theory.*
                              Wiley, New York, 1962.
                              Reprinted by Dover, New
                              York, 1980.

1964 BOREVICH, Z.I. &   *Number Theory.*
      SHAFAREVICH, I.R.   Moscow, 1964.
                              English translation by N. Greenleaf,
                              published by Academic Press,
                              New York, 1966.

1964 WILLANS, C.P.   On formulae for the nth prime.
                              Math. Gaz., 48, 1964, 413-415.

1966 BAKER, A.   Linear forms in the logarithms of
                              algebraic numbers.
                              Mathematika, 13, 1966, 204-216.

1967 GOODSTEIN, R.L.   Formulae for primes.
      & WORMELL, C.P.   Mat. Gaz., 51, 1967, 35-38.

1967 STARK, H.M.   A complete determination of the
                              complex quadratic fields of
                              class-number one.
                              Michigan Math. J., 14, 1967, 1-27.

1968 DEURING, M.   Imaginäre quadratische Zahlkörper
                              mit der Klassenzahl Eins.
                              Invent. Math., 5, 1968, 169-179.

1969 BAKER, A.   A remark on the class number of
                                   quadratic fields.
                                   Bull. London Math. Soc., 1, 1969,
                                   98-102.

1969 DUDLEY, U.   History of a formula for primes.
                                   Amer. Math. Monthly, 76,
                                   1969, 23-28.

1969 STARK, H.M.   A historical note on complex
                                   quadratic fields with class-number one.
                                   Proc. Amer. Math. Soc., 21, 1969,
                                   254-255.

1971 BAKER, A.   Imaginary quadratic fields with class
                                   number 2.
                                   Annals of Math., 94, 1971, 139-152.

1971 BAKER, A.   On the class number of imaginary
                                   quadratic fields.
                                   Bull. Amer. Math. Soc., 77, 1971,
                                   678-684.

1971 GANDHI, J.M.   Formulae for the nth prime.
                                   Proc. Washington State Univ.
                                   Conf. on Number Theory, 96-106.
                                   Wash. St. Univ., Pullman,
                                   Wash., 1971.

1971 MATIJASEVIČ, Yu.V.   Diophantine representation of the set
                                   of prime numbers (in Russian).
                                   Dokl. Akad. Nauk SSSR, 196,
                                   1971, 770-773.
                                   English translation by R. N. Goss,
                                   in Soviet Math. Dokl. 11, 1970, 354-358.

1972 SIEGEL, C.L.   Zur Theorie der quadratischen
                                   Formen.
                                   Nachr. Akad. Wiss. Göttingen,

Math. Phys. Kl., 1972, No. 3, 21-46.
Reprinted in *Gesammelte Abhandlungen*
(edited by K. Chandrasekharan &
H. Maaß), Vol. IV, 224-249.
Springer-Verlag, New York, 1979.

1972   VANDEN EYNDEN,   A proof of Gandhi's formula
       C.   for the nth prime.
                            Amer. Math. Monthly, 79, 1972, p. 625.

1973   DAVIS, M.   Hilbert's tenth problem is unsolvable.
                            Amer. Math. Monthly, 80,
                            1973, 233-269.

1973   KARST, E.   New quadratic forms with high
                            density of primes.
                            Elem. d. Math., 28, 1973, 116-118.

1974   GOLOMB, S.W.   A direct interpretation of
                            Gandhi's formula.
                            Amer. Math. Monthly, 81, 1974,
                            752-754.

1974   HENDY, M.D.   Prime quadratics associated with
                            complex quadratic fields of class
                            number two.
                            Proc. Amer. Math. Soc., 43, 1974,
                            253-260.

1974   MONTGOMERY, H.L.   Notes on small class numbers
       & WEINBERGER, P.J.   Acta Arith., 24, 1974, 529-542.

1974   SZEKERES, G.   On the number of divisors of
                            $x^2 + x + A$.
                            J. Nb. Th., 6, 1974, 434-442.

1975   BAKER, A.   *Transcendental Number Theory.*
                            Cambridge Univ. Press, Cambridge,
                            1975.

1975   ERNVALL, R.   A formula for the least prime
                                      greater than a given integer.
                                      Elem. d. Math., 30, 1975, 13-14.

1975   JONES, J.P.   Diophantine representation of the
                                      Fibonacci numbers.
                                      Fibonacci Quart., 13, 1975, 84-88.

1975   MATIJASEVIČ, Yu.V.   Reduction of an arbitrary
                                      diophantine equation to one in 13
                                      unknowns.
                                      Acta Arithm., 27, 1975, 521-553.

1975   STARK, H.M.   On complex quadratic fields with
                                      class-number two.
                                      Math. Comp., 29, 1975, 289-302.

1976   GOLDFELD, D.   The class number of quadratic
                                      fields and conjectures of Birch
                                      and Swinnerton-Dyer.
                                      Ann. Scuola Norm. Sup. Pisa,
                                      Ser. 4, 3, 1976, 623-663.

1976   JONES, J.P.,   Diophantine representation
       SATO, D.,   of the set of prime numbers.
       WADA, H. &   Amer. Math. Monthly, 83,
       WIENS, D.   1976, 449-464.

1977   MATIJASEVIČ, Yu.V.   Primes are nonnegative values of a
                                      polynomial in 10 variables.
                                      Zapiski Sem. Leningrad Mat. Inst.
                                      Steklov, 68, 1977, 62-82.
                                      English translation by L. Guy &
                                      J.P. Jones, J. Soviet Math., 15,
                                      1981, 33-44.

1977   MATIJASEVIČ, Yu.V.   Some purely mathematical results
                                      inspired by mathematical logic,
                                      in *Logic, Foundations of Mathematics
                                      and Computability Theory*

(Proc. 5th Intern. Congress Logic,
Methodology and Philosophy of
Science, Univ. Western Ont., London,
Ont., 1975), 121-127.
Reidel, Dordrecht, 1977.

1979	JONES, J.P.	Diophantine representation of Mersenne and Fermat primes. Acta Arithm., 35, 1979, 209-221.
1981	AYOUB, R.G. & CHOWLA, S.	On Euler's polynomial. J. Nb. Th., 13, 1981, 443-445.
1981	HATCHER, W.S. & HODGSON, B.R.	Complexity bounds on proof. J. Symb. Logic, 46, 1981, 255-258.
1982	JONES, J.P.	Universal diophantine equation. J. Symb. Logic, 47, 1982, 549-571.
1983	GROSS, B. & ZAGIER, D.	Points de Heegner et dérivées de fonctions L. C.R. Acad. Sci. Paris, 297, 1983, 85-87.
1984	OESTERLÉ, J.	Nombres de classes des corps quadratiques imaginaires. Séminaire Bourbaki, exposé No. 631, 1983/4, Paris. Astérisque, 121/2, 1985, 213-224.
1985	GOLDFELD, D.	Gauss' class number problem for imaginary quadratic fields. Bull. Amer. Math. Soc., 13, 1985, 23-37.
1987	RIBENBOIM, P.	Euler's famous prime generating polynomial and the class number of imaginary quadratic fields. L'Enseign. Math. (to appear). Same in Spanish translation by V. Albis Gonzalez, in Rev. Colombiana de Matematicas (to appear).

Chapter 4

1815   CAUCHY, A.L.   Démonstration générale du
                             théorème de Fermat sur les
                             nombres polygônes.
                             Bull. Soc. Philomatique, 1815,
                             196-197.
                             Reprinted in *Oeuvres*, (2),
                             Vol. 2, 202-204.
                             Gauthier-Villars, Paris, 1903.

1837   DIRICHLET, G.L.   Beweis des Satzes, dass jede
                             unbegrenzte arithmetische Progression,
                             deren erstes Glied und Differenz
                             ganze Zahlen ohne gemeinschaftlichen
                             Factor sind, unendlich viele
                             Primzahlen enthält.
                             Abh. d. Königl. Akad. d. Wiss.,
                             1837, 45-81.
                             Reprinted in *Werke*, Vol. I,
                             315-350.
                             G. Reimer, Berlin, 1889.

1885   MEISSEL, E.D.F.   Berechnung der Menge von Primzahlen,
                             welche innerhalb der ersten Milliarde
                             natürlicher Zahlen vorkommen.
                             Math. Ann., 25, 1885, 251-257.

1892   SYLVESTER, J.J.   On arithmetical series.
                             Messenger of Math., 21, 1892, 1-19
                             and 87-120.
                             Reprinted in *Gesammelte
                             Abhandlungen*, Vol. III, 573-587.
                             Springer-Verlag, New York, 1968.

1895   WENDT, E.   Elementarer Beweis des Satzes dass
                              in jeder unbegrenzter arithmetischen
                              Progression $my + 1$ unendlich viele
                              Primzahlen vorkommen.
                              Journal f. d. reine u. angew. Math.,
                              115, 1895, 85-88.

1901   VON KOCH, H.   Sur la distribution des nombres
                              premiers.
                              Acta Math., 24, 1901, 159-182.

1901   WOLFSKEHL, P.   Ueber eine Aufgabe der elementaren
                              Arithmetik.
                              Math. Ann., 54, 1901, 503-504.

1902   TORELLI, G.   Sulla totalità dei numeri primi
                              fino a un limite assegnato.
                              Atti d. Reale Acad. d. Sci. Fis. e
                              Mat. di Napoli, (2), 11, 1902,
                              1-222.

1903   LANDAU, E.   Neuer Beweis des Primzahlsatzes und
                              Beweis des Primidealsatzes.
                              Math. Ann., 56, 1903, 645-670.

1909   HILBERT, D.   Beweis für die Darstellbarkeit
                              der ganzen Zahlen durch eine feste
                              Anzahl n-ter Potenzen (Waringsches
                              Problem).
                              Math. Ann., 67, 1909, 281-300.
                              Reprinted in *Gesammelte Abhandlungen*,
                              Vol. I, 510-527, 2nd edition,
                              Chelsea, Bronx, N.Y., 1965.

1909   LANDAU, E.   *Handbuch der Lehre von der Verteilung
                              der Primzahlen.*
                              Teubner, Leipzig, 1909.
                              Reprinted by Chelsea, Bronx, N.Y.,
                              1974.

1909   LEHMER, D.N.   *List of Prime Numbers from 1 to*
                             10,006,721.
                             Reprinted by Hafner, New York, 1956.

1909   LEHMER, D.N.   *Factor Table for the First Ten Millions*
                             Containing the Smallest Factor of Every
                             Number not Divisible by 2, 3, 5, or 7,
                             Between the Limits 0 and 10,017,000.
                             Reprinted by Hafner, New York, 1956.

1909   WIEFERICH, A.   Beweis des Satzes, dass sich eine
                             jede ganze Zahl als Summe von
                             höchstens neun positiven Kuben
                             darstellen lässt.
                             Math. Ann., 66, 1909, 95-101.

1912   SCHUR, I.   Über die Existenz unendlich vieler
                             Primzahlen in einiger speziellen
                             arithmetischen Progressionen.
                             Sitzungsber. Berliner Math. Ges. 11,
                             1912, 40-50.

1912   STRIDSBERG, E.   Sur la démonstration de M. Hilbert
                             du théorème de Waring.
                             Math. Ann., 72, 1912, 145-152.

1913   COBLYN   Sur les couples de nombres premiers.
                             Soc. Math. France, C.R., 1913, 55-56.

1914   LITTLEWOOD, J.E.   Sur la distribution des nombres
                             premiers.
                             C.R. Acad. Sci. Paris, 158, 1914,
                             1869-1872.

1916   WEYL, H.   Über die Gleichverteilung von
                             Zahlen mod. Eins.
                             Math. Ann., 77, 1916, 313-352.

1919   BRUN, V.   Le crible d'Eratosthène et le
                                    théorème de Goldbach.
                                    C.R. Acad. Sci. Paris, 168, 1919,
                                    544-546.

1919   BRUN, V.   La série $\frac{1}{5} + \frac{1}{7} + \frac{1}{11} + \frac{1}{13} + \frac{1}{17} + \frac{1}{19}$
                                    $+ \frac{1}{29} + \frac{1}{31} + \frac{1}{41} + \frac{1}{43} + \frac{1}{59} + \frac{1}{61} + \cdots$
                                    où les dénominateurs sont "nombres
                                    premiers jumeaux" est convergente ou
                                    finie.
                                    Bull. Sci. Math., (2), 43, 1919,
                                    100-104 and 124-128.

1919   RAMANUJAN, S.   A proof of Bertrand's postulate.
                                    J. Indian Math. Soc., 11, 1919,
                                    181-182.
                                    Reprinted in *Collected Papers*
                                    (edited by G.H. Hardy, P.V. Seshu
                                    Aiyar, & B.M. Wilson), 208-209.
                                    Cambridge Univ. Press, Cambridge,
                                    1927.
                                    Reprinted by Chelsea, Bronx, N.Y.,
                                    1962.

1920   BRUN, V.   Le crible d'Erathostène et la
                                    théorème de Goldbach.
                                    Videnskapsselskapets Skrifter
                                    Kristiania, Mat.-nat. Kl. 1920,
                                    No. 3, 36 pages.

1920   HARDY, G.H. &   A new solution of Waring's problem.
       LITTLEWOOD, J.E.   Quart. J. Math., 1920, 48, 272-293.
                                    Reprinted in *Collected Papers of
                                    G.H. Hardy*, Vol. I, 382-393.
                                    Clarendon Press, Oxford, 1966.

1920   HARDY, G.H. &   Some problems of "Partitio
       LITTLEWOOD, J.E.   Numerorum", I: A new solution of
                              Waring's Problem.
                              Nachr. Königl. Ges. d. Wiss. zu
                              Göttingen, 1920, 33-54.
                              Reprinted in *Collected Papers of*
                              G.H. Hardy, Vol. I, 405-426.
                              Clarendon Press, Oxford, 1966.

1921   HARDY, G.H. &   Some problems of "Partitio
       LITTLEWOOD, J.E.   Numerorum", II: Proof that every
                              number is the sum of at most 21
                              biquadrates.
                              Math. Zeits., 9, 1921, 14-27.
                              Reprinted in *Collected Papers of*
                              G.H. Hardy, Vol. I, 427-440.
                              Clarendon Press, Oxford, 1966.

1921   KAMKE, E.   Verallgemeinerung des Waring-
                              Hilbertschen Satzes.
                              Math. Ann., 83, 1921, 85-112.

1922   HARDY, G.H. &   Some problems of "Partitio
       LITTLEWOOD, J.E.   Numerorum", IV: The singular
                              series in Waring's Problem and
                              the value of the number $G(k)$.
                              Math. Zeits., 12, 1922, 161-188.
                              Reprinted in *Collected Papers of*
                              G.H. Hardy, Vol. I, 441-468.
                              Clarendon Press, Oxford, 1966.

1923   HARDY, G.H. &   Some problems of "Partitio
       LITTLEWOOD, J.E.   Numerorum", III: On the expression
                              of a number as a sum of primes.
                              Acta Math., 44, 1923, 1-70.
                              Reprinted in *Collected Papers of*
                              G.H. Hardy, Vol. I, 441-468.
                              Clarendon Press, Oxford, 1966.

1923   HARDY, G. H. &   Some problems of "Partitio
       LITTLEWOOD, J.E.   Numerorum", III: On the
                                expression of a number as a sum
                                of primes.
                                Acta Math., 44, 1923, 1-70.
                                Reprinted in *Collected Papers of*
                                G.H. Hardy, Vol. I, 561-630.
                                Clarendon Press, Oxford, 1966.

1923   NAGELL, T.   Zahlentheoretische Notizen I, Ein
                                Beitrag zur Theorie der höheren
                                Kongruenzen.
                                Vidensk. Skrifter, ser. I, Math. Nat.
                                Kl., No. 13, Kristiania, 1923, 3-6.

1924   RADEMACHER, H.   Beiträge zur Viggo Brunschen
                                Methode in der Zahlentheorie.
                                Abh. Math. Sem. Univ. Hamburg,
                                3, 1924, 12-30.
                                Reprinted in *Collected Papers*
                                (edited by E. Grosswald),
                                Vol. I, 280-288.
                                M.I.T. Press, Cambridge, 1974.

1928   SCHOENBERG, I.J.   Über die asymptotische Verteilung
                                reeller Zahlen (mod 1).
                                Math. Zeits., 28, 1928, 171-199.

1928   WINTNER, A.   Über den Konvergenzbegriff der
                                mathematischen Statistik.
                                Math. Zeits., 28, 1928, 476-480.

1930   ERDÖS, P.   Beweis eines Satzes von
                                Tschebychef.
                                Acta Sci. Math. Szeged, 5, 1930,
                                194-198.

1930   HOHEISEL, G.   Primzahlprobleme in der Analysis.
                              Sitzungsberichte Berliner Akad. d.
                              Wiss., 1930, 580-588.

1930   SCHNIRELMANN, L.   Über additive Eigenschaften
                              von Zahlen.
                              Ann. Inst. Polytechn. Novocerkask,
                              14, 1930, 3-28 and Math. Ann., 107,
                              1933, 649-690.

1930   TITCHMARSH, E.C.   A divisor problem.
                              Rend. Cir. Mat. Palermo, 54, 1930,
                              414-429.

1931   WESTZYNTHIUS, E.   Über die Verteilung der Zahlen,
                              die zu den n ersten Primzahlen
                              teilerfremd sind.
                              Comm. Phys. Math. Helsingfors,
                              (5), 25, 1931, 1-37.

1932   LANDAU, E.   Über den Wienerschen neuen Weg zum
                              Primidealsatz.
                              Sitzungsber. Berliner Akad. d. Wiss.,
                              1932, 514-521.

1933   SKEWES, S.   On the difference $\pi(x) - \mathrm{li}(x)$.
                              J. London Math. Soc., 8, 1933,
                              277-283.

1934   CHOWLA, S.   On the least prime in an arithmetic
                              progression.
                              J. Indian Math. Soc., (2), 1, 1934,
                              1-3.

1934   ISHIKAWA, H.   Über die Verteilung der Primzahlen.
                              Sci. Rep. Tokyo Bunrika Daigaku,
                              A, 2, 1934, 27-40.

1935   ERDÖS, P.   On the density of some sequences
                                   of numbers, I.
                                   J. London Math. Soc., 10, 1935,
                                   120-125.

1935   ERDÖS, P.   On the difference of consecutive
                                   primes.
                                   Quart. J. Pure & Appl. Math.,
                                   Oxford, 6, 1935, 124-128.

1935   ERDÖS, P.   On the normal number of prime
                                   factors of $p-1$ and some related
                                   problems concerning Euler's
                                   ϕ-function.
                                   Quart. J. Pure & Appl. Math.,
                                   Oxford, 6, 1935, 205-213.

1935   VINOGRADOV, I.M.   On Waring's problem.
                                   Annals of Math., 36, 1935, 395-405.

1936   DICKSON, L.E.   Solution of Waring's problem.
                                   Amer. J. Math., 58, 1936, 530-535.
                                   Reprinted in *The Collected
                                   Mathematical Papers* (edited by
                                   A.A. Albert), Vol. III,
                                   290-295.
                                   Chelsea, Bronx, N.Y., 1975.

1936   HEILBRONN, H.   Über das Waringsche Problem.
                                   Acta Arith., 1, 1936, 212-221.

1937   CRAMÉR, H.   On the order of magnitude of the
                                   difference between consecutive
                                   prime numbers.
                                   Acta Arith., 2, 1937, 23-46.

1937   ERDÖS, P.   On the density of some
                                   sequences of numbers, II.
                                   J. London Math. Soc., 12, 1937, 7-11.

1937   INGHAM, A.E.   On the difference between
                              consecutive primes.
                              Quart. J. Pure & Appl. Math.,
                              Oxford, Ser. 2, 8, 1937,
                              255-266.

1937   LANDAU, E.   *Über einige neuere Fortschritte*
                              der additiven Zahlentheorie.
                              Cambridge Univ. Press, Cambridge,
                              1937.
                              Reprinted by Stechert-Hafner,
                              New York, 1964.

1937   TURÁN, P.   Über die Primzahlen der
                              arithmetischen Progressionen.
                              Acta Sci. Math. Szeged, 8, 1937,
                              226-235.

1937   VAN DER CORPUT,   Sur l'hypothèse de Goldbach pour
       J.G.   presque tous les nombres pairs.
                              Acta Arithm., 2, 1937, 266-290.

1937   VINOGRADOV, I.M.   Representation of an odd number
                              as the sum of three primes (in
                              Russian).
                              Dokl. Akad. Nauk SSSR, 15,
                              1937, 169-172.

1937   VINOGRADOV, I.M.   Some new problems in the theory
                              of primes (in Russian).
                              Doklady Akad. Nauk SSSR, 16,
                              1937, 131-132.

1937   VINOGRADOV, I.M.   Some theorems concerning the
                              theory of primes (in Russian).
                              Mat. Sbornik, N.S., 2, (44),
                              1937, 179-195.

1938 ERDÖS, P.   On the density of some sequences
                               of numbers, III.
                               J. London Math. Soc., 13, 1938,
                               119-127.

1938 ESTERMANN, T.   Proof that almost all even positive
                               integers are sums of two primes.
                               Proc. London Math. Soc., 44, 1938,
                               307-314.

1938 HARDY, G.H. &   Edmund Landau.
      HEILBRONN, H.   J. London Math. Soc., 13, 1938,
                               302-310.
                               Reprinted in *Collected Papers*
                               of G.H. Hardy, Vol. VII,
                               762-770.
                               Clarendon Press, Oxford, 1979.

1938 POULET, P.   Table des nombres composés
                               vérifiant le théorème de
                               Fermat pour le module 2, jusqu'
                               à 100.000.000.
                               Sphinx, 8, 1938, 52-52
                               Corrections: Math. Comp., 25,
                               1971, 944-945 and 26, 1972, p. 814.

1938 RANKIN, R.A.   The difference between consecutive
                               prime numbers.
                               J. London Math. Soc., 13,
                               1938, 242-247.

1938 ROSSER, J.B.   The nth prime is greater than
                               $n \log n$.
                               Proc. London Math. Soc. 45, 1938,
                               21-44.

1938   TSCHUDAKOFF, N.G.   On the density of the set of even
                          integers which are not representable
                          as a sum of two odd primes (in
                          Russian).
                          Izv. Akad. Nauk SSSR, Ser. Mat., 1,
                          1938, 25-40.

1938   VINOGRADOV, I.M.   Some general theorems about primes
                          (in Russian).
                          Trav. Inst. Math. Tbilissi, 3,
                          1938, 35-67.

1939   DAVENPORT, H.   On Waring's problem for cubes.
                          Acta Math., 71, 1939, 123-143.

1939   DAVENPORT, H.   On Waring's problem for fourth powers.
                          Annals of Math., 40, 1939, 731-747.

1939   DICKSON, L.E.   All integers, except 23 and 239, are
                          sums of 8 cubes.
                          Bull. Amer. Math. Soc., 45, 1939,
                          588-591.
                          Reprinted in *The Collected
                          Mathematical Papers* (edited by
                          A.A. Albert), Vol. V, 66-69.
                          Chelsea, Bronx, N.Y., 1975.

1939   VAN DER CORPUT,   Über Summen von Primzahlen
       J.G.   und Primzahlquadraten.
                          Math. Ann., 116, 1939, 1-50.

1940   ERDÖS, P.   The difference of consecutive primes.
                          Duke Math. J., 6, 1940, 438-441.

1940   INGHAM, A.E.   On the estimation of $N(\sigma,T)$.
                          Quart. J. Math., Oxford, 11, 1940,
                          291-292.

1940   PILLAI, S.S.   On Waring's problem: $g(6) = 73$.
                                Proc. Indian Acad. Sci., A, 12,
                                1940, 30-40.

1941   NARASIMHAMURTI,   On Waring's problem for 8th, 9th
       V.   and 10th powers.
                                J. Indian Math. Soc., 5, 1941,
                                p. 122.

1942   DAVENPORT, H.   On Waring's problem for fifth and
                                sixth powers.
                                Amer. J. Math., 64, 1942, 199-207.

1942   MANN, H.B.   A proof of the fundamental theorem
                                of sums of sets of positive integers.
                                Annals of Math., 43, 1942, 523-527.

1942   RUBUGUNDAY, R.   On $g(k)$ in Waring's problem.
                                J. Indian Math. Soc., N.S., 6,
                                1942, 192-198.

1943   ARTIN, E. &   On the sums of two sets of
       SCHERK, P.   integers.
                                Annals of Math., 44, 1943, 138-142.
                                Reprinted in Collected Papers
                                (edited by S. Lang & J.T. Tate), 346-350.
                                Addison-Wesley, Reading, Mass., 1965.

1943   LINNIK, Yu.V.   An elementary solution of a problem
                                of Waring by Schnirelmann's method.
                                Mat. Sbornik, 12 (54), 1943,
                                225-230.

1944   CHOWLA, S.   There exists an infinity of
                                3-combinations of primes in A. P.
                                Proc. Lahore Phil. Soc., 6,
                                1944, 15-16.

1944 LINNIK, Yu.V.   On the least prime in an
                             arithmetic progression I. The
                             basic theorem (in Russian).
                             Mat. Sbornik, 15 (57), 1944,
                             139-178.

1944 NIVEN, I.   An unsolved case of the Waring
                             problem.
                             Amer. J. Math., 66, 1944, 137-143.

1945 ERDÖS, R.   Some remarks on Euler's ϕ-function
                             and some related problems.
                             Bull. Amer. Math. Soc., 51, 1945,
                             540-544.

1946 BRAUER, A.   On the exact number of primes below a
                             given limit.
                             Amer. Math. Monthly, 9, 1946,
                             521-523.

1946 JOFFE, S.A.   Review of Kulik's "Magnus Canon
                             Divisorum ..."
                             Math. Comp., 2, 1946/7, 139-140.

1947 KHINCHIN, A.Ya.   Three Pearls of Number Theory.
                             Original Russian edition in OGIZ,
                             Moscow, 1947.
                             Translation into English published
                             by Graylock Press, Baltimore, 1952.

1947 RÉNYI, A.   On the representation of even
                             numbers as the sum of a prime
                             and an almost prime.
                             Dokl. Akad. Nauk SSSR, 56,
                             1947, 455-458.

1947   VINOGRADOV, I.M.   The method of trigonometrical
                                 sums in the theory of numbers
                                 (in Russian).
                                 Trav. Inst. Math. Steklov, 23,
                                 1947, 109 pages.

1948   ERDÖS, P. &   On some new questions on the
        TURÁN, P.   distribution of prime numbers.
                                 Bull. Amer. Math. Soc., 54, 1948,
                                 371-378.

1948   VINOGRADOV, I.M.   Über die Abschätzung
                                 trigonometrischer Summen mit
                                 Primzahlen.
                                 Izv. Akad. Nauk SSSR, Ser. Mat.,
                                 12, 1948, 225-248.

1949   CLEMENT, P.A.   Congruences for sets of primes.
                                 Amer. Math. Monthly, 56,
                                 1949, 23-25.

1949   ERDÖS, P.   Problems and results on the
                                 difference of consecutive primes.
                                 Publ. Math. Debrecen, 1, 1949,
                                 33-37.

1949   ERDÖS, P.   On a new method in elementary
                                 number theory which leads to an
                                 elementary proof of the prime
                                 number theorem.
                                 Proc. Nat. Acad. Sci. U.S.A.,
                                 35, 1949, 374-384.

1949   ERDÖS, P.   On the converse of Fermat's
                                 theorem.
                                 Amer. Math. Monthly, 56, 1949,
                                 623-624.

1949	MOSER, L.	A theorem on the distribution of primes. Amer. Math. Monthly, 56, 1949, 624-625.
1949	RICHERT, H.E.	Über Zerfällungen in ungleiche Primzahlen. Math. Zeits., 52, 1949, 342-343.
1949	SELBERG, A.	An elementary proof of the prime number theorem. Annals of Math., 50, 1949, 305-313.
1949	SELBERG, A.	An elementary proof of Dirichlet's theorem about primes in an arithmetic progression. Annals of Math., 50, 1949, 297-304.
1949	SELBERG, A.	An elementary proof of the prime number theorem for arithmetic progressions. Can. J. Math., 2, 1950, 66-78.
1950	ERDÖS, P.	On some application of Brun's method. Acta Sci. Math. Szeged, 13, 1950, 57-63.
1950	ERDÖS, P.	On almost primes. Amer. Math. Monthly, 57, 1950, 404-407.
1950	HASSE, H.	*Vorlesungen über Zahlentheorie.* Springer-Verlag, Berlin, 1950.
1950	SELBERG, A.	The general sieve method and its place in prime number theory. Proc. Int. Congr. Math., Cambridge,

1950   SHAPIRO, H.N.   On a theorem of Selberg and
                              generalization.
                              Annals of Math., (2), 51, 1950,
                              485-497.

1951   NAGELL, T.   *Introduction to Number Theory.*
                              Almqvist & Wiksell, Stockholm, 1964.
                              Reprinted by Chelsea, Bronx, N.Y.,
                              1964.

1951   TITCHMARSH, E.C.   *The Theory of the Riemann Zeta-
                              Function.*
                              Clarendon Press, Oxford, 1951.

1951   WATSON, G.L.   A proof of the seven cube theorem.
                              J. London Math. Soc., 26, 1951,
                              153-156.

1952   NAGURA, J.   On the interval containing at least one
                              prime number.
                              Proc. Japan Acad., 28, 1952, 177-181.

1953   PYATETSKII-   On the distribution of prime numbers
       SHAPIRO, I.I.   in sequence of the form $[f(n)]$
                              (in Russian).
                              Mat. Sbornik, N.S. 33 (75), 1953,
                              559-566.

1953   RIEGER, G.J.   Zur Hilbertschen Lösung der
                              Waringschen Problems: Abschätzung
                              von $g(n)$.
                              Arch. d. Math., 4, 1953, 275-281.

1954   BREUSCH, R.   Another proof of the prime number
                              theorem.
                              Duke M.J., 21, 1954, 49-53.

1954 RIEGER, G.J. Zu Linniks Lösung des Waringsches
Problems: Abschätzung von $g(n)$.
Math. Zeits., 60, 1954, 213-234.

1955 FJELLSTEDT, L. Bemerkungen über gleichzeitige
Lösbarkeit von Kongruenzen.
Arkiv Mat., 3, 1955, 193-198.

1955 RICCI, G. Recherches sur l'allure de la suite
$\{(p_{n+1} - p_n)/\log p_n\}$.
Coll. Th. Nombres Bruxelles 1955,
93-106.
G. Thone, Liège, 1956.

1955 SKEWES, S. On the difference $\pi(x) - \mathrm{li}(x)$, II.
Proc. London Math. Soc., 5, 1955, 48-70.

1956 AMITSUR, S.A. On arithmetic functions.
J. Anal. Math., 5, 1956/7, 273-314.

1956 ERDÖS, P. On pseudo-primes and
Carmichael numbers.
Publ. Math. Debrecen, 4,
1956, 201-206.

1956 NIVEN, I. *Irrational Numbers.*
Carus Math. Monographs, No. 11.
Math. Assoc. of America, Washington,
1956.

1956 OSTMANN, H.H. *Additive Zahlentheorie* (2 volumes).
Springer-Verlag, Berlin, 1956
(2nd edition 1969).

1957 HUA, L.K. *Additive Theory of Prime Numbers.*
Inst. Math. Chinese Acad. Sciences,
Peking, 1957.
Translated into English by H.H. Ng.
Amer. Math. Soc., Providence, 1965.

1957   LEECH, J.   Note on the distribution of prime
                                    numbers.
                                    J. London Math. Soc., 32, 1957,
                                    56-58.

1957   MAHLER, K.   On the fractional parts of powers
                                    of real numbers.
                                    Matematika, 4, 1957, 122-124.

1958   BAKER, C.L. &   *Primes in the Thousandth Million.*
       GRUENBERGER, E.J.   The Rand Corp., Santa Monica, 1958.

1958   CHEN, J.R.   On Waring's problem for nth powers.
                                    Acta Math. Sinica, 8, 1958, 253-257;
                                    Chinese Math. Acta, 8, 1966/7,
                                    849-853.

1958   SCHINZEL, A. &   Sur certaines hypothèses concernant
       SIERPIŃSKI, W.   les nombres premiers.
                                    Acta Arithm., 4, 1958, 185-208;
                                    Erratum, 5, 1959, p. 259.

1959   BAKER, C.L. &   *The First Six Million Prime Numbers.*
       GRUENBERGER, F.J.   Microcard Found., Madison, 1959.

1959   HUA, L.K.   Abschätzungen von
                                    Exponentialsummen und ihre
                                    Anwendung in der Zahlentheorie.
                                    Enzykl. d. Math. Wiss., I2, Heft 13,
                                    t.1.
                                    Teubner, Leipzig, 1959.

1959   KAC, M.   *Statistical Independence in*
                                    Probability, Analysis and Number
                                    Theory.
                                    Carus Math. Monographs, No. 12,
                                    Math. Assoc. of America,
                                    Washington, 1959.

1959   LEHMER, D.H.   On the exact number of primes less
                                 than a given limit.
                                 Illinois J. Math., 3, 1959,
                                 381-388.
                                 Reprinted in *Selected Papers*
                                 (edited by D. McCarthy),
                                 Vol. III, 1104-1111.
                                 Ch. Babbage Res. Centre, St. Pierre,
                                 Manitoba, Canada, 1981.

1959   SHANKS, D.   Quadratic residues and the
                                 distribution of primes.
                                 Math. Comp., 13, 1959, 272-284.

1959   SCHINZEL, A.   Démonstration d'une conséquence
                                 de l'hypothèse de Goldbach.
                                 Compositio Math., 14, 1959, 74-76.

1959   SCHINZEL, A.   Sur une conséquence de l'hypothèse
                                 de Goldbach.
                                 Izvestija Mat. Inst., Bulgarian
                                 Acad. Sci., 4, 1959, 35-38.

1959   VINOGRADOV, I.M.   On an upper bound for $G(n)$
                                 (in Russian).
                                 Izv. Akad. Nauk SSSR, Ser.
                                 Mat., 23, 1959, 637-642.

1960   JACOBSTHAL, E.   Über Sequenzen ganzer Zahlen, von
                                 denen keine zu n teilerfremd ist,
                                 I, II, III.
                                 Norske Videnskabsselskab Forhdl.,
                                 33, 1960, 117-139.

1960   NEWMAN, D.J.   A simplified proof of Waring's
                                 conjecture.
                                 Michigan Math. J., 7, 1960,
                                 291-295.

1961   JURKAT, W.B.   Eine Bemerkung zur Vermutung von
                                 Mertens.
                                 Machr. der Österr. Math. Ges.,
                                 Sondernummer Ber. V. Osterr.
                                 Math.-Kongress, Vienna, 1961, 11.

1961   PRACHAR, K.   Über die kleinste Primzahl einer
                                 arithmetischen Reihe.
                                 Journal f. d. reine u. angew. Math.,
                                 206, 1961, 3-4.

1961   ROTKIEWICZ, A.   Démonstration arithmétique
                                 d'existence d'une infinité de
                                 nombres premiers de la forme $nk + 1$.
                                 L'Enseign. Math., (2), 7,
                                 1962, 277-280.

1961   WRENCH, J.W.   Evaluation of Artin's constant and
                                 the twin-prime constant.
                                 Math. Comp., 15, 1961, 396-398.

1962   ERDÖS, P.   On the integers relatively prime to
                                 n and on a number theoretic function
                                 considered by Jacobsthal.
                                 Math. Scand., 10, 1962, 163-170.

1962   ROSSER, J.B. &   Approximate formulas for some
        SCHOENFELD, L.   functions of prime numbers.
                                 Illinois J. Math., 6, 1962, 64-94.

1962   SCHINZEL, A.   Remark on a paper of K. Prachar,
                                 "Über die kleinste Primzahl einer
                                 arithmetischen Reihe".
                                 Journal f. d. reine u. angew. Math.,
                                 210, 1962, 121-122.

1962   SEGAL, S.   On $\pi(x+y) \leqslant \pi(x) + \pi(y)$.
                                 Trans. Amer. Math. Soc., 104, 1962,
                                 523-527.

1963   AYOUB,   R.G.   *An Introduction to the Theory of Numbers.*
                              Amer. Math. Soc., Providence, R.I.,
                              1963.

1963   ESTERMANN, T.   Note on a paper of A. Rotkiewicz.
                              Acta Arithm., 8, 1963, 465-467.

1963   KANOLD, H.J.   Elementare Betrachtungen zur
                              Primzahltheorie.
                              Arch. Math., 14, 1963, 147-151.

1963   NEUBAUER, G.   Eine empirische Untersuchung zur
                              Mertenssche Funktion.
                              Numer. Math., 5, 1963, 1-13.

1963   RANKIN, R.A.   The difference between consecutive
                              prime numbers, V.
                              Proc. Edinburgh Math. Soc.,
                              (2), 13, 1963, 331-332.

1963   ROTKIEWICZ, A.   Sur les nombres pseudo-premiers
                              de la forme $ax + b$.
                              C.R. Acad. Sci. Paris, 257, 1963,
                              2601-2604.

1963   WALFISZ, A.Z.   *Weylsche Exponentialsummen in der*
                              neueren Zahlentheorie.
                              VEB Deutscher Verlag d. Wiss.,
                              Berlin, 1963.

1964   CHEN, J.R.   Waring's problem for $g(5) = 37$.
                              Sci. Sinica, 13, 1964, 1547-1568.
                              Reprinted in Chinese Mathematics,
                              6, 1965, 105-127.

1964/ KANOLD, H.J.   Über Primzahlen in arithmetischen
1965   Folgen.
                              Math. Ann., 156, 1964, 393-395;
                              157, 1965, 358-362.

1964   KAPFERER, H.   Verifizierung des symmetrischen
                             Teils der Fermatschen Vermutung
                             für unendlich viele paarweise
                             teilerfremde Exponenten E.
                             Journal f. d. reine u. angew. Math.,
                             214/5, 1964, 360-372.

1964   ROHRBACH, H. &   Zum finiten Fall des
       WEIS, J.   Bertrandschen Postulats.
                             Journal f. d. reine u. angew. Math.,
                             214/5, 1964, 432-440.

1964   SHEN, M.K.   On checking the Goldbach
                             conjecture.
                             Nordisk Tidskr., 4, 1964, 243-245.

1964   SIERPIŃSKI, W.   Elementary Theory of Numbers.
                             Hafner, New York, 1964.

1964   STEMMLER, R.M.   The ideal Waring theorem for
                             exponents 401-200,000.
                             Math. Comp., 18, 1964, 144-146.

1965   BATEMAN, P.T. &   Prime numbers in arithmetic
       LOW, M.E.   progression with difference 24.
                             Amer. Math. Monthly, 72, 1965,
                             139-143.

1965   GELFOND, A.O. &   Elementary Methods in Analytic
       LINNIK, Yu.V.   Number Theory.
                             Translated by A. Feinstein,
                             revised and edited by L.J. Mordell.
                             Rand McNally, Chicago, 1965.

1965   PAN, C.D.   On the least prime in an arithmetic
                             progression.
                             Sci. Record (N.S.), 1, 1957, 311-313.

1965   ROTKIEWICZ, A.   Les intervalles contenant les
                              nombres pseudo premiers.
                              Rend. Circ. Mat. Palermo (2), 14,
                              1965, 278-280.

1965   STEIN, M.L. &   New experimental results on the
       STEIN, P.R.   Goldbach conjecture.
                              Math. Mag., 38, 1965, 72-80.

1965   STEIN, M.L. &   Experimental results on
       STEIN, P.R.   additive 2-bases.
                              Math. Comp., 19, 1965, 427-434.

1966   BEILER, A.H.   *Recreations in the Theory of Numbers*
                              (*The Queen of Mathematics Entertains*).
                              Dover, New York, 1966.

1966   CHEN, J.R.   On the representation of a large
                              even integer as the sum of a prime
                              and the product of at most two primes.
                              Kexue Tongbao, 17, 1966, 385-386.

1966   BOMBIERI, E. &   Small differences between prime
       DAVENPORT, H.   numbers.
                              Proc. Roy. Soc., A, 293, 1966, 1-18.

1966   LEHMAN, R.S.   On the difference $\pi(x) - \mathrm{li}(x)$.
                              Acta Arithm., 11, 1966, 397-410.

1966   JARDEN, D.   Existence of arbitrarily long sequences
                              of consecutive numbers in arithmetic
                              progressions divisible by arbitrarily
                              many different primes.
                              Fibonacci Quart., 5, 1967, p. 287.

1967   JONES, M.F.,   Statistics on certain large primes.
       LAL, M. &   Math. Comp., 21, 1967, 103-107.
       BLUNDON, W.J.

1967 KOLESNIK, G.A.   The distribution of primes in
                              sequences of the form $[n^c]$
                              (in Russian).
                              Mat. Zametki, 2, 1967, 117-128.

1967 LANDER, L.J. &   Consecutive primes in
      PARKIN, T.R.   arithmetic progression.
                              Math. Comp., 21, 1967,
                              p. 489.

1967 ROTKIEWICZ, A.   On the pseudo-primes of the form
                              $ax + b$.
                              Proc. Cambridge Phil. Soc., 63,
                              1967, 389-392.

1967 SZYMICZEK, K.   On pseudo-primes which are
                              products of distinct primes.
                              Amer. Math. Monthly, 74,
                              1967, 35-37.

1968 HALBERSTAM, H. &   A gap theorem for pseudoprimes in
      ROTKIEWICZ, A.   arithmetic progressions.
                              Acta Arithm., 13, 1968, 395-404.

1969 GRÖLZ, W.   Primteiler von Polynomen.
                              Math. Ann., 181, 1969, 134-136.

1969 MONTGOMERY, H.L.   Zeros of L-functions.
                              Invent. Math., 8, 1969, 346-354.

1969 NAGELL, T.   Sur les diviseurs premiers des
                              polynômes.
                              Acta Arithm., 15, 1969, 235-244.

1969 RICHERT, H.E.   Selberg's sieve with weights.
                              Mathematika, 16, 1969, 1-22.

1969   ROSSER, J.B.,   Rigorous computation of the
        YOHE, J.M. &   zeros of the Riemann zeta-function
        SCHOENFELD, L.   (with discussion).
                                   Inform. Processing 68 (Proc.
                                   IFIP Congress, Edinburgh, 1968),
                                   Vol. I, 70-76.
                                   North-Holland, Amsterdam, 1969.

1969   SCHOENFELD, L.   An improved estimate for the
                                   summatory function of the Möbius
                                   function.
                                   Acta Arithm., 15, 1969, 221-233.

1969   WÓJCYK, J.   A refinement of a theorem of Schur
                                   on primes in arithmetic
                                   progressions, III.
                                   Acta Arithm. 15, 1969, 193-197.

1970   DRESSLER, R.E.   A density which counts multiplicity.
                                   Pacific J. Math., 34, 1970,
                                   371-378.

1970   HORNFECK, B.   Primteiler von Polynomen.
                                   Journal f. d. reine u. angew.
                                   Math., 243, 1970, p. 120.

1970   MOTOHASHI, Y.   A note on the least prime in an
                                   arithmetic progression with a
                                   prime difference.
                                   Acta Arithm., 17, 1970,
                                   283-285.

1970   SERRE, J.P.   Cours d'Arithmétique.
                                   Presses Univ. France, Paris,
                                   1970.
                                   English translation published
                                   by Springer-Verlag, New York, 1973.

1971 ELLIOTT, P.D.T.A. &   The least prime in arithmetic
      HALBERSTAM, H.   progression.
                              Studies in Pure Mathematics
                              (edited by R. Radò), 59-61.
                              Academic Press, London, 1971.

1971 ELLISON, W.J.   Waring's problem.
                              Amer. Math. Monthly, 78, 1971,
                              10-36.

1971 GERST, I. &   On prime divisors of polynomials.
      BRILLHART, J.   Amer. Math. Monthly, 78, 1971,
                              250-266.

1971 MONTGOMERY, H.L.   *Topics in Multiplicative
                              Number Theory.*
                              Lecture Notes in Math., #227.
                              Springer-Verlag, New York, 1971.

1971 SERGUSOV, I.S.A.   On the problem of prime-twins
                              (in Russian).
                              Jaroslav. Gos. Ped. Inst. Ucen.
                              Zap., 82, 1971, 85-86.

1971 TITCHMARSH, E.C.   *The Theory of the Riemann Zeta
                              Function.*
                              Clarendon Press, Oxford, 1951.

1971 TURÁN, P.   On some recent results in the
                              analytical theory of numbers.
                              Proc. Symp. Pure Mathematics (1969
                              Number Theory Institute), vol. 20,
                              359-374.
                              Amer. Math. Soc., Providence, R.I.,
                              1971.

1972 BATEMAN, P.T.   The distribution of values of
                              Euler function.
                              Acta Arithm., 21, 1972, 329-345.

1972 DESHOUILLERS, J.M.   Nombres premiers de la forme [n^c].
                            C.R. Acad. Sci. Paris, Sér. A,
                            282, 1976, 131-133.

1972 HUXLEY, M.N.   On the difference between
                            consecutive primes.
                            Invent. Math., 15, 1972, 164-170.

1972 HUXLEY, M.N.   *The Distribution of Prime Numbers.*
                            Oxford Univ. Press, Oxford, 1972.

1972 ROTKIEWICZ, A.   On a problem of W. Sierpiński.
                            Elem. d. Math., 27, 1972, 83-85.

1972 WALL, C.R.   Density bounds for Euler's function.
                            Math. Comp., 26, 1972, 779-783.

1973 APOSTOL, T.M.   Another elementary proof of Euler's
                            formula for $\zeta(2n)$.
                            Amer. Math. Monthly, 80, 1973,
                            425-431.

1973 BRENT, R.P.   The first occurrence of certain
                            large prime gaps.
                            Math. Comp., 35, 1980, 1435-1436.

1973/ CHEN, J.R.   On the representation of a large
1978   even integer as the sum of a
                            prime and the product of at most
                            two primes, I and II.
                            Sci. Sinica, 16, 1973, 157-176;
                            and 21, 1978, 421-430.

1973 HENSLEY, D. &   Primes in intervals.
      RICHARDS, I.   Acta Arithm., 25, 1973/4, 375-391.

1973   MONTGOMERY, H.L.   The pair correlation of zeros of the
                          zeta function.
                          Analytic Number Theory (Proc. Symp.
                          Pure Math., Vol. XXIV, St. Louis,
                          1972), 181-193.
                          Amer. Math. Soc., Providence, R.I.,
                          1973.

1973   WUNDERLICH, M.C.   On the Gaussian primes on the line
                          $\text{Im}(x) = 1$.
                          Math. Comp., 27, 1973, 399-400.

1974   AYOUB, R.G.   Euler and the zeta function.
                          Amer. Math. Monthly, 81, 1974,
                          1067-1086.

1974   BRENT, R.P.   The distribution of small gaps
                          between successive primes.
                          Math. Comp., 28, 1974, 315-324.

1974   EDWARDS, H.M.   *Riemann's Zeta Function.*
                          Academic Press, New York, 1974.

1974   HALBERSTAM, H. &   *Sieve Methods.*
       RICHERT, H.E.   Academic Press, New York, 1974.

1974   LEVINSON, N.   More than one third of zeros of
                          Riemann's zeta function are on
                          $\sigma = 1/2$.
                          Adv. in Math., 13, 1984,
                          383-436.

1974   MĄKOWSKI, A.   On a problem of Rotkiewicz on
                          pseudo-primes.
                          Elem. d. Math., 29, 1974, p. 13.

1974   SHANKS, D. &   Brun's constant.
       WRENCH, J.W.   Math. Comp., 28, 1974,
                          293-299.

1975   BRENT, R.P.   Irregularities in the distribution
                                      of primes and twin primes.
                                      Math. Comp., 29, 1975, 43-56.

1975   MONTGOMERY, H.L.   The exceptional set in Goldbach's
       & VAUGHAN, R.C.   problem.
                                      Acta Arithm., 27, 1975, 353-370.

1975   RAM MURTY, P.M.   *On the Existence of "Euclidean
                                      Proofs" of Dirichlet's Theorem on
                                      Primes in Arithmetic Progressions.*
                                      B.Sc. Thesis, Carleton University,
                                      Ottawa, 1975, 39 pages.

1975   ROSS, P.M.   On Chen's theorem that each
                                      large even number has the form
                                      $p_1 + p_2$ or $p_1 + p_2 p_3$.
                                      J. London Math. Soc., (2), 10,
                                      1975, 500-506.

1975   ROSSER, J.B. &   Sharper bounds for Chebyshev
       SCHOENFELD, L.   functions $\theta(x)$ and $\psi(x)$.
                                      Math. Comp., 29, 1975, 243-269.

1975   SWIFT, J.D.   Table of Carmichael numbers to 10^9.
                                      Math. Comp., 29, 1975, 338-339.

1976   APOSTOL, T.M.   *Introduction to Analytic Number
                                      Theory.*
                                      Springer-Verlag, New York, 1976.

1976   BRENT, R.P.   Tables concerning irregularities
                                      in the distribution of primes and
                                      twin primes to 10^{11}.
                                      Math. Comp., 30, 1976, p. 379.

1976   GERIG, S.   A simple proof of the prime number
                                      theorem.
                                      J. Nb. Th., 8, 1976, 131-136.

1976   NIVEN, I. &   Primes in certain arithmetic
       POWELL, B.   progressions.
                             Amer. Math. Monthly, 83, 1976,
                             467-469.

1976   SCHOENFELD, L.   Sharper bounds for Chebyshev
                             functions $\theta(x)$ and $\psi(x)$, II.
                             Math. Comp., 30, 1976, 337-360.

1976   VAUGHAN, R.C.   A note on Schnirelmann's approach
                             to Goldbach's problem.
                             Bull. London Math. Soc., 8, 1976,
                             245-250.

1977   DESHOUILLERS, J.M.   Sur la constante de Schnirelmann.
                             Sém. Delange-Pisot-Poitou, 17^e
                             année, 1975/6, fasc. 2, exp.
                             No. G16, 6 p., Paris, 1977.

1977   HUDSON, R.H.   A formula for the exact number of
                             primes below a given bound in any
                             arithmetic progression.
                             Bull. Austral. Math. Soc., 16,
                             1977, 67-73.

1977   HUDSON, R.H. &   On the exact number of primes in the
       BRAUER, A.   arithmetic progressions $4n \pm 1$
                             and $6n \pm 1$.
                             Journal f. d. reine u. angew. Math.,
                             291, 1977, 23-29.

1977   HUXLEY, M.N.   Small differences between
                             consecutive primes, II.
                             Mathematika, 24, 1977, 142-152.

1977   JUTILA, M.   On Linnik's constant.
                             Math. Scand., 41, 1977, 45-62.

1977   JUTILA, M.   Zero-density estimates for L-functions.
                             Acta Arithm., 32, 1977, 52-62.

1977 KUMAR MURTY, V. *The Least Prime in an Arithmetical Progression and an Estimate of Linnik's Constant.* B.Sc. Thesis, Carleton Univ., Ottawa, 1977, 45 pages.

1977 LANGEVIN, M. Methodes élémentaires en vue du théorème de Sylvester. Sém. Delange-Pisot-Poitou, 17e année, 1975/76, fasc. 1, exp. No. G12, 9 pages, Paris, 1977.

1977 POWELL, B. Proof of a special case of Dirichlet's theorem. Fibonacci Quart., 15, 1977, 167-169.

1977 SMALL, C. Waring's problem. Math. Mag., 50, 1977, 12-16.

1977 WEINTRAUB, S. Seventeen primes in arithmetic progression. Math. Comp., 31, 1977, p. 1030.

1977 ZAGIER, D. The first 50 million prime numbers. Math. Intelligencer, Vol. 0, 1977, 7-19. Reprinted in German in *Lebendige Zahlen*, by W. Borho, J.C. Jantzen, H. Kraft, J. Rohlfs, D. Zagier. Birkhäuser, Basel, 1981.

1978 BAYS, C. & HUDSON, R.H. Details of the first region of integers x with $\pi_{3,2}(x) < \pi_{3,1}(x)$. Math. Comp., 32, 1978, 571-576.

1978 BAYS, C. & HUDSON, R.H. On the fluctuations of Littlewood for primes of the form $4n \pm 1$. Math. Comp., 32, 141, 281-286.

1978 ELLISON, W.J. &   Théorie des Nombres.
      ELLISON, F.   *Abrégé d'Histoire des*
                              Mathématiques, Vol. I, Chapter V,
                              §VI (edited by J. Dieudonné).
                              Hermann, Paris, 1978.

1978 HEATH-BROWN, D.R.   Almost-primes in arithmetic
                              progressions and short intervals.
                              Math. Proc. Cambridge Phil.
                              Soc., 83, 1978, 357-375.

1978 IWANIEC, H.   On the problem of Jacobsthal.
                              Demo. Math., 11, 1978, 225-231.

1978 WAGSTAFF, Jr., S.S.   The least prime in arithmetic
                              progression with prime difference.
                              Journal f. d. reine u. angew.
                              Math., 301, 1978, 114-115.

1979 ATKIN, A.O.L. &   On a larger pair of twin primes.
      RICKERT, N.W.   Abstract 79T-A132, Notices
                              Amer. Math. Soc., 26, 1979, A-373.

1979 BALASUBRAMANIAN,   On Waring's problem: $g(4) \leqslant 21$.
      R.   Hardy & Ramanujan J., 2, 1979,
                              31 pages.

1979 BAYS, C. &   Numerical and graphical descrip-
      HUDSON, R.H.   tion of all axis crossing regions for
                              the moduli 4 and 8 which occur
                              before 10^{12}.
                              Intern. J. Math. & Math. Sci., 2,
                              1979, 111-119.

1979 CHEN, J.R.   On the least prime in an
                              arithmetical progression and
                              theorems concerning the zeros
                              of Dirichlet's L-functions, II.
                              Sci. Sinica, 22, 1979, 859-889.

1979   ELLIOTT, P.D.T.A.   *Probabilistic Number Theory* (in
                                2 volumes).
                                Springer-Verlag, New York, 1979.

1979   GROSSWALD, E. &   Arithmetic progressions consisting
       HAGIS, Jr., P.   only of primes.
                                Math. Comp., 33, 1979, 1343-1352.

1979   HEATH-BROWN, D.R.   On the difference between
       & IWANIEC, H.   consecutive powers.
                                Bull. Amer. Math. Soc.,
                                N.S., 1, 1979, 758-760.

1979   HLAWKA, E.   *Theorie der Gleichverteilung.*
                                Bibliographisches Institut,
                                Zürich, 1979.

1979   IWANIEC, H. &   Primes in short intervals.
       JUTILA, M.   Arkiv f. Mat., 17, 1979, 167-176.

1979   POMERANCE, C.   The prime number graph.
                                Math. Comp., 33, 1979,
                                399-408.

1979   ROTKIEWICZ, A.   On a number-theoretical
       & WASÉN, R.   series.
                                Publ. Math. Debrecen, 26,
                                1979, 1-4.

1979   WAGSTAFF, Jr., S.S.   Greatest of the least primes in
                                arithmetic progressions having a
                                given modulus.
                                Math. Comp., 33, 1979, 1073-1080.

1979   WOOLDRIDGE, K.   Values taken many times by Euler's
                                phi-function.
                                Proc. Amer. Math. Soc., 76, 1979,
                                229-234.

1980   BRENT, R.P.   The first occurrence of certain
                               large prime gaps.
                               Math. Comp., 35, 1980, 1435-1436.

1980   CHEN, J.R. &   The exceptional set of Goldbach
       PAN, C.D.   numbers, I.
                               Sci. Sinica, 23, 1980, 416-430.

1980   ERDÖS, P. &   Remarks on the difference between
       STRAUS, E.G.   consecutive primes.
                               Elem. d. Math., 35, 1980, 115-118.

1980   IVIĆ, A.   Exponent pairs and the zeta-function
                               of Riemann.
                               Studia Sci. Math. Hung., 15, 1980,
                               157-181.

1980   KUTSUNA, M.   On a criterion for the class
                               number of a quadratic number
                               field to be one.
                               Nagoya Math. J., 79, 1980, 123-129.

1980   LIGHT, W.A.,   A note on Goldbach's conjecture.
       FORREST, J.,   BIT, 20, 1980, p. 525.
       HAMMOND, N., &
       ROE, S.

1980   NEWMAN, D.J.   Simple analytic proof of the prime
                               number theorem.
                               Amer. Math. Monthly, 87, 1980,
                               693-696.

1980   PINTZ, J.   On Legendre's prime number formula.
                               Amer. Math. Monthly, 87, 1980,
                               733-735.

1980   POMERANCE, C.   Popular values of Euler's function.
                               Mathematika, 27, 1980, 84-89.

1980 POMERANCE, C.   A note on the least prime in an
                             arithmetic progression.
                             J. Nb. Th., 12, 1980, 218-223.

1980 POMERANCE, C.,   The pseudo primes to
      SELFRIDGE, J.L., & $25 \cdot 10^9$.
      WAGSTAFF, Jr., S.S.   Math. Comp., 35, 1980,
                             1003-1026.

1980 POWELL, B.   Problem E2844 (Difference between
                             consecutive primes).
                             Amer. Math. Monthly, 87, 1980,
                             p. 577; 90, 1983, p. 286.

1980 VANDEN EYNDEN,   Proofs that $\Sigma \, 1/p$ diverges.
      C.   Amer. Math. Monthly, 87, 1980,
                             394-397.

1980 VAN DER POORTEN,   On strong pseudoprimes in
      A.J. &   arithmetic progressions.
      ROTKIEWICZ, A.   J. Austral. Math. Soc., A, 29,
                             1980, 316-321.

1980 WAGSTAFF, Jr., S.S.   Greatest of the least primes in
                             arithmetic progressions having a
                             given modulus.
                             Math. Comp., 33, 1979, 1073-1080.

1981 BOHMAN, J. &   Numerical investigations of
      FRÖBERG, C.E.   Waring's problem for cubes.
                             BIT 21, 1981, 118-122.

1981 GRAHAM, S.   On Linnik's constant.
                             Acta Arithm., 39, 1981, 163-179.

1981 HEATH-BROWN, D.R.   Three primes and an almost
                             prime in arithmetic progression.
                             J. London Math. Soc., (2), 23,
                             1981, 396-414.

1981   LEAVITT, W.G. &   Primes differing by a fixed
       MULLIN, A.A.   integer.
                                Math. Comp., 37, 1981, 581-585.

1981   LEHMER, D.H.   On Fermat's quotient, base two.
                                Math. Comp., 36, 1981,
                                289-290.

1981   MAIER, H.   Chains of large gaps between
                                consecutive primes.
                                Adv. in Math., 39, 1981, 257-269.

1981   PINTZ, J.   On primes in short intervals, I.
                                Studia Sci. Math. Hung., 16, 1981,
                                395-414.

1981   POMERANCE, C.   On the distribution of
                                pseudo-primes.
                                Math. Comp., 37, 1981,
                                587-593.

1981   WEINTRAUB, S.   A large prime gap.
                                Math. Comp., 36, 1981,
                                p. 279.

1982   DIAMOND, H.G.   Elementary methods in the study of
                                the distribution of prime numbers.
                                Bull. Amer. Math. Soc., 7, 1982,
                                553-589.

1982   NAIR, M.   On Chebyshev type inequalities for
                                primes.
                                Amer. Math. Monthly, 81, 1982,
                                126-129.

1982   POMERANCE, C.   A new lower bound for the pseudo-
                                primes counting function.
                                Illinois J. Math., 26, 1982, 4-9.

1982   PRITCHARD, P.A.   18 primes in arithmetic
                              progression.
                              J. Recr. Math., 15, 1982/3,
                              p. 288.

1982   ROMANI, F.   Computations concerning Waring's
                              problem for cubes.
                              Calcolo, 19, 1982, 415-431.

1982   THANIGASALAM, K.   Some new estimates for $G(k)$ in
                              Waring's problem.
                              Acta Arithm., 42, 1982/3, 73-78.

1982   WILLIAMS, H.C.   A note on the Fibonacci quotient
                              $F_{p-\epsilon}/p$.
                              Can. Math. Bull., 25, 1982,
                              366-370.

1983   CHEN, J.R.   The exceptional value of Goldbach
                              numbers, II.
                              Sci. Sinica, Ser. A, 26, 1983,
                              714-731.

1983   CONREY, J.B.   Zeros of derivatives of Riemann's
                              xi-function on the critical line.
                              J. Nb. Th., 16, 1983, 49-74.

1983   FOUVRY, E. &   Primes in arithmetic progressions.
       IWANIEC, H.   Acta Arithm., 42, 1983, 197-218.

1983   IVIĆ, A.   Topics in Recent Zeta Function
                              Theory.
                              Publ. Math. d'Orsay, Univ. Paris-Sud.
                              1983.

1983   KELLER, W.   Large twin prime pairs related to
                              Mersenne numbers.
                              Abstracts Amer. Math. Soc., 4,
                              1983, p. 490.

1983 NICOLAS, J.L.   Petites valeurs de la fonction
                             d'Euler.
                             J. Nb. Th., 17, 1983, 375-388.

1983 NICOLAS, J.L.   Distribution de valeurs de la
                             fonction d'Euler.
                             In *Algorithmique, Calcul Formel
                             Arithmétique*, exposé 24,
                             7 pages.
                             Univ. Saint-Etienne, 1983.
                             Reprinted in L'Enseign. Math.
                             30, 1984, 331-338.

1983 POWELL, B.   Problem 6429 (Difference between
                             consecutive primes).
                             Amer. Math. Monthly, 90, 1983, p. 338.

1983 RIESEL, H. &   On sums of primes.
      VAUGHAN, R.C.   Arkiv f. Mat., 21, 1983, 45-74.

1983 ROBIN, G.   Estimation de la fonction de
                             Tschebychef θ sur le k-ième nombre
                             premier et grandes valeurs de la
                             fonction $\omega(n)$, nombre de diviseurs
                             premiers de n.
                             Acta Arithm., 42, 1983, 367-389.

1984 BALASUBRAMANIAN,   An improved upper bound for $G(k)$
      R. & MOZZOCHI, C.J.   in Waring's problem for relatively
                             small k.
                             Acta Arithm., 63, 1984, 283-285.

1984 DABOUSSI, H.   Sur le théorème des nombres
                             premiers.
                             C.R. Acad. Sci. Paris, Sér. I, 298,
                             1984, 161-164.

1984 DAVIES, R.O.   Solution of problem 6429.
                             Amer. Math. Monthly, 91, 1984, p. 64.

1984   GUPTA, R. &   A remark on Artin's conjecture.
       RAM MURTY, P.M.   Invent. Math., 78, 1984, 127-130.

1984   IWANIEC, H. &   Primes in short intervals.
       PINTZ, J.   Monatsh. Math., 98, 1984, 115-143.

1984   PINTZ, J.   On primes in short intervals, II.
                                    Stud. Sci. Math. Hung., 19, 1984,
                                    89-96.

1984   POWELL, B. &   Solution of problem E 2844.
       SHAFER, R.E.   Amer. Math. Monthly, 91, 1984,
                                    310-311.

1984   SCHROEDER, M.R.   *Number Theory in Science and
                                    Communication.*
                                    Springer-Verlag, New York, 1984.

1984   WANG, Y.   *Goldbach Conjecture.*
                                    World Scientific Publ., Singapore,
                                    1984.

1985   BALASUBRAMANIAN,   Asymptotic mean square of the
       R., CONREY, J.B., &   product of the Riemann zeta-
       HEATH-BROWN, D.R.   function and a Dirichlet polynomial.
                                    Journal f. d. reine u. angew. Math.,
                                    357, 1985, 161-181.

1985   FOUVRY, E.   Théorème de Brun-Titchmarsh,
                                    application au théorème de Fermat.
                                    Invent. Math., 79, 1985, 383-407.

1985   HUDSON, R.H.   Averaging effect on irregularities
                                    in the distribution of primes in
                                    arithmetic progressions.
                                    Math. Comp., 44, 1985, 561-571.

1985   IVIĆ, A.   *The Riemann Zeta-Function.*
                                    J. Wiley & Sons, New York, 1985.

1985 LAGARIAS, J.C.,   Computing $\pi(x)$: The
      MILLER, V.S. &   Meissel-Lehmer method.
      ODLYZKO, A.M.   Math. Comp., 44, 1985, 537-560.

1985 MAIER, H.   Small differences between prime
                             numbers.
                             Michigan Math. J., 32, 1985, 221-225.

1985 ODLYZKO, A.M. &   Disproof of the Mertens conjecture.
      TE RIELE, H.J.J.   Journal f. d. reine u. angew. Math.,
                             357, 1985, 138-160.

1985 PINTZ, J.   An effective disproof of the
                             Mertens conjecture.
                             Preprint No. 55/1985, 9 pages.
                             Math. Inst. Hungarian Acad. Sci.,
                             Budapest.

1985 POWELL, B.   Problem 1207 (A generalized
                             weakened Goldbach theorem).
                             Math. Mag., 58, 1985, p. 46; 59, 1986,
                             48-49.

1985 PRITCHARD, P.A.   Long arithmetic progressions of
                             primes; some old, some new.
                             Math. Comp., 45, 1985, 263-267.

1985 THANIGASALAM, K.   Improvement on Davenport's
                             iterative method and new results
                             in additive number theory, I and II
                             (Proof that $G(5) \leqslant 22$).
                             Acta Arithm., 46, 1985, 1-31 and
                             91-112.

1986 BALASUBRAMANIAN,   Problème de Waring pour les
      R., DESHOUILLERS, J.M. bicarrés, 2: résultats
      & DRESS, F.   auxiliaires pour le théorème
                             asymptotique.
                             C.R. Acad. Sci. Paris, Sér. I, 303,
                             1986, 161-163.

1986 BOMBIERI, E.,   Primes in arithmetic
      FRIEDLANDER, J.B.   progression to large moduli, I.
      & IWANIEC, H.   Acta Math., 156, 1986, 203-251.

1986 COSTA PEREIRA, N.   Sharp elementary estimates for the
                              sequence of primes.
                              Port. Math., 43, 1986, 399-406.

1986 FINN, M.V. &   Solution of problem 1207.
      FROHLIGER, J.A.   Math. Mag., 59, 1986, 48-49.

1986 MOZZOCHI, C.J.   On the difference between
                              consecutive primes.
                              J. Nb. Th., 24, 1986, 181-187.

1986 PINTZ, J.   A note on the exceptional set in
                              Goldbach's problem.
                              Math. Institute Hungarian Acad.
                              Sci., Preprint No. 14/1986.

1986 TE RIELE, H.J.J.   On the sign of the difference
                              $\pi(x) - \text{l}i(x)$.
                              Report NM-R8609, Centre for Math.
                              and Comp. Science, Amsterdam, 1986;
                              Math. Comp., 48, 1987, 323-328.

1986 VAN DE LUNE, J.,   On the zeros of the Riemann zeta
      TE RIELE, H.J.J., &   function in the critical strip, IV.
      WINTER, D.T.   Math. Comp., 47, 1986, 667-681.

1986 VAUGHAN, R.C.   On Waring's problem for small
                              exponents.
                              Proc. London Math. Soc., 52, 1986,
                              445-463.

1986 VAUGHAN, R.C.   On Waring's problem for sixth
                              powers.
                              J. London Math. Soc., (2), 33,
                              1986, 227-236.

1986 WAGON, S.   Where are the zeros of zeta of s?
                             Math. Intelligencer 8, No. 4, 1986,
                             57-62.

1987 BOMBIERI, E.,   Primes in arithmetic
      FRIEDLANDER, J.B.   progressions to large moduli, II.
      & IWANIEC, H.   Math. Ann. 277, 1987, 361-393.

1987 BOMBIERI, E.,   On the order of $\zeta(\frac{1}{2} + it)$.
      IWANIEC, H.   Ann. Scuola Norm. Sup. Pisa
                             (to appear).

1987 ODLYZKO, A.M.   On the distribution of spacings
                             between zeros of the zeta function.
                             Math. Comp., 48, 1987, 273-308.

1987 THANIGASALAM, K.   Improvement on Davenport's iterative
                             method and new results in additive
                             number theory, III.
                             Acta Arithm., 48, 1987, 97-116.

Chapter 5

1878 LUCAS, E.   Théorie des fonctions numériques
                             simplement périodiques.
                             Amer. J. Math., 1, 1878, 185-240
                             and 289-321.

1902 BACHMANN, P.   Niedere Zahlentheorie, Vol. II.
                             Teubner, Leipzig, 1902.
                             Reprinted by Chelsea, Bronx, N.Y.,
                             1968.

1917 POLLACZEK, F.   Über den grossen Fermat'schen Satz.
                             Sitzungsber. Akad. d. Wiss. Wien,
                             Abt. IIa, 126, 1917, 45-59.

1937   HALL, Jr., M.   Divisors of second-order sequences.
                                       Bull. Amer. Math. Soc., 43, 1937, 78-80.

1940   KRASNER, M.   À propos du critère de Sophie
                                       Germain - Furtwängler pour le
                                       premier cas du théorème de
                                       Fermat.
                                       Mathematica Cluj, 16, 1940,
                                       109-114.

1948   GUNDERSON, N.G.   *Derivation of Criteria for the First*
                                       Case of Fermat's Last Theorem and the
                                       Combination of these Criteria to
                                       Produce a New Lower Bound for the
                                       Exponent.
                                       Ph.D. Thesis, Cornell University,
                                       1948, 111 pages.

1951   DÉNES, P.   An extension of Legendre's criterion
                                       in connection with the first case
                                       of Fermat's last theorem.
                                       Publ. Math. Debrecen, 2, 1951, 115-120.

1952   ERDÖS, P. &   The distribution of values of the
       MIRSKY, L.   divisor function $d(n)$.
                                       Proc. London Math. Soc., (3), 2,
                                       1952, 257-271.

1953   GOLDBERG, K.   A table of Wilson quotients and the
                                       third Wilson prime.
                                       J. London Math. Soc., 28,
                                       1953, 252-256.

1954   WARD, M.   Prime divisors of second order
                                       recurring sequences.
                                       Duke Math. J., 21, 1954, 607-614.

1954 WARD, M. Prime divisors of second order recurring sequences. Duke Math. J., 21, 1954, 607-614.

1956 OBLÁTH, R. Une propriété des puissances parfaites. Mathesis, 65, 1956, 356-364.

1959 SIERPIŃSKI, W. Sur les nombres premiers ayant des chiffres initiaux et finals donnés. Acta Arithm., 5, 1959, 265-266.

1969 SIERPIŃSKI, W. Sur un problème concernant les nombres $k \cdot 2^n + 1$. Elem. d. Math., 15, 1960, 73-74.

1960 WALL, D.D. Fibonacci series modulo m. Amer. Math. Monthly, 67, 1960, 525-532.

1961 AIGNER, A. Folgen der Art $ar^n + b$, welche nur teilbare Zahlen liefern. Math. Nachr., 23, 1961, 259-264.

1961 WARD, M. The prime divisors of Fibonacci numbers. Pacific J. Math., 11, 1961, 379-386.

1963 BRILLHART, J. Some miscellaneous factorizations. Math. Comp., 17, 1963, 447-450.

1963 PEARSON, E.H. On the congruences $(p-1)! \equiv -1$ and $2^{p-1} \equiv 1 \pmod{p^2}$. Math. Comp., 17, 1963, 194-195.

1963 SELFRIDGE, J.L. Solution to problem 4995 (proposed by O. Ore). Amer. Math. Monthly, 70, 1963, p. 101.

1963   VOROB'EV, N.N.   *Fibonacci Numbers.*
                               Heath & Co., Boston, 1963.

1964   GRAHAM, R.L.   A Fibonacci-like sequence of
                               composite numbers.
                               Math. Mag., 37, 1964, 322-324.

1965   KLOSS, K.E.   Some number theoretic calculations.
                               J. Res. Nat. Bureau of Stand., B, 69,
                               1965, 335-336.

1965   ROTKIEWICZ, A.   Sur les nombres de Mersenne dépourvus
                               de diviseurs carrés et sur les nombres
                               naturels n tels que $n^2 \mid 2^n - 2$.
                               Mathem. Vecnik, (2), 17, 1965, 78-80.

1966   HASSE, H.   Über die Dichte der Primzahlen p,
                               fur die eine vorgegebene ganzra-
                               tionale Zahl $a \neq 0$ von gerader bzw.
                               ungerader Ordnung mod p ist.
                               Math. Ann., 168, 1966, 19-23.

1966   KRUYSWIJK, D.   On the congruence $u^{p-1} \equiv 1 \pmod{p^2}$
                               (in Dutch).
                               Math. Centrum Amsterdam, 1966,
                               7 pages.

1967   WARREN, L.J. &   On the square-freeness of Fermat and
       BRAY, H.   Mersenne numbers.
                               Pacific J. Math., 22, 1967, 563-564.

1968   PUCCIONI, S.   Un teorema per una resoluzione
                               parziale del famoso teorema di Fermat.
                               Archimede 20, 1968, 219-220.

1970   GOLOMB, S.W.   Powerful numbers.
                               Amer. Math. Monthly, 77, 1970,
                               848-852.

| 1971 | BRILLHART, J., TONASCIA, J., & WEINBERGER, P.J. | On the Fermat quotient. *Computers in Number Theory* (edited by A.L. Atkin & B.J. Birch), 213-222. Academic Press, New York, 1971. |

1971 UCHIDA, K.

Class numbers of imaginary abelian number fields, III.
Tôhoku Math. J., 23, 1971, 573-580.

1972 IWASAWA, K.

Lectures on p-adic L-functions.
Annals of Math. Studies, Princeton Univ. Press, Princeton, 1972.

1972 MASLEY, J.M.

On the Class Number of Cyclotomic Fields.
Ph.D. Thesis, Princeton Univ., 1972, 51 pages.

1973 VAUGHAN, R.C.

A remark on the divisor function $d(n)$.
Glasgow Math. J., 14, 1973, 54-55.

1974 ANGELL, I.O. & GODWIN, H.J.

Some factorizations of $10^n \pm 1$.
Math. Comp., 28, 1974, 307-308.

1974 BORUCKI, L.J. & DIAZ, J.B.

A note on primes, with arbitrary initial or terminal decimal ciphers in Dirichlet arithmetic progressions.
Amer. Math. Monthly, 81, 1974, 1001-1002.

1975 ERDÖS, P.

Problems and results on consecutive integers.
Eureka, 38, 1975/6, 3-8.

1975 JOHNSON, W.   Irregular primes and cyclotomic
                                   invariants.
                                   Math. Comp., 29, 1975, 113-120.

1975 METSÄNKYLÄ, T.   On the cyclotomic invariants of
                                   Iwasawa.
                                   Math. Scand., 37, 1975, 61-75.

1976 ERDÖS, P.   Problems and results on consecutive
                                   integers.
                                   Publ. Math. Debrecen, 23, 1976,
                                   271-282.

1976 MASLEY, J.M. &   Unique factorization in cyclotomic
      MONTGOMERY, H.L.   fields.
                                   Journal f. d. reine u. angew. Math.,
                                   286/7, 1976, 248-256.

1976 MENDELSOHN, N.S.   The equation $\phi(x) = k$.
                                   Math. Mag., 49, 1976, 37-39.

1976 STEPHENS, P.J.   Prime divisors of second order
                                   linear recurrences, I, II.
                                   J. Nb. Th., 8, 1976, 313-345.

1977 JOHNSON, W.   On the non-vanishing of Fermat's
                                   quotient (mod p).
                                   Journal f. d. reine u. angew. Math.,
                                   292, 1977, 196-200.

1977 POMERANCE, C.   On composite n for which
                                   $\phi(n) \mid n-1$, II.
                                   Pacific J. Math., 69, 1977, 177-186.

1977 POWELL, B.   Problem E 2631 (Prime satisfying
                                   Mirimanoff's condition).
                                   Amer. Math. Monthly, 84, 1977, p. 57.

1978   DE LEON, M.J.   Solution of problem E 2631.
                              Amer. Math. Monthly, 85, 1978, 279-280.

1978   METSÄNKYLÄ, T.   Iwasawa invariants and Kummer
                              congruences.
                              J. Nb. Th., 10, 1978, 510-522.

1978   WAGSTAFF, Jr., S.S.   The irregular primes to 125000.
                              Math. Comp., 32, 1978, 583-591.

1978   WILLIAMS, H.C.   Some primes with interesting
                              digit patterns.
                              Math. Comp., 32, 1978, 1306-1310.

1979   BÁYER, P.   Sobre el indice de irregularidad de
                              los números primos.
                              Collect. Math., 30, 1979, 11-20.

1979   ERDÖS, P. &   On the density of odd integers of
       ODLYZKO, A.M.   the form $(p-1)2^{-n}$ and related
                              questions.
                              J. Nb. Th., 11, 1979, 257-263.

1979   FERRERO, B. &   The Iwasawa invariant μ_p vanishes
       WASHINGTON, L.C.   for abelian number fields.
                              Ann. Math., 109, 1979, 377-395.

1979   RIBENBOIM, P.   *13 Lectures on Fermat's*
                              Last Theorem.
                              Springer-Verlag, New York, 1979.

1979   WILLIAMS, H.C.   Some primes of the form
       & SEAH, E. $(a^n-1)/(a-1)$.
                              Math. Comp., 33, 1979, 1337-1342.

1980   NEWMAN, M.,   Simple groups of square order and
       SHANKS, D. &   an interesting sequence of primes.
       WILLIAMS, H.C.   Acta Arithm., 38, 1980, 129-140.

1980   POWELL, B.   Primitive densities of certain sets
                              of primes.
                              J. Nb. Th., 12, 1980, 210-217.

1980   SKULA, L.   Index of irregularity of a prime.
                              Journal f. d. reine u. angew. Math.,
                              315, 1980, 92-106.

1980   WASHINGTON, L.C.   *Introduction to Cyclotomic Fields.*
                              Springer-Verlag, New York, 1980.

1981   LEHMER, D.H.   On Fermat's quotient, base two.
                              Math. Comp., 36, 1981, 289-290.

1981   SHANKS, D. &   Gunderson's function in Fermat's
       WILLIAMS, H.C.   last theorem.
                              Math. Comp., 36, 1981, 291-295.

1981   SPIRO, C.A.   *The Frequency with which an
                              Integral-Valued, Prime-Independent,
                              Multiplicative or Additive Function
                              of n Divides a Polynomial Function of n.*
                              Ph.D. Thesis, University of Illinois,
                              Urbana-Champaign, 1981, 179 pages.

1982   POWELL, B.   Problem E 2948 ($p^e \parallel x^{p-1} - y^{p-1}$,
                              $2 \nmid pe$, p prime occurs frequently).
                              Amer. Math. Monthly, 89, 1982, p. 334.

1982   POWELL, B.   Problem E 2956 (The existence of small
                              prime solutions of $x^{p-1} \not\equiv 1 \pmod{p^2}$).
                              Amer. Math. Monthly, 89, 1982,
                              p. 498.

1982   WILLIAMS, H.C.   The influence of computers in the
                              development of number theory.
                              Comp. & Maths. with Appl., 8, 1982,
                              75-93.

1982 YATES, S. *Repunits and Repetends.*
Star Publ. Co., Boynton
Beach, Florida, 1982.

1983 JAESCHKE, G. On the smallest k such that
$k \cdot 2^N + 1$ are composite. Corrigendum.
Math. Comp., 40, 1983, 381-384; 45,
1985, p. 637.

1983 KELLER, W. Factors of Fermat numbers and
large primes of the form $k \cdot 2^n + 1$.
Math. Comp., 41, 1983, 661-673.

1983 RIBENBOIM, P. 1093.
Math. Intelligencer, 5, No. 2,
1983, 28-34.

1984 DAVIS, J.A. & Most wanted factorization using the
HOLDRIDGE, D.B. quadratic sieve.
Sandia Nat. Lab. Report SAND
84-1658, 1984.

1984 HEATH-BROWN, D.R. The divisor function at consecutive
integers.
Mathematika, 31, 1984, 141-149.

1984 MATTICS, L.E. Solution of problem E 2948.
Amer. Math. Monthly, 91, 1984,
650-651.

1985 ADLEMAN, L.M. & The first case of Fermat's last
HEATH-BROWN, D.R. theorem.
Invent. Math., 79, 1985, 409-416.

1985 FOUVRY, E. Théorème de Brun-Titchmarsh;
application au théorème de Fermat.
Invent. Math., 79, 1985, 383-407.

1985   GRANVILLE, A.J.   Refining the conditions on the Fermat
                              quotient.
                              Math. Proc. Cambridge Phil. Soc., 98,
                              1985, 5-8.

1985   KELLER, W.   The 17th prime of the form
                              $5 \cdot 2^n + 1$.
                              Abstracts Amer. Math. Soc. 6, No. 1,
                              1985, p. 121.

1985   LAGARIAS, J.C.   The set of primes dividing the
                              Lucas numbers has density $\frac{2}{3}$.
                              Pacific J. Math., 118, 1985, 19-23.

1985   RIBENBOIM, P.   An extension of Sophie Germain's
                              method to a wide class of
                              diophantine equations.
                              Journal f. d. reine u. angew. Math.,
                              356, 1985, 49-66.

1986   GRANVILLE, A.J.   Powerful numbers and Fermat's last
                              theorem.
                              C.R. Math. Rep. Acad. Sci. Canada, 8,
                              1986, 215-218.

1986   MOLLIN, R.A. &   On powerful numbers.
       WALSH, P.G.   Intern. J. Math. & Math. Sci., 9,
                              1986, 801-806.

1986   SKULA, L.   A note on the index of irregularity.
                              J. Nb. Th., 22, 1986, 125-138.

1986   SPIRO, C.A.   An iteration problem involving the
                              divisor function.
                              Acta Arithm., 46, 1986, 17-27.

1986   TZANAKIS, N.   Solution to problem E 2956.
                              Amer. Math. Monthly, 93, 1986, p. 569.

1986   WILLIAMS, H.C. &   The primality of R1031.
       DUBNER, H.   Math. Comp., 47, 1986, 703-712.

1987   GRANVILLE, A.J.   *Diophantine Equations with Variable*
                            Exponents with Special Reference to
                            Fermat's Last Theorem.
                            Ph.D. Thesis, Queen's University,
                            Kingston, 1987, 207 pages.

1987   RIBENBOIM, P.   Impuissants devant les puissances.
                            Expo. Math., 5, 1987 (to appear).

1987   TANNER, J.W. &   New congruences for the Bernoulli
       WAGSTAFF, Jr., S.S. numbers.
                            Math. Comp., 48, 1987, 341-350.

Chapter 6

1857   BOUNIAKOWSKY, V.   Nouveaux théorèmes relatifs à
                            la distribution des nombres premiers
                            et à la décomposition des
                            entiers en facteurs.
                            Mém. Acad. Sci. St. Petersbourg,
                            (6), Sci. Math. Phys., 6, 1857,
                            305-329.

1904   DICKSON, L.E.   A new extension of Dirichlet's
                            theorem on prime numbers.
                            Messenger of Math., 33, 1904,
                            155-161.

1922   NAGELL, T.   Zur Arithmetik der Polynome.
                            Abhandl. Math. Sem. Univ.
                            Hamburg, 1, 1922, 179-194.

1923 HARDY, G.H. &   Some problems in "Partitio
      LITTLEWOOD, J.E.   Numerorum," III: On the expres-
                            sion of a number as a sum of
                            primes.
                            Acta Math., 44, 1923, 1-70.
                            Reprinted in *Collected Papers of
                            G.H. Hardy*, Vol. I, 561-630.
                            Clarendon Press, Oxford, 1966.

1931 HEILBRONN, H.   Über die Verteilung der Primzahlen
                            in Polynomen.
                            Math. Ann., 104, 1931, 794-799.

1937 BILHARZ, H.   Primdivisoren mit vorgegebener
                            Primitivwurzel.
                            Math. Ann., 114, 1937, 476-492.

1937 RICCI, G.   Su la congettura di Goldbach e la
                            constante de Schnirelmann.
                            Annali Scuola Norm. Sup. Pisa, 6
                            (2), 1937, 71-116.

1948 LIÉNARD, R.   Tables fondamentales à 50 décimales
                            des sommes S_n, u_n, Σ_n.
                            Centre de Docum. Univ., Paris, 1948.

1948 WEIL, A.   *Sur les Courbes Algébriques et
                            les Variétés qui s'en Déduisent.*
                            Hermann, Paris, 1948.

1950 HASSE, H.   *Vorlesungen über Zahlentheorie.*
                            Springer-Verlag, Berlin, 1950.

1954 KUHN, P.   Über die Primteiler eines
                            Polynoms.
                            Proc. Intern. Congress Math.,
                            Amsterdam, 2, 1954, 35-37.

1958   SCHINZEL, A. &   Sur certaines hypothèses
        SIERPIŃSKI, W.   concernant les nombres
                                premiers. Remarque.
                                Acta Arithm., 4, 1958,
                                185-208 and 5, 1959, p. 259.

1960   SHANKS, D.   On the conjecture of Hardy &
                                Littlewood concerning the number of
                                primes of the form $n^2 + a$.
                                Math. Comp., 14, 1960, 321-332.

1961   SCHINZEL, A.   Remarks on the paper "Sur
                                certaines hypothèses
                                concernant les nombres premiers."
                                Acta Arithm., 7, 1961, 1-8.

1961   WRENCH, J.W.   Evaluation of Artin's constant and
                                the twin prime constant.
                                Math. Comp., 15, 1961, 396-398.

1962   BATEMAN, P.T. &   A heuristic asymptotic formula
        HORN, R.A.   concerning the distribution of
                                prime numbers.
                                Math. Comp., 16, 1962, 363-367.

1964   ANKENY, N.C. &   The general sieve.
        ONISHI, H.   Acta Arithm., 10, 1964, 31-62.

1964   GILLIES, D.B.   Three new Mersenne primes and a
                                statistical theory.
                                Math. Comp., 18, 1964, 93-98.

1964   SHANKS, D.   An analytic criterion for the
                                existence of infinitely many primes
                                of the form $\frac{1}{2}(n^2 + 1)$.
                                Illinois J. Math., 8, 1964, 377-379.

1964   SIEGEL, C.L.   Zu zwei Bemerkungen Kummers.
                                   Nachr. Akad. d. Wiss. Göttingen,
                                   Math. Phys. Kl., II, 1964, 51-62.
                                   Reprinted in *Gesammelte Abhandlungen*
                                   (edited by K. Chandrasekharan &
                                   H. Maaß), Vol. III, 436-442.
                                   Springer-Verlag, Berlin, 1966.

1965   BATEMAN, P.T. &   Primes represented by irreducible
       HORN, R.A.   polynomials in one variable.
                                   Theory of Numbers (Proc. Symp. Pure
                                   Math., Vol. VIII), 119-132.
                                   Amer. Math. Soc., Providence, RI, 1965.

1966   GROSSWALD, E.   *Topics from the Theory of Numbers.*
                                   Macmillan, New York, 1966; second
                                   edition Birkhäuser, Boston, 1984.

1967   HOOLEY, C.   On Artin's conjecture.
                                   Journal f. d. reine u. angew. Math.,
                                   225, 1967, 209-220.

1967   SHANKS, D. &   On the distribution of Mersenne
       KRAVITZ, S.   divisors.
                                   Math. Comp., 21, 1967, 97-101.

1969   RIEGER, G.J.   On polynomials and almost-primes.
                                   Bull. Amer. Math. Soc., 75, 1969,
                                   100-103.

1971   SHANKS, D.   A low density of primes.
                                   J. Recr. Math., 4, 1971/2, 272-275.

1972   RIBENBOIM, P.   *Algebraic Numbers.*
                                   Wiley-Interscience, New York, 1972.

1973   RADEMACHER, H.   *Topics in Analytic Number Theory.*
                                   Springer-Verlag, New York, 1975.

1973   WUNDERLICH, M.C.   On the Gaussian primes on the line
                             $\text{Im}(x) = 1$.
                             Math. Comp., 27, 1973, 399-400.

1975   SHANKS, D.   Calculation and applications of
                             Epstein zeta functions.
                             Math. Comp., 29, 1975, 271-287.

1976   CHOWLA, S. &   Class numbers and quadratic residues.
       FRIEDLANDER, J.B.   Glasgow Math. J., 17, 1976, 47-52.

1978   IWANIEC, H.   Almost-primes represented by
                             quadratic polynomials.
                             Invent. Math., 47, 1978, 171-188.

1978   SLOWINSKI, D.   Searching for the 27th Mersenne prime.
                             J. Recr. Math., 11, 1978/9, 258-261.

1980   FELTHI, C.   Non-nullité des fonctions zeta des
                             corps quadratiques réels pour $0 < s < 1$.
                             C.R. Acad. Sci. Paris, Sér. A, 291,
                             1980, 623-625.

1980   LENSTRA, Jr., H.W.   Primality testing.
                             Studieweek Getaltheorie en Computers.
                             Stichting Math. Centrum, Amsterdam,
                             1980.

1981   POWELL, B.   Problem 6384 (Numbers of the form
                             $m^p - n$).
                             Amer. Math. Monthly, 89, 1982, p. 278.

1982   ISRAEL, R.B.   Solution of problem 6384.
                             Amer. Math. Monthly, 90, 1983, p. 650.

1982   SCHINZEL, A.   *Selected Topics on Polynomials.*
                             Univ. of Michigan Press, Ann Arbor,
                             1982.

1983   ADLEMAN, L.M. &   Irreducibility testing and factorization
       ODLYZKO, A.M.   of polynomials.
                               Math. Comp., 41, 1983, 699-709.

1983   SCHROEDER, M.R.   Where is the next Mersenne prime
                               hiding?
                               Math. Intelligencer, 5, No. 3, 1983,
                               31-33.

1983   WAGSTAFF, Jr., S.S.   Divisors of Mersenne numbers.
                               Math. Comp., 40, 1983, 385-397.

1984   GUPTA, R. &   A remark on Artin's conjecture.
       RAM MURTY, P.M.   Invent. Math., 78, 1984, 127-130.

1984   McCURLEY, K.S.   Prime values of polynomials and
                               irreducibility testing.
                               Bull. Amer. Math. Soc., 11, 1984,
                               155-158.

1986   McCURLEY, K.S.   The smallest prime value of $x^n + a$.
                               Can. J. Math., 38, 1986, 925-936.

1986   McCURLEY, K.S.   Polynomials with no small prime
                               values.
                               Proc. Amer. Math. Soc., 97, 1986,
                               393-395.

1987   MOLLIN, R.A.   Necessary and sufficient conditions
                               for the class number of a real
                               quadratic field to be 1, and a
                               conjecture of Chowla.
                               Proc. Amer. Math. Soc., 101, 1987
                               (to appear).

1987   MOLLIN, R.A. &   A conjecture of S. Chowla via the
       WILLIAMS, H.C.   generalized Riemann hypothesis.
                               Proc. Amer. Math. Soc., 101, 1987
                               (to appear).

1987   MOLLIN, R.A. &   On prime valued polynomials
        WILLIAMS, H.C.   and class numbers of real
                                  quadratic fields.
                                  Preprint, University of Calgary,
                                  1987.

1987   RAM MURTY, P.M. &   Some remarks on Artin's conjecture.
        SRINIVASAN, S.   Can. Math. Bull. 30, 1987, 80-85.

Conclusion

1981   BATEMAN, P.T.   Major figures in the history of the
                                  prime number theorem.
                                  Abstracts Amer. Math. Soc. (87th
                                  annual meeting, San Francisco),
                                  1981, p. 2.

        BASHŌ   *Narrow Road to a Far Province.*
                                  Translated by Dorothy Britton.
                                  Kodanshu Publ., Tokyo, 1974.

PRIMES UP TO 10,000

2	3	5	7	11	13	17	19	23	29	31
79	83	89	97	101	103	107	109	113	127	131
191	193	197	199	211	223	227	229	233	239	241
311	313	317	331	337	347	349	353	359	367	373
439	443	449	457	461	463	467	479	487	491	499
577	587	593	599	601	607	613	617	619	631	641
709	719	727	733	739	743	751	757	761	769	773
857	859	863	877	881	883	887	907	911	919	929
1009	1013	1019	1021	1031	1033	1039	1049	1051	1061	1063
1151	1153	1163	1171	1181	1187	1193	1201	1213	1217	1223
1297	1301	1303	1307	1319	1321	1327	1361	1367	1373	1381
1459	1471	1481	1483	1487	1489	1493	1499	1511	1523	1531
1607	1609	1613	1619	1621	1627	1637	1657	1663	1667	1669
1759	1777	1783	1787	1789	1801	1811	1823	1831	1847	1861
1933	1949	1951	1973	1979	1987	1993	1997	1999	2003	2011
2089	2099	2111	2113	2129	2131	2137	2141	2143	2153	2161
2269	2273	2281	2287	2293	2297	2309	2311	2333	2339	2341
2411	2417	2423	2437	2441	2447	2459	2467	2473	2477	2503
2609	2617	2621	2633	2647	2657	2659	2663	2671	2677	2683
2741	2749	2753	2767	2777	2789	2791	2797	2801	2803	2819
2909	2917	2927	2939	2953	2957	2963	2969	2971	2999	3001
3089	3109	3119	3121	3137	3163	3167	3169	3181	3187	3191
3299	3301	3307	3313	3319	3323	3329	3331	3343	3347	3359
3461	3463	3467	3469	3491	3499	3511	3517	3527	3529	3533
3613	3617	3623	3631	3637	3643	3659	3671	3673	3677	3691
3779	3793	3797	3803	3821	3823	3833	3847	3851	3853	3863
3943	3947	3967	3989	4001	4003	4007	4013	4019	4021	4027
4129	4133	4139	4153	4157	4159	4177	4201	4211	4217	4219
4289	4297	4327	4337	4339	4349	4357	4363	4373	4391	4397

37	41	43	47	53	59	61	67	71	73
137	139	149	151	157	163	167	173	179	181
251	257	263	269	271	277	281	283	293	307
379	383	389	397	401	409	419	421	431	433
503	509	521	523	541	547	557	563	569	571
643	647	653	659	661	673	677	683	691	701
787	797	809	811	821	823	827	829	839	853
937	941	947	953	967	971	977	983	991	997
1069	1087	1091	1093	1097	1103	1109	1117	1123	1129
1229	1231	1237	1249	1259	1277	1279	1283	1289	1291
1399	1409	1423	1427	1429	1433	1439	1447	1451	1453
1543	1549	1553	1559	1567	1571	1579	1583	1597	1601
1693	1697	1699	1709	1721	1723	1733	1741	1747	1753
1867	1871	1873	1877	1879	1889	1901	1907	1913	1931
2017	2027	2029	2039	2053	2063	2069	2081	2083	2087
2179	2203	2207	2213	2221	2237	2239	2243	2251	2267
2347	2351	2357	2371	2377	2381	2383	2389	2393	2399
2521	2531	2539	2543	2549	2551	2557	2579	2591	2593
2687	2689	2693	2699	2707	2711	2713	2719	2729	2731
2833	2837	2843	2851	2857	2861	2879	2887	2897	2903
3011	3019	3023	3037	3041	3049	3061	3067	3079	3083
3203	3209	3217	3221	3229	3251	3253	3257	3259	3271
3361	3371	3373	3389	3391	3407	3413	3433	3449	3457
3539	3541	3547	3557	3559	3571	3581	3583	3593	3607
3697	3701	3709	3719	3727	3733	3739	3761	3767	3769
3877	3881	3889	3907	3911	3917	3919	3923	3929	3931
4049	4051	4057	4073	4079	4091	4093	4099	4111	4127
4229	4231	4241	4243	4253	4259	4261	4271	4273	4283
4409	4421	4423	4441	4447	4451	4457	4463	4481	4483

4493	4507	4513	4517	4519	4523	4547	4549	4561	4567	4583
4663	4673	4679	4691	4703	4721	4723	4729	4733	4751	4759
4871	4877	4889	4903	4909	4919	4931	4933	4937	4943	4951
5021	5023	5039	5051	5059	5077	5081	5087	5099	5101	5107
5227	5231	5233	5237	5261	5273	5279	5281	5297	5303	5309
5417	5419	5431	5437	5441	5443	5449	5471	5477	5479	5483
5573	5581	5591	5623	5639	5641	5647	5651	5653	5657	5659
5749	5779	5783	5791	5801	5807	5813	5821	5827	5839	5843
5923	5927	5939	5953	5981	5987	6007	6011	6029	6037	6043
6131	6133	6143	6151	6163	6173	6197	6199	6203	6211	6217
6301	6311	6317	6323	6329	6337	6343	6353	6359	6361	6367
6481	6491	6521	6529	6547	6551	6553	6563	6569	6571	6577
6689	6691	6701	6703	6709	6719	6733	6737	6761	6763	6779
6863	6869	6871	6883	6899	6907	6911	6917	6947	6949	6959
7027	7039	7043	7057	7069	7079	7103	7109	7121	7127	7129
7237	7243	7247	7253	7283	7297	7307	7309	7321	7331	7333
7477	7481	7487	7489	7499	7507	7517	7523	7529	7537	7541
7607	7621	7639	7643	7649	7669	7673	7681	7687	7691	7699
7817	7823	7829	7841	7853	7867	7873	7877	7879	7883	7901
8011	8017	8039	8053	8059	8069	8081	8087	8089	8093	8101
8219	8221	8231	8233	8237	8243	8263	8269	8273	8287	8291
8389	8419	8423	8429	8431	8443	8447	8461	8467	8501	8513
8609	8623	8627	8629	8641	8647	8663	8669	8677	8681	8689
8761	8779	8783	8803	8807	8819	8821	8831	8837	8839	8849
8963	8969	8971	8999	9001	9007	9011	9013	9029	9041	9043
9157	9161	9173	9181	9187	9199	9203	9209	9221	9227	9239
9341	9343	9349	9371	9377	9391	9397	9403	9413	9419	9421
9497	9511	9521	9533	9539	9547	9551	9587	9601	9613	9619
9719	9721	9733	9739	9743	9749	9767	9769	9781	9787	9791
9883	9887	9901	9907	9923	9929	9931	9941	9949	9967	9973

4591	4597	4603	4621	4637	4639	4643	4649	4651	4657
4783	4787	4789	4793	4799	4801	4813	4817	4831	4861
4957	4967	4969	4973	4987	4993	4999	5003	5009	5011
5113	5119	5147	5153	5167	5171	5179	5189	5197	5209
5323	5333	5347	5351	5381	5387	5393	5399	5407	5413
5501	5503	5507	5519	5521	5527	5531	5557	5563	5569
5669	5683	5689	5693	5701	5711	5717	5737	5741	5743
5849	5851	5857	5861	5867	5869	5879	5881	5897	5903
6047	6053	6067	6073	6079	6089	6091	6101	6113	6121
6221	6229	6247	6257	6263	6269	6271	6277	6287	6299
6373	6379	6389	6397	6421	6427	6449	6451	6469	6473
6581	6599	6607	6619	6637	6653	6659	6661	6673	6679
6781	6791	6793	6803	6823	6827	6829	6833	6841	6857
6961	6967	6971	6977	6983	6991	6997	7001	7013	7019
7151	7159	7177	7187	7193	7207	7211	7213	7219	7229
7349	7351	7369	7393	7411	7417	7433	7451	7457	7459
7547	7549	7559	7561	7573	7577	7583	7589	7591	7603
7703	7717	7723	7727	7741	7753	7757	7759	7789	7793
7907	7919	7927	7933	7937	7949	7951	7963	7993	8009
8111	8117	8123	8147	8161	8167	8171	8179	8191	8209
8293	8297	8311	8317	8329	8353	8363	8369	8377	8387
8521	8527	8537	8539	8543	8563	8573	8581	8597	8599
8693	8699	8707	8713	8719	8731	8737	8741	8747	8753
8861	8863	8867	8887	8893	8923	8929	8933	8941	8951
9049	9059	9067	9091	9103	9109	9127	9133	9137	9151
9241	9257	9277	9281	9283	9293	9311	9319	9323	9337
9431	9433	9437	9439	9461	9463	9467	9473	9479	9491
9623	9629	9631	9643	9649	9661	9677	9679	9689	9697
9803	9811	9817	9829	9833	9839	9851	9857	9859	9871
10007	10009	10037	10039	10061	10067	10069	10079	10091	10093

INDEX OF NAMES

GALLIMAWFRIES

SUBJECT INDEX